CONSTITUTIVE MODELING OF **GEOMATERIALS**

PRINCIPLES AND APPLICATIONS

CONSTITUTIVE MODELING OF GEOMATERIALS

PRINCIPLES AND APPLICATIONS

TERUO NAKAI

CRC Press is an imprint of the
Taylor & Francis Group, an **informa** business
A SPON PRESS BOOK

CRC Press
Taylor & Francis Group
6000 Broken Sound Parkway NW, Suite 300
Boca Raton, FL 33487-2742

Version Date: 20120522

International Standard Book Number: 978-0-415-55726-9 (Hardback)

Library of Congress Cataloging-in-Publication Data

Nakai, Teruo.
Constitutive modeling of geomaterials : principles and applications / Teruo Nakai.
p. cm.
Includes bibliographical references and index.
ISBN 978-0-415-55726-9 (hardback)
1. Soil mechanics. 2. Soils--Mathematical models. 3. Foundations. I. Title.

TA710.N258 2012
624.1'5136--dc23 2012016097

Visit the Taylor & Francis Web site at
http://www.taylorandfrancis.com

and the CRC Press Web site at
http://www.crcpress.com

Contents

Preface

When I was a student (almost 40 years ago), my supervisor, Sakuro Murayama, often told us that the most important challenge in the field of soil mechanics was to establish the stress–strain–time–temperature relation of soils. Since the beginning of his academic career, he had pursued research on a constitutive model for soils, and he summarized his experience in a thick book of almost 800 pages (Murayama 1990) when he was almost 80 years old. In his book, the elastoplasticity theory was not used in a straightforward manner, but he discussed soil behavior, focusing his attention not on the plane where shear stress is maximized, called the τ_{max} plane or 45° plane, but rather on the plane where the shear–normal stress ratio is maximized, called the $(\tau/\sigma)_{max}$ plane or mobilized plane, because the soil behavior is essentially governed by a frictional law.

In retrospect, I realize how sharp was his vision to pay attention to the mobilized plane at a time when most people looked at the τ_{max} plane. Now, in three-dimensional (3D) conditions in which the intermediate principal stress must be considered, the plane corresponding to the τ_{max} plane in two-dimensional (2D) conditions is the commonly used octahedral plane because the shear stress on the octahedral plane is the quadratic mean of maximum shear stresses between two respective principal stresses. For 3D constitutive modeling in this book, attention is paid to the so-called spatially mobilized plane (SMP) on which the shear–normal stress ratio is the quadratic mean of maximum shear–normal stress ratios between two respective principal stresses.

The Cam clay model, proposed in the 1960s by Roscoe and collaborators, was the first model to unify relevant aspects of soil behavior under shear or isotropic compression. However, this model had some limitations, mostly stemming from the explicit adoption of stress invariants (p, q), which represent the normal and shear stresses acting on the octahedral plane. Although many constitutive models have been proposed since the Cam clay model was developed, most of them are formulated for describing some specific features of soil behavior, and their applicability becomes restricted—for example, models for sand, models for clay, models under cyclic loading, models for describing time-dependent behavior, models for the analysis of embankment, models for the analysis of excavation, and others. Then, even though the ground is the same, the adopted models and/or the values of their material parameters become dependent on the problem to be solved. The principal quest during my research activities summarized in this book was the formulation of a constitutive model capable of simulating the main features of soil behavior, using a unified and rational framework leading to a unified set of material parameters. I hope to have achieved or at least to have contributed to this ambitious goal.

Several books and state-of-the-art reports on constitutive models for geomaterials have given general overviews and explanations of past models. This book is not intended to give such explanations of various models developed in the world. Rather, the aim of this book is to present a framework of simple and rational modeling of various soil features based on the unified concepts developed by me and my colleagues, mostly at the Nagoya Institute of Technology (NIT).

The starting point of my research was the discovery of the spatially mobilized plane (SMP; Matsuoka and Nakai 1974), when I was studying the stress–strain behavior of soils in 3D stress conditions under the supervision of Hajime Matsuoka as a master student of Murayama's laboratory in Kyoto University. Then, the concept of SMP was developed into the extended concept of spatially mobilized plane (SMP*; Nakai and Matsuoka 1980, 1983) to describe more comprehensively the stress–strain behavior of soils in 3D conditions; this was the main content of my PhD thesis. The concept of modified stress t_{ij} (Nakai and Mihara 1984), which is one of the main parts of this book, is the generalized procedure for applying the concept of SMP* to the models of geomaterials based on conventional elastoplasticity.

In this book, the ideas and formulations to model various soil features are firstly explained in one-dimensional (1D) conditions for easy understanding, and then these 1D models are extended to 3D ones by using the t_{ij} concept. In the modeling approach described, the influence of density and/or confining pressure is considered by introducing and revising the subloading surface concept of Hashiguchi (1980). The 3D model considering this influence is called the subloading t_{ij} model (Nakai and Hinokio 2004). To describe the behavior of structured soils such as natural clays, the influence of bonding as well as of density is considered. For this modeling, the super/subloading surface concept by Asaoka, Nakano, and Noda (2000a) is of much help. The methods to consider the time effect, temperature effect, suction effect, and others were recently developed with my colleagues (Hossain Md. Shahin, Mamoru Kikumoto, Hiroyuki Kyokawa, and others).

In order to confirm the applicability of the constitutive models to practical problems, the numerical simulations of typical geotechnical problems (numerical modeling) and the corresponding model tests (physical modeling) are also presented in this book. Hossain Md. Shahin contributed much to these numerical analyses.

The cooperative research with Marcio M. Farias (University of Brasilia) has continued since he worked in my laboratory on his sabbatical leave in 1998. He visited NIT several times and offered many valuable comments and helpful discussions on our research and also kindly checked the contents and English of the draft. Feng Zhang (NIT), who joined us in 1998, gave me helpful and critical comments on our research. Richard Wan (University of Calgary), who was collaborating with us on a research project in my laboratory at the time of the Great East Japan Earthquake in 2011, checked the draft and gave me valuable comments.

During my academic life in Nagoya, I often discussed current and future geotechnical problems and other subjects with my senior, Akira Asaoka (Nagoya University). Our discussions and his comments were interesting, exciting, and helpful for me.

I would like to express my sincere gratitude to all of these individuals for their cooperation and support on this work. Also, I wish to express my thanks to former and present students in my laboratory for their help in experiments and analyses. Finally, I give my thanks to my family for supporting my research life.

Teruo Nakai

List of symbols

PART I

a	material parameter to describe the influence of density and/or confining pressure
a_1, a_2, and a_3	principal values of a_{ij} (i.e., direction cosines of normal to the spatially mobilized plane, SMP)
a_{ij}	symmetric tensor whose principal values are given by a_1, a_2, and a_3
b	material parameter to describe the influence of bonding
b_σ	intermediate principal stress parameter (= $(\sigma_2 - \sigma_3)/(\sigma_1 - \sigma_3)$)
$d\varepsilon_N^*$	strain increment invariant in t_{ij} concept (i.e., normal component of $d\varepsilon_{ij}$ with respect to the SMP (= $d\varepsilon_{ij}a_{ij}$)
$d\varepsilon_S^*$	strain increment invariant in t_{ij} concept (i.e., in-plane component of $d\varepsilon_{ij}$ with respect to the SMP (= $\sqrt{d\varepsilon'_{ij}d\varepsilon'_{ij}}$)
$d\varepsilon'_{ij}$	deviatoric strain increment tensor based on t_{ij} concept (= $d\varepsilon_{ij} - d\varepsilon^*_N a_{ij}$))
e	void ratio
e_{ij}	deviatoric strain tensor (= $\varepsilon_{ij} - \varepsilon_v\delta_{ij}/3$)
e_{ijk}	permutation tensor
e_N	void ratio on the normal consolidation line (NCL)
$f = 0$	yield function
h^p	plastic modulus
k	coefficient of permeability
l	material parameter for describing suction effect
p	stress invariant used in ordinary models (i.e., mean principal stress (= $\sigma_{ij}\delta_{ij}/3$))
p_1	value of p on p-axis for the current yield surface

q	stress invariant used in ordinary models (i.e., deviatoric stress $(=\sqrt{(3/2)s_{ij}s_{ij}})$)
r_{ij}	square root of σ_{ij} (i.e., $r_{ik}r_{kj}=\sigma_{ij}$)
s	suction
s_{ij}	deviatoric stress tensor $(=\sigma_{ij}-p\delta_{ij})$
t	time
t_1, t_2, and t_3	principal value of t_{ij}
t_{ij}	modified stress tensor based on the t_{ij} concept $(=a_{ik}\sigma_{kj})$
t'_{ij} :	deviatoric stress tensor based on t_{ij} concept $(=t_{ij}-t_N a_{ij})$
t_N	stress invariant in t_{ij} concept (i.e., normal component of t_{ij} with respect to the SMP $(=t_{ij}a_{ij})$)
t_{N1}	value of t_N on t_N-axis for the current yield surface
t_S	stress invariant in the t_{ij} concept (i.e., in-plane component of t_{ij} with respect to the SMP $(=\sqrt{t'_{ij}t'_{ij}})$)
u, u_w	pore water pressure
u_a	pore air pressure
w	water content
x_{ij}	stress ratio tensor based on t_{ij} concept $(=t'_{ij}/t_N)$
D	Shibata's dilatancy coefficient
D_{ijkl}	stiffness tensor
E, E_e	tangential Young's modulus of the elastic component
F	stress term in the yield function $(=(\lambda-\kappa)\ln\frac{\sigma}{\sigma_0}$, $(\lambda-\kappa)\ln\frac{p_1}{p_0}$ or $(\lambda-\kappa)\ln\frac{t_{N1}}{t_{N0}})$
G	shear modulus
$G(\rho)$	increasing function of ρ that satisfies $G(0)=0$
H	plastic strain term in yield function $(=(-\Delta e)^p=(1+e_0)\varepsilon_v^p)$
I_1, I_2, and I_3	first, second, and third invariants of σ_{ij}
I_{r1}, I_{r2}, and I_{r3}	first, second, and third invariants of r_{ij} (where $r_{ik}r_{kj}=\sigma_{ij}$)
K	bulk modulus
N	void ratio at the NCL for $p=98$ kPa (at $(-\dot{e})^p=(-\dot{e})^p_{ref}$ in case of time-dependent model)
$Q(\omega)$	increasing function of ω that satisfies $Q(0)=0$
R_{CS}	principal stress ratio at critical state in triaxial compression $(=(\sigma_1/\sigma_3)_{CS(comp)})$
S_r	degree of saturation
T	temperature
X	stress ratio $(=t_S/t_N=\sqrt{x_{ij}x_{ij}})$
Y	plastic strain increment ratio $(=d\varepsilon_N^{*p}/d\varepsilon_S^{*p})$

α	angle between the coordinate axes and principal stress axes in 2D and axisymmetric conditions
α_T	coefficient of thermal expansion
β	material parameter to determine the shape of the yield surface
δ_{ij}	unit tensor (Kronecker's delta)
ε	strain in one-dimensional condition
ε_1, ε_2, and ε_3	three principal strains
ε_{ij}	strain tensor
ε_d	deviatoric strain ($=\sqrt{(2/3)e_{ij}e_{ij}}$)
ε_v	volumetric strain ($=\varepsilon_{ij}\delta_{ij}$)
ϕ_{moij}	mobilized angle between two principal stresses (σ_i and σ_j)
η	stress ratio ($= q/p = \sqrt{(3/2)\eta_{ij}\eta_{ij}}$)
η_{ij}	stress ratio tensor by Sekiguchi and Ohta ($= s_{ij}/p$)
φ_1 and φ_2	principal values of the fabric tensor
κ	swelling index
λ	compression index
λ_k	coefficient for considering influence of void ratio on permeability
λ_T	material parameter for describing thermal effect
λ_α	coefficient of secondary consolidation
ν, ν_e	Poisson's ratio of elastic component
θ	angle between σ_1-axis and radial stress path on the octahedral plane
ρ	state variable representing density; difference between the current void ratio and the void ratio on the NCL at the same stress level
σ	stress in one-dimensional condition
σ_y	yield stress in one-dimensional condition
σ_1, σ_2, and σ_3	three principal stresses
σ_{ij}	stress tensor
σ_{ij}^*	modified stress tensor defined by Satake
σ''	Bishop's effective stress ($\cong \sigma - u_a + S_r s$)
σ^{net}	net stress for unsaturated soil ($= \sigma - u_a$)
ω	state variable considering the bonding effect as an imaginary increase of density
$\xi(\eta)$	increasing function of η that satisfies $\xi(0) = 0$
ψ	state variable for shifting the NCL on the e–$\ln\sigma$ plane or the e–$\ln t_{N1}$ plane

$\zeta(\eta)$, $\zeta(X)$	increasing function of the stress ratio (η or X) that satisfies $\zeta(0) = 0$ (i.e., function to determine the shape of yield surface)
Λ	proportionality constant ($= \frac{dF}{h^p}$ or $\frac{dF+d\psi}{h^p}$)
M	stress ratio η at the critical state
M*	intercept with X-axis in stress–dilatancy relations based on the t_{ij} concept; determined from X_{CS} and Y_{CS}
superscript e	elastic component
superscript p	plastic component
superscript (AF)	component that satisfies the associated flow rule
superscript (IC)	component under increasing mean stress
subscript 0	initial value
subscript f	value at failure
subscript ref	value at reference state
subscript sat	value at saturated state
subscript (equ)	equivalent value
subscript CS	value at critical state
subscript NC	value at normally consolidated state
overhead dot ($\dot{}$)	rate of quantities
prefix d	infinitesimal increment of quantities
prefix Δ	finite change in quantities

PART 2

d	imposed displacement of trapdoor
d_c	amount of downward translation of the tunnel center
d_r	amount of shrinkage of circular tunnel
e	eccentricity of load from the center of foundation
h	horizontal displacement of foundation
k_n	unit elastic stiffness along the joint element
k_s	unit elastic stiffness across the joint element
p_n	normal stress on joint element
p_s	shear stress on joint element
q_v	vertical pressure on foundation
s	mean stress in 2D condition ($= (\sigma_1 + \sigma_3)/2$)
t	deviatoric stress in 2D condition ($= (\sigma_1 - \sigma_3)/2$))
v	vertical displacement of foundation
w_n	normal relative displacement on joint element

w_s	tangential relative displacement on joint element
B	width of tunnel
B, B_f	width of foundation
D	depth of tunnel
D	installation depth of reinforcement
$[D_J]$	stiffness matrix of joint element
D_p	vertical distance between the pile tip and the tunnel crown
EA	axial stiffness
EI	bending stiffness
E_p	elastic modulus of pile
K	earth pressure coefficient
L	length of reinforcement
L_w	distance between the retaining wall and the edge of raft of the foundation
L_p	pile length
P_h	horizontal total thrust against wall
Q, Q_v	vertical load
α	loading angle from vertical direction
β	reinforcement placement angle from horizontal direction
γ	unit weight of soil
δ	friction angle on joint
ε_a	axial strain
θ	rotation angle of foundation
ξ	multiplier for representing slip behavior on the joint element

Chapter 1

Introduction

1.1 BACKGROUND

The framework of soil mechanics (geomechanics) was established by Karl Terzaghi (Terzaghi 1943), and two famous English textbooks of soil mechanics (Terzaghi and Peck 1948; Taylor 1948) were published in 1948. Later, soil mechanics from that time was named the "1948 model of soil mechanics" by my supervisor, Toru Shibata (coincidentally, 1948 is my birth year). Practical designs of most geotechnical problems have been done more or less based on the ideas of these books.

In this old framework, the same soil is assumed to be different kinds of materials when solving different problems (e.g., porous rigid material in seepage problems, nonlinear elastoplastic material in settlement problems, porous linear elastic material in consolidation problems, or rigid plastic material in earth pressure and stability problems). The "1948 model of soil mechanics" has been adapted to various practical geotechnical problems in the epoch when there was no constitutive model of geomaterials and computers were not available for practical use. Most practical designs of earth structures, foundations, and countermeasures against earth disasters have been conceived without consideration of the important deformation characteristics of geomaterials, such as soil dilatancy.

On the other hand, for example, designs of steel structures have adopted consistent methods to predict their deformation and failure because there are unified constitutive models of metals to describe their behavior from deformation to failure as elastoplastic materials. Thus the most important task for researchers of geomechanics is to establish a general constitutive model for geomaterials.

The Cam clay model, which was developed at Cambridge University (Roscoe, Schofield, and Thurairajah 1963; Schofield and Wroth 1968; Roscoe and Burland 1968), is certainly the first elastoplastic model applicable to the practical deformation analysis of ground. The model at least describes the behavior of soils under shear loading and under consolidation using the same framework. This model is very simple; it has few material parameters and the physical meaning of each parameter is clear.

Since the initial development of the Cam clay model, nonlinear elastoplastic analyses using this model have been carried out to solve boundary value problems. However, applications to practical design have been limited, in part because the Cam clay model has problems in describing the following important features of soil behavior:

1. Influence of intermediate principal stress on the deformation and strength of geomaterials
2. Dependency of the direction of plastic flow on the stress path
3. Positive dilatancy during strain hardening
4. Stress-induced anisotropy and cyclic loading
5. Inherent anisotropy
6. Influence of density and/or confining pressure on deformation and strength
7. Behavior of structured soils such as naturally deposited clay
8. Time-dependent behavior and rheological characteristics
9. Temperature-dependent behavior
10. Behavior of unsaturated soils
11. Influence of particle crushing

Although many constitutive models have been proposed to overcome the limitations of the Cam clay model, most of them are more complex than the model and/or the conditions to which they can be applied are still restricted.

At the 16th International Conference on Soil Mechanics and Geotechnical Engineering, which was held in Osaka in 2005, there was a practitioner/academic forum called "Which Is Better for Practical Geotechnical Engineering, Simplified Model or Complex Model?" In my opinion, the model should be simple and sophisticated (but not complicated) so as to be applicable and useful to the practice. This book was written considering this opinion.

Many books have been published on constitutive modeling. Some of them review and explain extensively previous works on soil modeling together with the relevant experimental results—for example, Muir Wood (1990), Hicher and Shao (2008), Nova (2010), and Yu (2006). Other books emphasize the understanding of soil behavior mostly from the experimental point of view (Mitchell and Soga 2005). There are also books with emphasis on the formulation of recent plasticity theory (Hashiguchi 2009). This book is intended to present the key concepts to model various features of geomaterials in a uniform and rational manner.

1.2 CONTENTS OF THIS BOOK

As mentioned, the Cam clay model is not the most appropriate to be applied to many practical problems. However, it is a good model to study the fundamental framework of the constitutive models for geomaterials—particularly the concepts of critical state and state boundary surface used in the model. In **Part 1**, after explaining the general modeling of materials such as elastic and plastic materials, a simple and clear explanation of the Cam clay model is given for a better understanding of the subsequent contents. Every explanation about how to model the basic features of soil behavior starts using a one-dimensional (1D) condition for easy understanding of the idea and formulation, and then the models are extended to general three-dimensional (3D) conditions. Validations of the present models are achieved by comparing simulations of various kinds of element tests and the observed results reported in the literature.

Chapter 2 explains the fundamental formulations and the meaning of elasticity and plasticity, initially under 1D conditions; then the 3D formulation of plasticity theory is presented. Chapter 3 describes a simple and unified 1D framework in order to model typical features of normally consolidated and naturally deposited soils considering the effects of density, bonding, time, temperature, suction, and others. Chapter 4 explains the Cam clay model considering that the model is a 3D extension of the well-known linear e–$\ln\sigma$ relation of remolded normally consolidated clay. Also, the applicability and drawbacks of ordinary modeling using Cam clay for describing 3D soil behavior are discussed based on test results.

Chapter 5 describes a simple and unified method to extend 1D models to general 3D conditions, using the so-called t_{ij} concept; a detailed explanation of its physical meaning is presented. Using this concept, the Cam clay model is extended to one that is valid under three different principal stresses as an example. Chapter 6 presents 3D models in which various features of soil behavior are properly taken into consideration. The 1D models described in Chapter 3 are extended to 3D ones by using the t_{ij} concept. Also, a simple method to describe the dependency of the direction of plastic flow on the stress path in the constitutive modeling is presented. The present 3D models are verified by simulations of various element tests on clay and sand. Chapter 7 summarizes the main conclusions of the chapters presented in Part 1.

In Part 2, the present constitutive models are implemented into a finite element code applied to simulate various kinds of boundary value problems in geotechnical engineering: the tunneling problem, retaining wall problem, braced open excavation problem, bearing capacity problem, reinforced soil problem, and shear band development. The applicability of the constitutive model to the practical geotechnical problems is confirmed by comparing the results of the numerical simulations (numerical modeling) with the observed results of the corresponding model tests (physical modeling) in laboratory scale. The same materials are used to represent the ground in all model tests and numerical simulations: aluminum rod mass for two-dimensional (2D) problems and alumina balls for 3D problems. The structure of Part 2 will be described in Chapter 8.

Figure 1.1 shows the relation between a field problem, numerical modeling, and physical modeling in a geotechnical investigation. The final target of geotechnical engineering is to predict the behavior of the grounds and structures in the field with sufficient accuracy and to propose efficient countermeasures if problems are identified. For this purpose, centrifuge tests, by which the same stress level as in the field can be reproduced in a small-scale physical model, are sometimes carried out. The qualitative and/or quantitative ground behavior in the field is predicted directly from the model tests. However, it is impossible to adjust all the similarity aspects of the physical quantities and to create the same model ground as the real natural ground

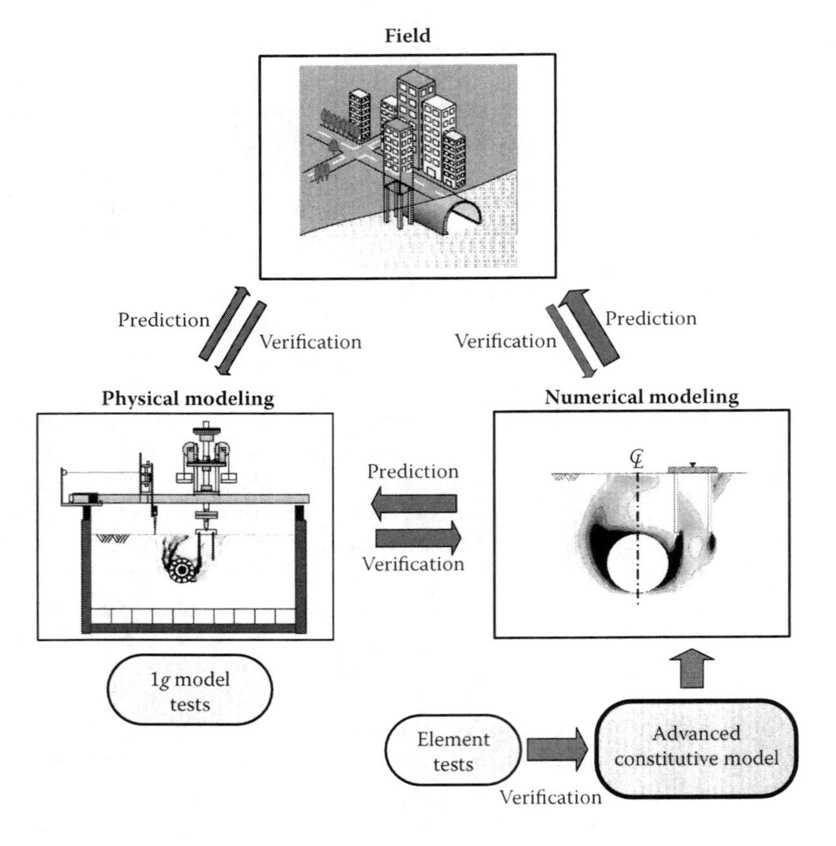

Figure 1.1 Procedure for the prediction of ground behavior in the field.

with sedimentation effects. Furthermore, it is difficult to simulate construction processes precisely (particularly in 3D conditions) and to know the overall information of the ground (distributions of deformations, stresses, strains, and others).

On the other hand, numerical modeling, by means of nonlinear finite element analysis, can provide the overall information of the ground, but the reliability of the results depends on the quality of the constitutive model used in the analysis. In this type of approach, the designer should consider the following aspects: What soil features can the model describe? What stress levels and stress conditions can the model handle? What are the material parameters of the constitutive model? Are they unified (not depending on density, confining pressure, and stress path)? Therefore it is necessary to check the validity of the constitutive model comprehensively by element tests under various stress paths and stress conditions.

Also, to obtain better reliability of the numerical modeling for the prediction of the real ground behavior, applicability of the numerical modeling should be confirmed by some physical modeling. The thick arrows in Figure 1.1 show the proposed procedure for the prediction of real ground behavior by the numerical modeling. If the simulations of the numerical model are confirmed by the experimental results obtained from physical model tests, then the numerical model can be applied with confidence to the prediction of real ground behavior. Here, the physical model tests were carried out under normal gravity conditions (1 *g* model tests) with the ground material represented by aluminum rod mass or alumina balls). To achieve representative model test validations, the constitutive model used in the numerical simulations should be able to describe various soil features in 3D condition comprehensively and to represent the behavior of various kinds of materials (including real soils or a mass of aluminum balls/rods) under a wide range of stress levels (from the stress condition in the 1 *g* model test to real ground stress level in the field).

Part I

Modeling of geomaterials

Chapter 2

Fundamentals of conventional elastoplasticity

2.1 INTRODUCTION

In this chapter, for easy understanding of elastoplasticity, the fundamentals of elastoplasticity—that is, yield stress (function), hardening function, and loading condition—are first explained in one-dimensional (1D) conditions because the formulation is given using one stress (σ) and one strain (ε). In addition to these fundamentals in 1D elastoplasticity, the other requirements in multidimensional plasticity (i.e., coaxiality and flow rule) are explained. Then, the formulation of conventional elastoplastic models is described.

2.2 ONE-DIMENSIONAL MODELING OF ELASTOPLASTIC MATERIALS

We start with typical 1D stress–strain behaviors of various solid materials as shown schematically in Figure 2.1. Figure 2.1(a) refers to an elastic material in which the following relation holds between stress (σ) and strain (ε), irrespective of loading or unloading:

$$\varepsilon = \frac{1}{E}\sigma \quad \text{or} \quad d\varepsilon = d\varepsilon^e = \frac{1}{E}d\sigma \qquad (E\text{: elastic modulus}) \tag{2.1}$$

Here, the material with constant E is called linear elastic, whereas a material in which E varies according to the magnitude of stress and/or strain is called nonlinear elastic. The stiffness E calculated as the slope of the line connecting the origin to a point on the stress–strain curve is called secant elastic modulus. When the slope is calculated at any specified point on the stress–strain curve, the tangent elastic modulus is obtained as the ratio of the local stress increment to the corresponding strain increment. Therefore, the secant modulus and tangential modulus are different for nonlinear elastic materials.

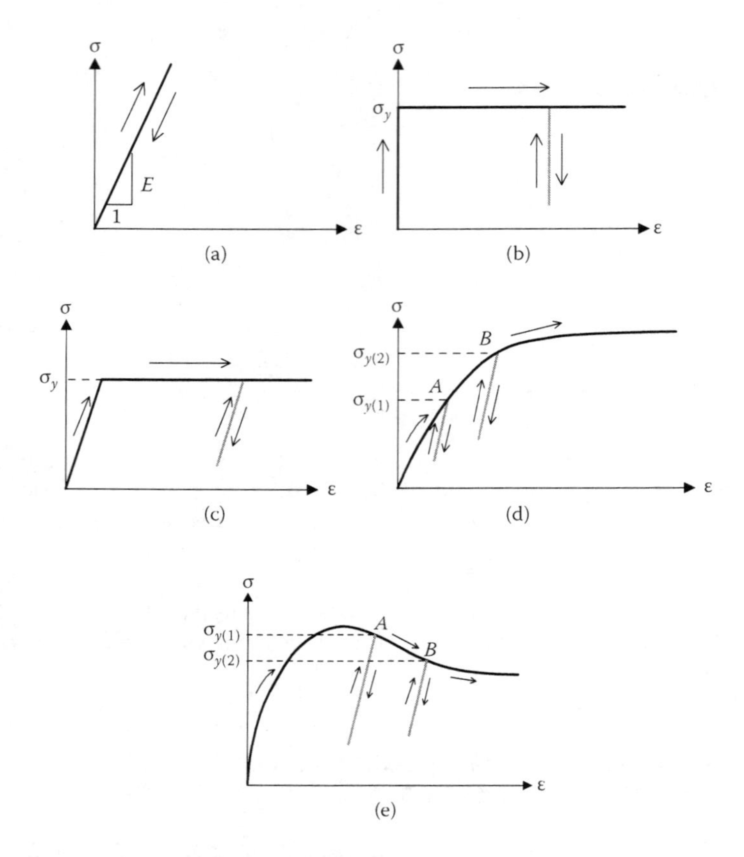

Figure 2.1 Typical one-dimensional stress–strain behaviors of solid materials.

Figure 2.1(b) shows the stress–strain behavior of rigid-plastic materials. No strain occurs until the stress reaches a yield stress value σ_y, where infinite strain occurs while keeping the stress constant at $\sigma = \sigma_y$. This strain is irrecoverable under unloading and reloading processes, as shown in the figure. Such an irrecoverable strain is called plastic strain, whereas recoverable strain as described in Figure 2.1(a) is called elastic strain. It can be seen that since there is not a unique relationship between stress and plastic strain, plastic deformation is expressed as a function of stress increment and strain increment as follows:

$$\begin{cases} d\varepsilon = d\varepsilon^p = \dfrac{d\sigma}{h^p} = \infty & \text{if } f = \sigma - \sigma_y = 0 \quad \& \quad df = 0 \\ d\varepsilon = 0 & \text{if } f = \sigma - \sigma_y < 0 \quad \text{or} \quad df < 0 \end{cases} \tag{2.2}$$

As shown in eq. (2.2), the criterion that determines whether plastic strain occurs or not refers to a "loading condition," with the function f being called a "yield function." In eq. (2.2), h^p which represents the stiffness of the plastic component, is called plastic modulus in analogy with the elastic modulus for the elastic component in eq. (2.1). The material in which infinite plastic strains develop without any change in stress (σ_y = const.), as shown in Figure 2.1(b), is called "perfectly plastic"; h^p inevitably becomes zero for perfectly plastic materials. Superscripts e and p denote elastic and plastic components, respectively, throughout this book.

Materials in general have a characteristic strength with an initial elastic behavior, so they can be described as being elastoplastic. The stress–strain behavior in such a case is described in Figure 2.1(c), where the total strain increment is expressed as the summation of elastic and plastic components given by eqs. (2.1) and (2.2), respectively. This material is called "elastic-perfectly plastic." It can be easily understood that the following relation between the total strain increment and stress increment holds:

$$\begin{cases} d\varepsilon = d\varepsilon^e + d\varepsilon^p = \dfrac{1}{E}d\sigma + \dfrac{d\sigma}{h^p} = \infty & \text{if } f = \sigma - \sigma_y = 0 \quad \& \quad df = 0 \\ d\varepsilon = d\varepsilon^e = \dfrac{1}{E}d\sigma & \text{if } f = \sigma - \sigma_y < 0 \quad \text{or} \quad df < 0 \end{cases} \tag{2.3}$$

As shown in these equations, the stress–strain behavior of 1D linearly elastic-perfectly plastic material is described by determining the elastic modulus E and the yield stress σ_y alone. However, geomaterials (even metals, in a strict sense) display elastoplastic behavior, as indicated in Figure 2.1(d). As shown in this figure, though, after loading to point A ($\sigma = \sigma_{y(1)}$), only elastic behavior is observed under unloading–reloading to the same point; the material shows elastoplastic behavior just as in Figure 2.1(c) when increasing the stress σ under further loading processes. Then, when unloading from point B, a new elastic limit is reached, signifying that the yield stress has increased to the value of stress at point B ($\sigma = \sigma_{y(2)}$).

Such a behavior is described as "strain-hardening elastoplastic," whereby further plastic strain occurs with an increase of stress. As such, it is necessary to express the yield stress as a function of plastic strain such as $\sigma_y = \sigma_y(\varepsilon^p)$. Although the relation between the strain increment and the stress increment is fundamentally the same as in eq. (2.3), the yield stress increases with the development of the plastic strain. Thus, the complete relationship, including loading condition, is expressed as

$$\begin{cases} d\varepsilon = d\varepsilon^e + d\varepsilon^p = \dfrac{1}{E}d\sigma + \dfrac{d\sigma}{h^p} & \text{if } f = \sigma - \sigma_y = 0 \;\; \& \;\; df \geq 0 \\ d\varepsilon = d\varepsilon^e = \dfrac{1}{E}d\sigma & \text{if } f = \sigma - \sigma_y < 0 \;\; \text{or} \;\; df < 0 \end{cases} \tag{2.4}$$

It can be seen from the preceding explanation that the yield stress σ_y varies with the occurrence of plastic strains, satisfying $\sigma = \sigma_y$ (or $f = 0$); this is the so-called "subsequent loading condition." As seen before, during plastic loading and upon unloading–reloading, the behavior is always elastic so that the current stress becomes (follows) the new yield stress. Based on this observation, it is concluded that, mathematically, $d\sigma = d\sigma_y$ (or $df = 0$), which leads to the so-called "consistency condition." Noting strain hardening as the increase in yield stress with plastic strain—that is, $\sigma_y(\varepsilon^p)$—the plastic modulus h^p in eq. (2.4) can be determined by applying the consistency condition:

$$d\sigma = d\sigma_y = \frac{\partial \sigma_y}{\partial \varepsilon^p} d\varepsilon^p = h^p d\varepsilon^p \tag{2.5}$$

So far, the occurrence of plastic strains has been described only for conditions of constant or increasing stress. However, as shown in Figure 2.1(e), plastic strain can also occur with decreasing stress. This process is called "strain softening." In this case, the loading condition cannot be given by eq. (2.4) because the stress change is negative, $d\sigma < 0$, under both loading and unloading from points A and B. Therefore, the loading condition cannot be expressed in terms of stress only. The loading condition of an elastoplastic model with both hardening and softening is given by the following equation, considering that the plastic strain increment should always be positive whenever it occurs:

$$\begin{cases} d\varepsilon^p \neq 0 & \text{if } f = \sigma - \sigma_y = 0 \;\; \& \;\; d\varepsilon^p > 0 \\ d\varepsilon^p = 0 & \text{if } f = \sigma - \sigma_y < 0 \;\; \text{or} \;\; d\varepsilon^p \leq 0 \end{cases} \tag{2.6}$$

In this formulation of elastoplastic models with hardening and/or softening, the yield function is represented by $f = \sigma - \sigma_y = 0$, where the yield stress is directly related to the plastic strain. Alternatively, the following expression of the yield function for an elastoplastic model with hardening and softening is more general and can be easily extended to three-dimensional (3D) stress and strain conditions as described later:

$$f = F(\sigma) - H(\varepsilon^p) = 0 \tag{2.7}$$

where F and H are functions of stress and plastic strain, respectively. In such formulations, the consistency condition ($df = 0$) gives

$$dF = \frac{\partial F}{\partial \sigma} d\sigma = \frac{\partial H}{\partial \varepsilon^p} d\varepsilon^p \tag{2.8}$$

Therefore, the incremental stress–strain relation is expressed as

$$\begin{cases} d\varepsilon = d\varepsilon^e + d\varepsilon^p = \dfrac{1}{E} d\sigma + \dfrac{dF}{\partial H / \partial \varepsilon^p} & \text{if } f = F - H = 0 \quad \& \quad d\varepsilon^p > 0 \\ d\varepsilon = d\varepsilon^e = \dfrac{1}{E} d\sigma & \text{if } f = F - H < 0 \quad \text{or} \quad d\varepsilon^p \leq 0 \end{cases} \tag{2.9}$$

2.3 MULTIDIMENSIONAL MODELING OF ELASTOPLASTIC MATERIALS

In this section, fundamentals of elastoplasticity in multidimensional stress conditions are explained as an extension to the previously presented 1D case. Under a 1D setting, it was shown that elastic strains differ from plastic strains in that the former are recoverable and can take either positive or negative values. However, whenever plastic strains occur, they are irrecoverable and always take positive values. Additionally, in multidimensional stress conditions, the following features on plastic deformations are introduced.

Let us consider two-dimensional (2D) materials—one elastic and the other one plastic (isotropic) and both subjected to the same principal stresses σ_1 and σ_2 ($\sigma_1 > \sigma_2$), as shown in Figure 2.2(a). Here, when an infinitesimal stress increment, $\Delta\sigma$ (principal stress increment), is applied in the direction of the σ_1-axis, the same deformation mode is observed for both elastic and plastic materials as shown in Figure 2.2(b)—that is, compressive in the direction of the σ_1-axis and expansive in the direction of the σ_2-axis. The magnitude of strains and the ratio of strain increment ($d\varepsilon_1/d\varepsilon_2$) are not discussed here.

Next, applying an infinitesimal stress increment, $\Delta\sigma$ (shear stress increment), in the tangential direction to the principal stress planes, the elastic and plastic deformation modes are as shown in Figure 2.2(c). Although the principal axes of strain increment coincide with those of stress increment in the elastic case, they are found to coincide only with the principal stress direction in the plastic case. The agreement between the principal strain increment and stress axes is called "coaxiality." It is also seen from the plastic deformations in Figures 2.2(b) and (c) that although the

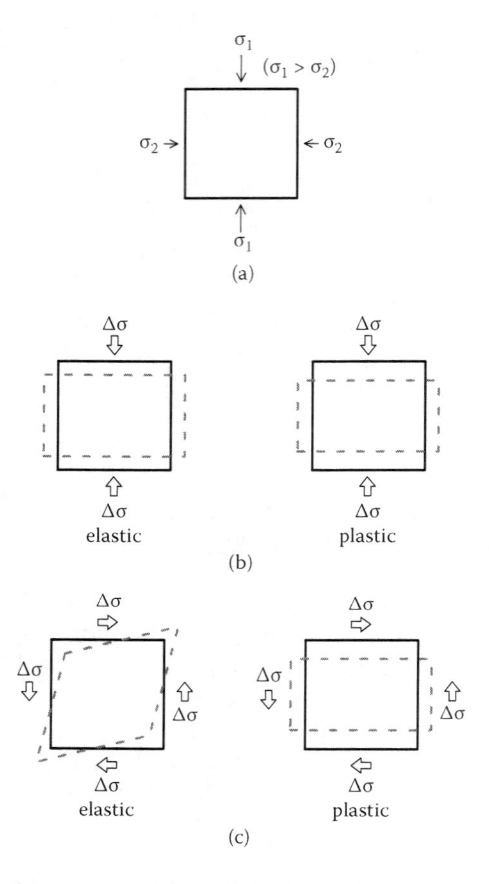

Figure 2.2 Modes of elastic and plastic deformations in multidimensional conditions.

current stresses are the same, the deformation modes, including the ratio of the plastic strain increments—for example, $(d\varepsilon_1^p/d\varepsilon_2^p)$—are independent of the applied stress increments and determined by the current stress condition alone.

This characteristic feature of incremental plastic deformations in plastic materials refers to the "flow rule" and can be interpreted as follows. Consider plastic deformations as a result of breakage of a material so that deformations are irrecoverable. Then, an old vase endowed with a pattern of cracks is likely to fail in the same mode, irrespective of the direction of loading applied to it. As such, this feature of plastic deformations is also valid in the multidimensional setting, so the elastoplastic behavior of materials in the multidimensional case can be interpreted in the same way as in the 1D case.

The formulation of 3D elastoplastic modeling is shown next. The yield function in 3D conditions takes the same form as in eq. (2.7) for the 1D condition, except that it is expressed in terms of 3D stresses and plastic strains:

$$\begin{cases} \text{in general form} \\ f = F(\sigma_{ij}) - H\left(\varepsilon_{ij}^{p}\right) = 0 \\ \text{in the case assuming an isotropic material} \\ \text{and expressed in terms of principal values} \\ f = F(\sigma_1, \sigma_2, \sigma_3) - H\left(\varepsilon_1^{p}, \varepsilon_2^{p}, \varepsilon_3^{p}\right) = 0 \end{cases} \tag{2.10}$$

where
σ_{ij} is the stress tensor
ε_{ij}^{p} is the plastic strain tensor
$(\sigma_1, \sigma_2, \sigma_3)$ are the principal stresses
$(\varepsilon_1^{p}, \varepsilon_2^{p}, \varepsilon_3^{p})$ are the principal plastic strains

The yield function is drawn as a surface in 3D stress space, when H is fixed, as shown in Figure 2.3. This surface is called "yield locus (yield surface)," which defines the limit of elastic region inside which no plastic strain is possible, in the same way as the yield stress σ_y in the 1D elastoplastic. As shown in Figure 2.3(a), when the stress condition moves from point A to point B, the yield locus expands with the development of plastic strains and satisfies eq. (2.10) in the same way as in the 1D condition (subsequent loading condition).

We next discuss how to determine the plastic strains in 3D conditions—that is, both the direction and magnitude of the ensuing plastic

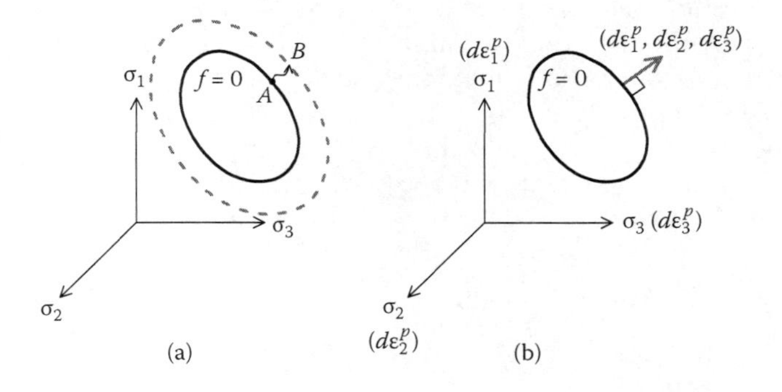

Figure 2.3 Yield surface and flow rule in principal stress space.

strain increment. From the principle of coaxiality in plasticity, the axes of the three principal plastic strain increments coincide with those of the three principal stresses, respectively, as shown in Figure 2.3(b). Then, it is assumed that the direction of plastic strain increments points outward to the yield locus, which has the same meaning as the plastic strain increment and is always positive in the 1D case.

Furthermore, this direction is not influenced by the stress increment (flow rule), but rather is normal to the yield locus. This is called the "normality rule" or an "associated flow rule." In fact, this condition can be derived from the principle of maximum plastic work by Hill (1948), in which it is assumed that, for a given plastic strain increment, the plastic work is maximized at the current state on the yield locus with respect to various possible stress increments. The principle of maximum plastic work not only leads to the "normality" condition but also ensures that the yield locus is convex (convexity condition). Detailed explanations of the principle of maximum plastic work can be found in most textbooks on plasticity (Hill 1950).

Since the direction of the plastic strain increment is normal and outward to the yield locus, the plastic strain increment is expressed as

$$\begin{cases} \text{in general form} \\ d\varepsilon_{ij}^p = \Lambda \dfrac{\partial F}{\partial \sigma_{ij}} \qquad\qquad (\Lambda : \text{positive scalar}) \\ \text{in the case assuming an isotropic material} \\ \text{and expressed in terms of principal values} \\ d\varepsilon_1^p = \Lambda \dfrac{\partial F}{\partial \sigma_1}, \quad d\varepsilon_2^p = \Lambda \dfrac{\partial F}{\partial \sigma_2}, \quad d\varepsilon_3^p = \Lambda \dfrac{\partial F}{\partial \sigma_3} \end{cases} \tag{2.11}$$

It is interesting to note that this formulation is analogous to seepage problems; that is, the direction of laminar flow is normal to an equipotential line in an isotropic porous material and expressed as the gradient of the associated scalar equipotential function.

The positive scalar introduced in eq. (2.11) represents the magnitude of the plastic strain increment, which can be determined from the consistency condition ($df = 0$) in the same way as in the 1D condition:

$$\begin{aligned} df &= dF - dH = dF - \frac{\partial H}{\partial \varepsilon_{ij}^p} d\varepsilon_{ij}^p \\ &= dF - \frac{\partial H}{\partial \varepsilon_{ij}^p} \frac{\partial F}{\partial \sigma_{ij}} \cdot \Lambda = 0 \end{aligned} \tag{2.12}$$

Therefore, the positive scalar (proportionality constant) Λ is given by the following equation:

$$
\left\{
\begin{array}{l}
\text{in general form} \\
\Lambda = \dfrac{dF}{\dfrac{\partial H}{\partial \varepsilon_{ij}^{p}} \dfrac{\partial F}{\partial \sigma_{ij}}} = \dfrac{dF}{h^{p}} \\
\text{in the case assuming an isotropic material} \\
\text{and expressed in terms of principal values} \\
\Lambda = \dfrac{dF}{\left(\dfrac{\partial H}{\partial \varepsilon_{1}^{p}} \dfrac{\partial F}{\partial \sigma_{1}} + \dfrac{\partial H}{\partial \varepsilon_{2}^{p}} \dfrac{\partial F}{\partial \sigma_{2}} + \dfrac{\partial H}{\partial \varepsilon_{3}^{p}} \dfrac{\partial F}{\partial \sigma_{3}} \right)} = \dfrac{dF}{h^{p}}
\end{array}
\right.
\tag{2.13}
$$

Here, since dF denotes the change in size of the yield surface (how much the yield surface expands, as in Figure 2.3), the dominator h^p in eq. (2.13) represents the plastic modulus in the same way as in the 1D model. Throughout this book, "summation convention"—indicial notation with repetition of indices (lowercase subscripts) meaning summation—is used, according to usual notation in tensor analysis. In conventional elastoplastic models, for a given yield stress state in the principal stress space, if the stress condition evolves so that the stress increment vector points inward to bring the stress state inside the yield surface, there is no incremental plastic strain induced and the yield surface does not change. Then, the loading condition is expressed as follows in the same way as that in the 1D model:

$$
\left\{
\begin{array}{lll}
d\varepsilon_{ij}^{p} \neq 0 & \text{if } f = F - H = 0 & \& \quad dF \geq 0 \\
d\varepsilon_{ij}^{p} = 0 & \text{if } f = F - H < 0 & \text{or} \quad dF < 0
\end{array}
\right.
\tag{2.14}
$$

As mentioned in 1D modeling, this condition cannot determine the occurrence of loading or unloading, as the stress evolves from its current state on the yield surface. Thus, according to eq. (2.9), in a 1D model, the following loading condition is defined in multidimensional conditions (Hashiguchi 2009):

$$
\left\{
\begin{array}{lll}
d\varepsilon_{ij}^{p} \neq 0 & \text{if } f = F - H = 0 & \& \quad \Lambda > 0 \\
d\varepsilon_{ij}^{p} = 0 & \text{if } f = F - H < 0 & \text{or} \quad \Lambda \leq 0
\end{array}
\right.
\tag{2.15}
$$

Turning to the elastic component of deformations, the elastic strain increment is more or less relatively small compared with the plastic strain increment. Therefore, it is assumed to be given by Hooke's law for isotropic elastic materials using Young's modulus E and Poisson's ratio ν or, alternatively, bulk elastic modulus K and shear elastic modulus G or others:

$$\begin{cases} d\varepsilon_{ij}^{e} = \dfrac{1+\upsilon}{E} d\sigma_{ij} - \dfrac{\upsilon}{E} d\sigma_{kk}\delta_{ij} = \dfrac{1}{2G} d\sigma_{ij} - \left(\dfrac{1}{6G} - \dfrac{1}{9K}\right) d\sigma_{kk}\delta_{ij} \\ \text{or} \\ d\sigma_{ij} = \left(K - \dfrac{2}{3}G\right) d\varepsilon_{kk}^{e}\delta_{ij} + 2G d\varepsilon_{ij}^{e} \end{cases} \tag{2.16}$$

Here, δ_{ij} is the unit tensor (or called the Kronecker delta) and implies that $\delta_{ij} = 1$ when $i = j$ and $\delta_{ij} = 0$ when $i \neq j$. Also, the bulk and shear elastic moduli are expressed using Young's modulus and Poisson's ratio as

$$K = \frac{E}{3(1-2\nu)}, \qquad G = \frac{E}{2(1+\nu)} \tag{2.17}$$

Using K and G, the following equations are also obtained:

$$\begin{cases} dp = \dfrac{d\sigma_{kk}}{3} = K d\varepsilon_{kk}^{e} = K d\varepsilon_{v}^{e} \\ ds_{ij} = 2G d\varepsilon_{ij}' \end{cases} \tag{2.18}$$

Here, $d\varepsilon_{v}^{e}$ is the elastic volumetric strain increment and is related with the mean stress increments dp alone. In the same way, $d\varepsilon_{ij}'$ is called the deviatoric strain increment tensor and is related with deviatoric stress increment ds_{ij}. These are defined as

$$\begin{cases} ds_{ij} = d\sigma_{ij} - \dfrac{d\sigma_{kk}}{3}\delta_{ij} = d\sigma_{ij} - dp\delta_{ij} \\ d\varepsilon_{ij}' = d\varepsilon_{ij} - \dfrac{d\varepsilon_{kk}}{3}\delta_{ij} = d\varepsilon_{ij} - \dfrac{d\varepsilon_{v}}{3}\delta_{ij} \end{cases} \tag{2.19}$$

Chapter 3

Modeling of one-dimensional soil behavior

3.1 INTRODUCTION

The well-known linear relations between void ratio (e) and stress in logarithmic scale ($\ln\sigma$) in one-dimensional (1D) consolidation under loading, unloading, and reloading are explained as a conventional elastoplastic behavior. However, it is experimentally known that even in the elastic region in the conventional model (e.g., under reloading and cyclic loading), the plastic strain develops. For instance, when vertical load increases on an overconsolidated clay in oedometer tests, the void ratio approaches gradually the normal consolidation line in the e–$\ln\sigma$ relation with development of the plastic strain. Also, the plastic strains develop during cyclic loadings with constant amplitude of stresses in shear tests. To model such behaviors, which cannot be described by conventional plasticity, some advanced elastoplastic models have been proposed: the subloading surface model (Hashiguchi and Ueno 1977; Hashiguchi 1980), the bounding surface model (Dafalias and Popov 1975), and the anisotropic hardening model (Mroz, Norris, and Zienkiewicz 1981), as well as others. Since these models have two or more yield surfaces, they are called multisurface models.

This chapter refers to the subloading surface concept by Hashiguchi and revises it to a 1D one; a simple 1D model to consider the influence of density and/or confining pressure is presented by employing a state variable and determining its evolution rule. Considering another state variable to represent bonding effect, this model is extended to one for describing the behavior of naturally deposited soils. Further, considering the experimental results that the normal consolidation line on the e–$\ln\sigma$ plane shifts depending on the strain rate, temperature, suction, and other factors, 1D models, which can take into account time, temperature, suction, and other effects, are developed (Nakai et al. 2011a).

3.2 MODELING OF NORMALLY CONSOLIDATED SOILS—CONVENTIONAL ELASTOPLASTIC MODELING

The well-known relation between stress and void ratio of soil under 1D conditions is formulated based on the conventional elastoplasticity. Figure 3.1 shows the typical relation between void ratio (e) and stress in logarithmic scale ($\ln\sigma$) for a normally consolidated clay, as described in many textbooks of soil mechanics. Here, point I ($\sigma = \sigma_0$, $e = e_0$) represents the initial state, and point P ($\sigma = \sigma$, $e = e$) represents the current state. Let e_{N0} and e_N denote the void ratios on the normal consolidation line (NCL) at the initial stress and at the current stress, respectively. The slopes λ and κ of the straight lines denote the compression and swelling indices, respectively. When the stress state evolves from σ_0 to σ, the total finite change in void ratio $(-\Delta e)$ of a normally consolidated clay is expressed as

$$(-\Delta e) = e_0 - e = e_{N0} - e_N = \lambda \ln \frac{\sigma}{\sigma_0} \tag{3.1}$$

where the elastic (recoverable) component $(-\Delta e)^e$ is computed using the swelling index κ as follows:

$$(-\Delta e)^e = \kappa \ln \frac{\sigma}{\sigma_0} \tag{3.2}$$

The plastic (irrecoverable) component $(-\Delta e)^p$ is then given by

$$(-\Delta e)^p = (-\Delta e) - (-\Delta e)^e = \lambda \ln \frac{\sigma}{\sigma_0} - \kappa \ln \frac{\sigma}{\sigma_0} \tag{3.3}$$

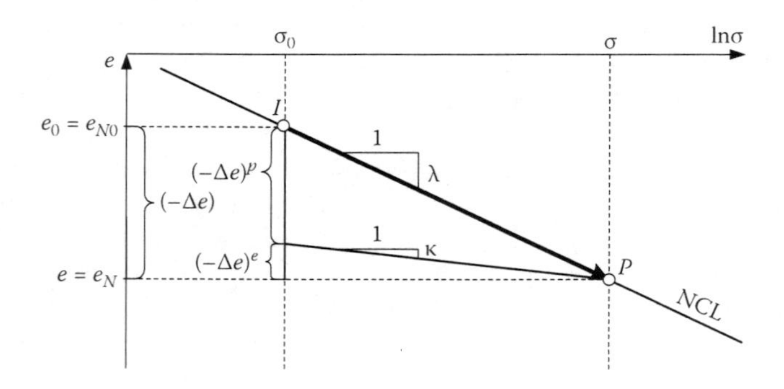

Figure 3.1 Change of void ratio in normally consolidated (NC) clay.

Now, let the F and H functions introduced in Chapter 2 be identified with the logarithmic change of stress and the change in plastic void ratio, respectively:

$$F = (\lambda - \kappa)\ln\frac{\sigma}{\sigma_0} \tag{3.4}$$

$$H = (-\Delta e)^p \tag{3.5}$$

According to the formulation of conventional 1D elastoplastic modeling described in Chapter 2 (see eq. 2.7), the term F is a scalar function of the stress, and the change in plastic void ratio $H = (-\Delta e)^p$ can be considered as the strain hardening parameter. Rewriting eq. (3.3) in terms of these variables, the yield function f for a normally consolidated soil in 1D condition can be expressed as

$$F = H \text{ or } f = F - H = 0 \tag{3.6}$$

Figure 3.2 shows the graphical representation of the relation between F and H given by eq. (3.6). This can be interpreted as a hardening rule. The change in plastic void ratio relates to the stress function F according to a straight line with unit slope, so the stress point is always on the yield surface to satisfy the consistency condition ($df = 0$):

$$df = dF - dH = (\lambda - \kappa)\frac{d\sigma}{\sigma} - d(-e)^p = 0 \tag{3.7}$$

Therefore, the infinitesimal increment of plastic void ratio is expressed as

$$d(-e)^p = (\lambda - \kappa)\cdot\frac{d\sigma}{\sigma} \tag{3.8}$$

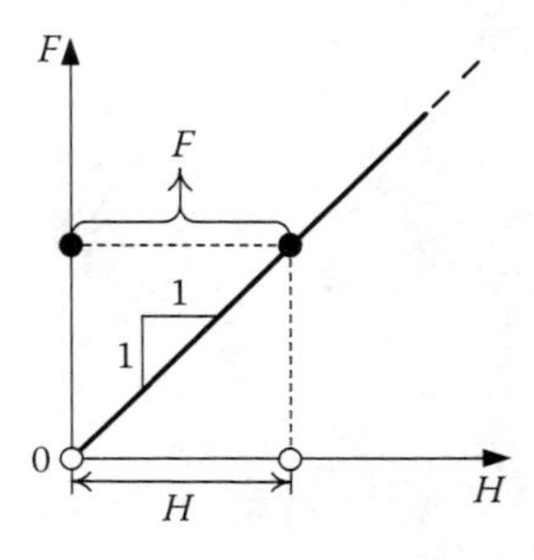

Figure 3.2 Evolution of F and H in NC soil.

This relation could also be obtained directly by differentiating eq. (3.3). On the other hand, differentiating eq. (3.2), the infinitesimal increment of elastic void ratio is expressed as follows:

$$d(-e)^e = \kappa \frac{d\sigma}{\sigma} \tag{3.9}$$

Referring to the loading condition for the conventional 1D elastoplastic model, the increment of total void ratio is given by the summation of eqs. (3.8) and (3.9) in the elastoplastic region and is expressed by eq. (3.9) alone in the elastic region:

$$\begin{cases} d(-e) = d(-e)^p + d(-e)^e = \{(\lambda - \kappa) + \kappa\} \dfrac{d\sigma}{\sigma} & \text{if } f = F - H = 0 \ \& \ dF \geq 0 \\ d(-e) = d(-e)^e = \kappa \dfrac{d\sigma}{\sigma} & \text{if } f = F - H < 0 \text{ or } dF < 0 \end{cases} \tag{3.10}$$

The approach described earlier is used in the following sections in order to introduce the influence of other relevant characteristics that affect the behavior of real soils.

3.3 MODELING OF OVERCONSOLIDATED SOILS—ADVANCED ELASTOPLASTIC MODELING AT STAGE I

The simple relation in eq. (3.10) describes well the behavior of remolded clays when they are normally consolidated under 1D or isotropic compression. A similar type of relation is assumed in the well known Cam clay model (Roscoe, Schofield, and Thurairajah 1963; Schofield and Wroth 1968; Roscoe and Burland 1968). Under those hypotheses, such a remolded clay in an overconsolidated state (i.e., for a current stress smaller than the yield stress) would behave according to a linear e–lnσ relation with constant slope κ, as described in eq. (3.10). In linear stress scale, this actually means a nonlinear elastic behavior under unloading and reloading. However, real clay shows elastoplastic behavior even in the overconsolidation region. Figure 3.3 shows schematically the e–lnσ relation for overconsolidated soils. Even in the overconsolidation region, elastoplastic deformation occurs, and the void ratio of the soil gradually approaches to the NCL with increasing stresses. In this section, an advanced elastoplastic modeling of the overconsolidated soils, called "advanced modeling at stage I," is presented.

Figure 3.4 shows the total or finite change of void ratio ($-\Delta e$) when the stress condition changes from the initial state I ($\sigma = \sigma_0$) to the current state

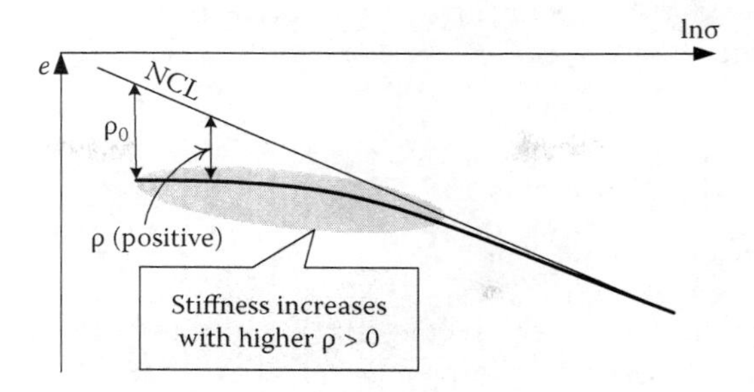

Figure 3.3 Void ratio (e)–lnσ relation in overconsolidated (OC) clay.

P ($\sigma = \sigma$). Here, e_0 and e are the initial and current void ratios of the overconsolidated soil, whereas e_{N0} and e_N denote the corresponding void ratios for virtual points I′ and P′ on the NCL. The difference between the void ratio (e_N) for the virtual point on the NCL and the actual void ratio (e), for the same stress level, is denoted as ρ. This variable represents a decrease in void ratio and therefore an increase in density for the overconsolidated remolded soil with respect to a normally consolidated condition. Hence, the value of ρ is always positive for unstructured overconsolidated soils; under this condition, the soil is denser and presents a higher stiffness as illustrated in Figure 3.3. The variable ρ also represents a measure of overconsolidation and can be easily related to the overconsolidation ratio (OCR) as

$$\rho = (\lambda - \kappa)\ln(\mathrm{OCR}) \tag{3.11}$$

When the actual stress level increases, the soil becomes less overconsolidated, and the value of ρ decreases and tends to zero as the e–lnσ relation approaches the normal consolidation line.

Referring back to Figure 3.4, when the stress condition changes from the initial state I to the current state P, the difference of void ratios between the normally consolidated and overconsolidated conditions for the same stress

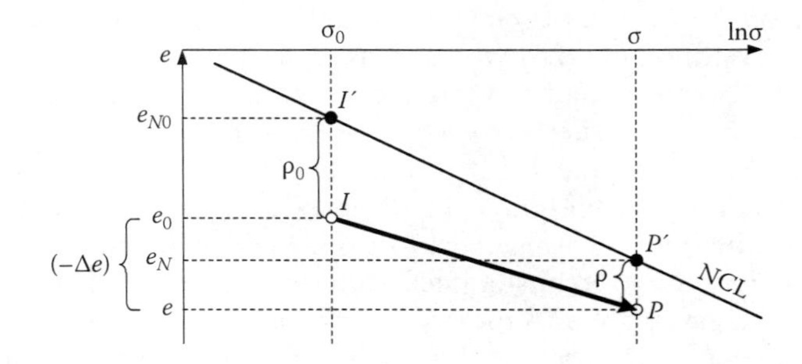

Figure 3.4 Change of void ratio in OC clay.

state is expressed as the change from ρ_0 $(= e_{N0} - e_0)$ to ρ $(= e_N - e)$. The total or finite change in void ratio $(-\Delta e)$ can be easily calculated as

$$
\begin{aligned}
(-\Delta e) &= e_0 - e \\
&= (e_{N0} - \rho_0) - (e_N - \rho) \\
&= (e_{N0} - e_N) - (\rho_0 - \rho)
\end{aligned}
\tag{3.12}
$$

Here, it is assumed that the recoverable change in void ratio $(-\Delta e)^e$ (elastic component) for overconsolidated soils is the same as that for normally consolidated soils, which is given by eq. (3.2). Then, referring to Figure 3.4 and eq. (3.12), the plastic change in void ratio $(-\Delta e)^p$ for overconsolidated soils is obtained as

$$
\begin{aligned}
(-\Delta e)^p &= (-\Delta e) - (-\Delta e)^e \\
&= \left\{(e_{N_0} - e_N) - (\rho_0 - \rho)\right\} - (-\Delta e)^e \\
&= \lambda \ln \frac{\sigma}{\sigma_0} - (\rho_0 - \rho) - \kappa \ln \frac{\sigma}{\sigma_0}
\end{aligned}
\tag{3.13}
$$

From this equation and using the definitions of F and H in eqs. (3.4) and (3.5), the yield function for overconsolidated soils is expressed as follows:

$$
F + \rho = H + \rho_0 \quad \text{or} \quad f = F - \{H + (\rho_0 - \rho)\} = 0 \tag{3.14}
$$

From the consistency condition ($df = 0$) during the occurrence of plastic deformation and satisfying eq. (3.14), the following equation is obtained:

$$
\begin{aligned}
df &= dF - \left\{dH - d\rho\right\} \\
&= (\lambda - \kappa)\frac{d\sigma}{\sigma} - \left\{d(-e)^p - d\rho\right\} = 0
\end{aligned}
\tag{3.15}
$$

Now there are two internal strain-like variables, H and ρ, that control the (isotropic) hardening of the model. This is schematically illustrated by the solid curve in Figure 3.5, which represents the relation between F and $(H + \rho_0)$ for an overconsolidated soil as expressed in eq. (3.14). Interestingly, this curve approaches the broken line ($F = H$) of normally consolidated soils with the development of plastic deformation. The solid curve is shifted by ρ_0 from the origin, which represents the accumulated plastic void ratio reduction, to bring the soil from a normally consolidated condition to the actual overconsolidated state. Here, the horizontal distance between the solid curve and the broken line indicates the current density variable ρ, which decreases with the development of plastic deformation and tends to zero as the soil approaches the normally consolidated state.

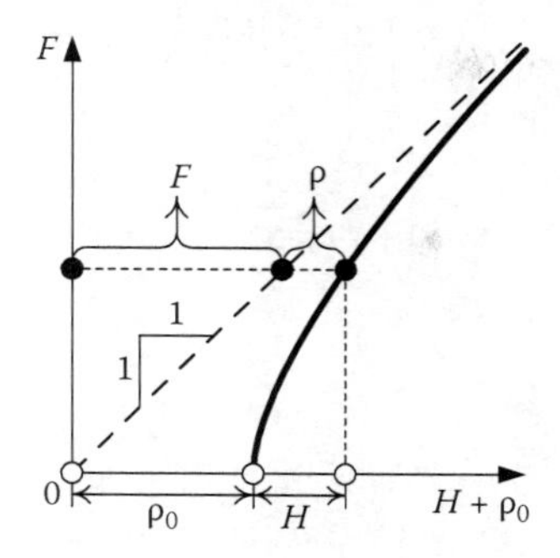

Figure 3.5 Evolution of F and H in OC soil.

The slope dF/dH of the solid line gives an idea of the stiffness with respect to plastic change in void ratio for overconsolidated soils. This can be compared with the stiffness of a normally consolidated soil given by the slope dF/dH of the broken line, which is always unity in this diagram.

Now it is necessary to formulate an evolution rule for the internal variable ρ, which represents the density. It can be assumed that the state variable ρ decreases ($d\rho < 0$) with the development of plastic deformation (volume contraction)—that is, $d\rho \propto d(-e)^p$—and finally becomes zero in the normally consolidated state. Furthermore, it can be considered that $d\rho$, the degree of degradation of ρ, decreases as the value of ρ becomes small. This can be expressed by means of function $G(\rho)$. In order to fulfill these conditions, the function $G(\rho)$ must decrease monotonically and satisfy $G(0) = 0$ in order to adhere to the NCL. Therefore, the evolution rule of ρ can be given in the following form:

$$d\rho = -G(\rho)\cdot d(-e)^p \tag{3.16}$$

Now, substituting eq. (3.16) into eq. (3.15), the infinitesimal increment of the plastic void ratio is given by

$$d(-e)^p = \frac{\lambda-\kappa}{1+G(\rho)}\cdot\frac{d\sigma}{\sigma} \tag{3.17}$$

The loading condition and the strain increment of normally consolidated soil are given by eq. (3.10) based on conventional elastoplasticity. However, according to conventional elastoplasticity, no plastic deformation occurs in the overconsolidation region because the current stress is less than the yield stress. Therefore, it is assumed that a plastic strain increment occurs under reloading as well, even if the current stress is smaller than the yield stress. That is, the current stress is considered to be the yield stress regardless of the loading condition (eq. 3.14 always holds). Then, the loading condition and the increment of total void ratio are expressed as follows from eqs. (3.9) and (3.17):

$$\begin{cases} d(-e) = d(-e)^p + d(-e)^e = \left\{ \dfrac{\lambda - \kappa}{1 + G(\rho)} + \kappa \right\} \dfrac{d\sigma}{\sigma} & \text{if } dF \geq 0 \\ d(-e) = d(-e)^e = \kappa \dfrac{d\sigma}{\sigma} & \text{if } dF < 0 \end{cases} \tag{3.18}$$

As seen from this equation, $G(\rho)$ has the effect of increasing the stiffness of the soil, and this effect becomes larger with the increase of the value of ρ. A value $G(\rho) = 0.2$, for instance, represents an increase of 20% in the plastic component of the stiffness, compared to that of a normally consolidated soil. After $G(\rho)$ becomes zero for vanishing values of ρ, then eq. (3.18) corresponds to eq. (3.10), which is the formulation for normally consolidated soil. The method of considering the influence of density presented here, in a sense, corresponds to an interpretation of the subloading surface concept by Hashiguchi (1980) in 1D conditions.

Figure 3.6 shows the calculated e–logσ relation of 1D compression for overconsolidated clays with different initial void ratios. Assuming Fujinomori clay, which has been often used in numerous experimental verifications of constitutive models (Nakai and Hinokio 2004; Nakai 2007), the following material parameters were employed in the numerical simulations: compression index $\lambda = 0.104$, swelling index $\kappa = 0.010$, and void ratio on the NCL at $\sigma = 98$ kPa (atmospheric pressure) $N = 0.83$. The function $G(\rho)$, which determines the evolution rule of ρ, is assumed to be linear:

$$G(\rho) = a\rho \tag{3.19}$$

where the parameter a is a proportionality constant.

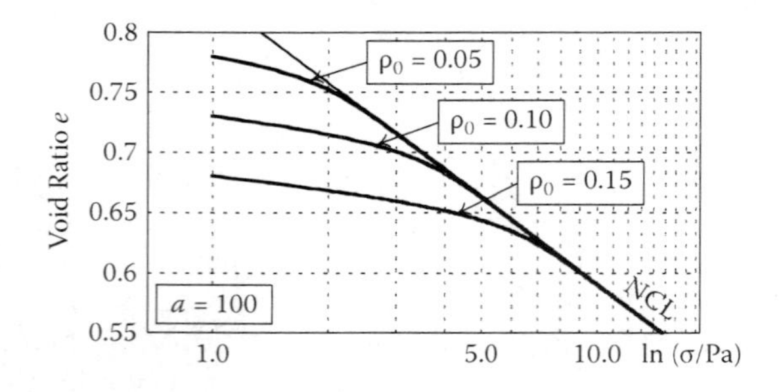

Figure 3.6 Calculated results of clay with different ρ_0, where Pa = atmospheric pressure (98 kPa).

Table 3.1 Values of material parameters for simulations of overconsolidated soils

λ	0.104
κ	0.010
N (e_N at σ = 98 kPa)	0.83
a	100

All the material parameters used for the simulations shown in Figure 3.6 are given in Table 3.1. It can be seen that the present model describes well the elastoplastic 1D behavior of overconsolidated clays with smooth transition from overconsolidated state to normally consolidated state.

Figures 3.7(a) and (b) show three types of function $G(\rho)$ used for the evolution rule of ρ and the corresponding numerical results obtained for 1D compression of the clay, respectively. Here, the initial values of the

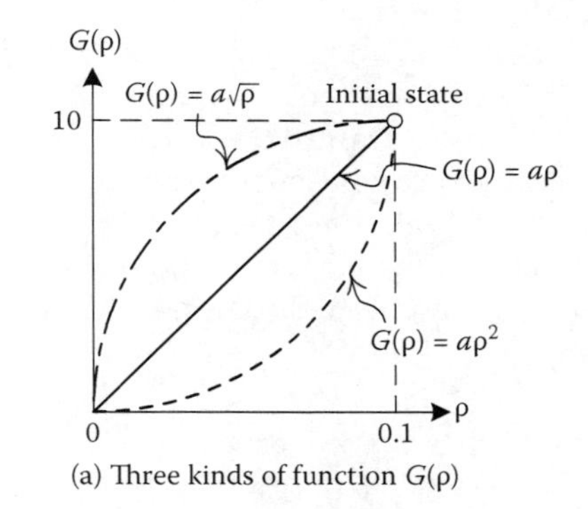

(a) Three kinds of function $G(\rho)$

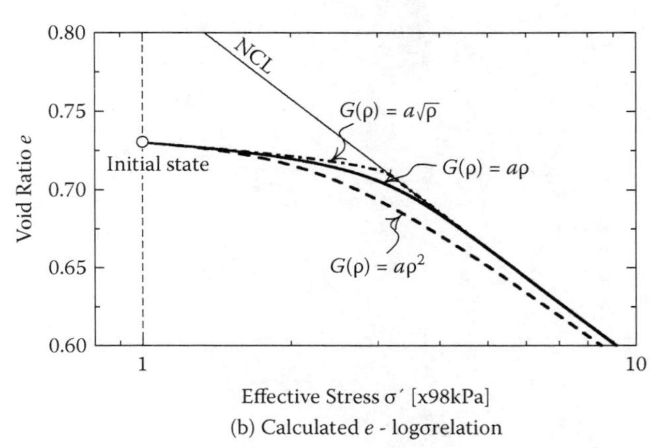

(b) Calculated e - logσrelation

Figure 3.7 Sensitivity of function $G(\rho)$.

state valuable ρ and the function $G(\rho)$ are the same; that is, $\rho_0 = 0.1$ and $G(0.1) = 10$. Indeed, Figure 3.7 shows the sensitivity of the function $G(\rho)$ on the calculated results in terms of how the compression curve approaches the NCL.

3.4 MODELING OF STRUCTURED SOILS (NATURALLY DEPOSITED SOILS)—ADVANCED ELASTOPLASTIC MODELING AT STAGE II

The model developed in the previous section can represent quite well the behavior of overconsolidated remolded clays. However, because natural clays develop a complex structure during the deposition process, they behave intricately compared with the remolded clays often used in laboratory tests. Such structured soils can exist naturally with a void ratio greater than that of a nonstructured normally consolidated soil under the same stress condition and yet the structured soil may have a higher stiffness due to natural cementation. Due to debonding effects, this type of structured soil shows a rather brittle and more compressive behavior than nonstructured soil after reaching a certain stress level.

This is illustrated by the solid curve in Figure 3.8, which shows schematically a typical e–lnσ relation for natural clays. Three regions may be identified in this figure: in region I, from point I to J, the stress level is low and the structured soil presents a denser ($\rho > 0$) and stiffer state than in the

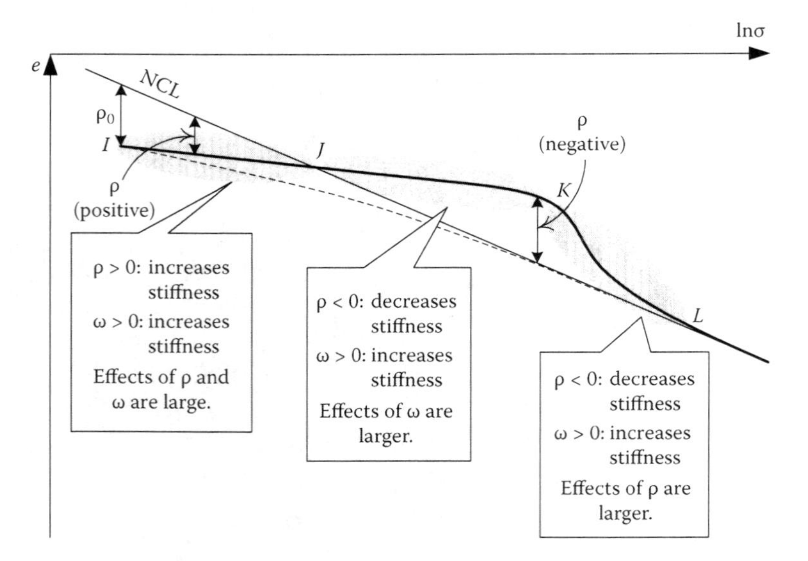

Figure 3.8 Void ratio (e)–lnσ relation in structured clay.

NCL; in region II, from point J to K, the structured soil is in a looser ($\rho < 0$) yet stiffer state than the NCL; in region III, from point K to L, the stress level is such that debonding effects prevail and the structure collapses fast with the corresponding changes in stiffness. Finally, the compression curve of the real soil approaches the NCL from above. Compare this with the thin broken curve of a nonstructured overconsolidated soil, which approaches the NCL from below in the same figure.

Asaoka, Nakano, and Noda (2000a) and Asaoka (2003) developed a model to describe such structured soils, introducing the concepts of subloading and superloading surfaces to the Cam clay model. In their modeling, a factor related to the overconsolidation ratio was introduced to increase the initial stiffness, and a factor related to the soil skeleton structure was introduced to decrease the stiffness as the stress state approached the normally consolidated condition. By controlling the evolution rules of these factors, they described various features of consolidation and shear behaviors of structured soils.

In this section, attention is focused not only on the real density, but also on bonding as the main factors that affect the behavior of a structured soil. This is because bonding effects can make the soil skeleton structure looser than that of a normally consolidated soil. From a behavioral point of view, the macroscopic effect of such bonding reflects in an overall increase of stiffness in very much the same way as an increase in overconsolidation ratio and/or in density does. Therefore, the global bonding effect can be mathematically simulated by means of an "imaginary" increase of density, here denoted by the Greek letter ω, despite the fact that the real structured soil may exist in looser states than a nonstructured soil.

Figure 3.9 gives a magnified view of part of region I of Figure 3.8, showing the total change (finite increment) of void ratio when the stress condition moves from the initial state I ($\sigma = \sigma_0$) to the current state P ($\sigma = \sigma$) in the same way as that in Figures 3.1 and 3.4. Here, e_0 and e are the initial and current void ratios of the structured soil, and e_{N0} and e_N are the

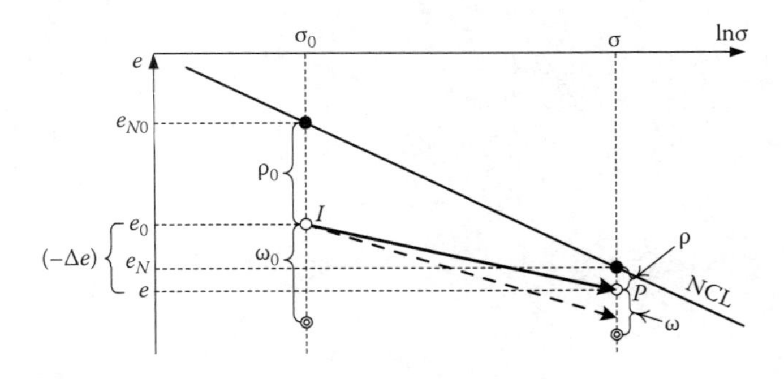

Figure 3.9 Change of void ratio in structured clay.

corresponding void ratios on the normal consolidation line as described before. The lower arrow in the broken line represents the same change of void ratio as that in Figure 3.4 for the overconsolidated unstructured soil.

Now it can be understood that the structured soil is stiffer than a nonstructured overconsolidated soil, even if the initial state variable ρ_0 is the same. Then, the change in void ratio for the structured soil indicated by the arrow in the solid line is smaller than that for a nonstructured overconsolidated soil (arrow in the broken line). Such an increase in stiffness is expressed by introducing an imaginary density ω, which represents the effect of the bonding, in addition to the real density ρ. Here, ω_0 is the initial value of ω. Concrete methods to determine the values of material parameters, including the initial value ω_0, are described later in Section 3.9.

Despite the fact that the structured soil shows a stiffer behavior up to certain stress levels, the total change in void ratio is computed in exactly the same way as that developed for the unstructured soil and formulated in eqs. (3.12) to (3.15). The main difference resides in the evolution law of the real density ρ, which now can also assume negative values, as illustrated in Figure 3.10, depending on the magnitude of the bonding effects represented by the imaginary density ω. The solid line in Figure 3.10 shows the relation between F and $(H + \rho_0)$ for a structured soil. When the degradation of the bonding effect ω is faster with the development of plastic deformation, the solid line monotonically approaches the broken line ($F = H$) of the normally consolidated (NC) soil in the same way as that for overconsolidated soil in Figure 3.5 (see Figure 3.10a).

On the other hand, when the degradation of ω is subdued, the solid line reaches the broken line before complete debonding ($\omega = 0$), so the solid line enters the region of $\rho < 0$. If it is assumed that a negative ρ has an

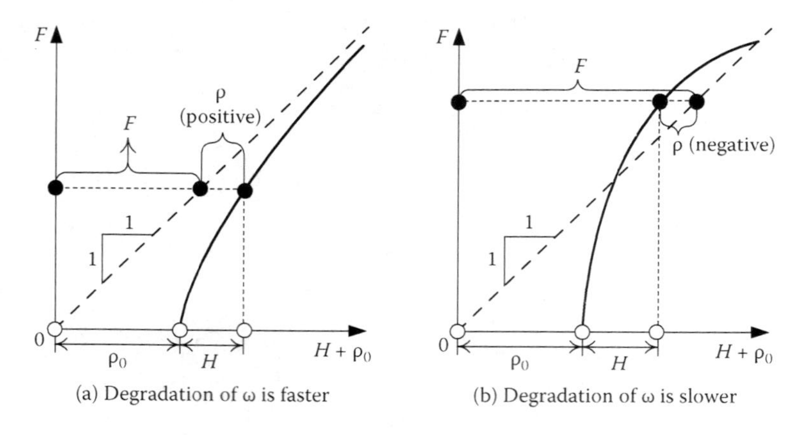

Figure 3.10 Evolution of F and H in structured soil.

effect to decrease the stiffness contrary to a positive ρ, the solid line finally approaches the broken line from the region of $\rho < 0$.

Now, it is necessary to account for the effect of bonding on the evolution rule of the density variable ρ. This should still be dependent on the development of plastic deformation for the structured soil so that $d\rho \propto d(-e)^p$. Furthermore, it is considered that the degree of degradation of ρ is determined not only by the state variable ρ related to the real density, but also by the state variable ω related to the imaginary increase of density due to bonding. This is introduced by an extra function, $Q(\omega)$, that is simply added to the already defined function $G(\rho)$. So, the evolution rule of ρ can be given in the following form:

$$d\rho = -\{G(\rho) + Q(\omega)\} \cdot d(-e)^p \tag{3.20}$$

An additional evolution rule must also be introduced for the imaginary density. Here, this evolution rule of ω is also given using the same function $Q(\omega)$ as follows:

$$d\omega = -Q(\omega) \cdot d(-e)^p \tag{3.21}$$

It is also possible to define the evolution rule of ω by another function that takes into account destructuring effects such as particle crushing due to increasing stress magnitude, decay of bonding due to chemical action and/or weathering, and so on. Both functions $G(\rho)$ and $Q(\omega)$ must be monotonically increasing (or decreasing) and satisfy the conditions $G(0) = 0$ and $Q(0) = 0$, so that the e–lnσ curve approaches the NCL when the soil becomes totally destructured ($\omega = 0$) and normally consolidated ($\rho = 0$).

The domain of function $G(\rho)$ can now assume positive or negative values of ρ, while ω is always positive. Therefore, $G(\rho)$ might be positive or negative, while $Q(\omega)$ is always positive. The simpler relations $G(\rho)$ and $Q(\omega)$ that satisfy these restrictions are given by linearly increasing functions, as illustrated in Figure 3.11. Here, the slopes a and b of these lines are material

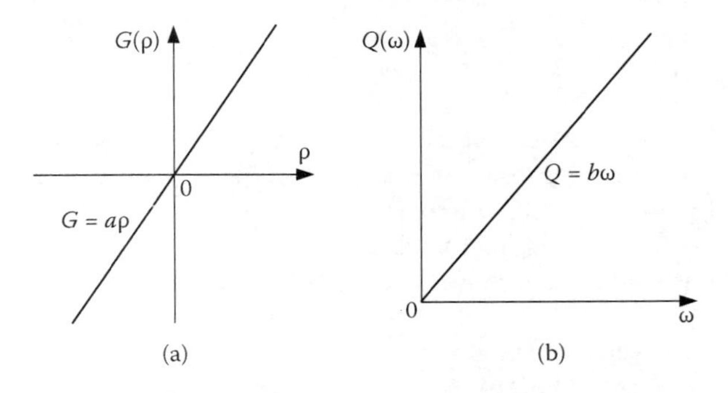

Figure 3.11 $G(\rho)$ and $Q(\omega)$ given by linear functions.

parameters that control the degradation rate of the state variables ρ and ω, respectively.

Equations (3.15) and (3.20) give the increment of the plastic void ratio as

$$d(-e)^p = \frac{\lambda - \kappa}{1 + G(\rho) + Q(\omega)} \cdot \frac{d\sigma}{\sigma} \tag{3.22}$$

The increment of total void ratio is expressed as follows from eqs. (3.9) and (3.22):

$$d(-e) = d(-e)^p + d(-e)^e = \left\{ \frac{\lambda - \kappa}{1 + G(\rho) + Q(\omega)} + \kappa \right\} \frac{d\sigma}{\sigma} \tag{3.23}$$

As shown in eq. (3.22), positive values of ρ and ω have the effects of increasing the stiffness of soils because $G(\rho)$ and $Q(\omega)$ are positive in the case when ρ and ω are positive. Now, although the same equation as eq. (3.32) is derived in the previous paper by Nakai, Kyokawa et al. (2009), the present derivation process of the model is more logical.

A brief discussion is presented now to describe how the formulation of eq. (3.22) can explain mathematically the conceptual behavior of structured soils under 1D consolidation, as depicted in Figure 3.8. Assume an initial state with positive ρ_0 and positive ω_0. At the first stage ($\rho > 0$ and $\omega > 0$), the stiffness of the soil is much larger than that of normally consolidated (NC) soil because of the positive values of $G(\rho)$ and $Q(\omega)$. When the current void ratio becomes the same as that on the NCL ($\rho = 0$), the stiffness of the soil is still larger than that of the NC soil because of $\omega > 0$. Thus it is possible for the structured soil to exist in a looser but stiffer state than that on the NCL. In this stage ($\rho < 0$ and $\omega > 0$), the effect of increasing stiffness due to the positive value of ω is larger than the effect of decreasing stiffness due to the negative value of ρ. After this stage, the effect of ω becomes smaller with the development of plastic deformation. On the other hand, the effect of ρ to decrease the stiffness becomes prominent because of the negative value of ρ. Finally, the void ratio approaches that on the NCL because both ρ and ω converge to zero.

In order to check the validity of the present model, numerical simulations of 1D compression tests are carried out. The expressions of $G(\rho)$ and $Q(\omega)$ are given by the simple linear functions of ρ and ω in Figure 3.11. The following material parameters are employed in the numerical simulations: compression index $\lambda = 0.104$, swelling index $\kappa = 0.010$, and void ratio on the NCL at $\sigma = 98$ kPa (atmospheric pressure) $N = 0.83$, which are the same as those in Table 3.1.

Figure 3.12(a) shows the calculated e–logσ relation of the 1D compression of remolded overconsolidated clays without bonding ($\omega_0 = 0$) for which the initial void ratios are the same ($e_0 = 0.73$), but the parameter a assumes

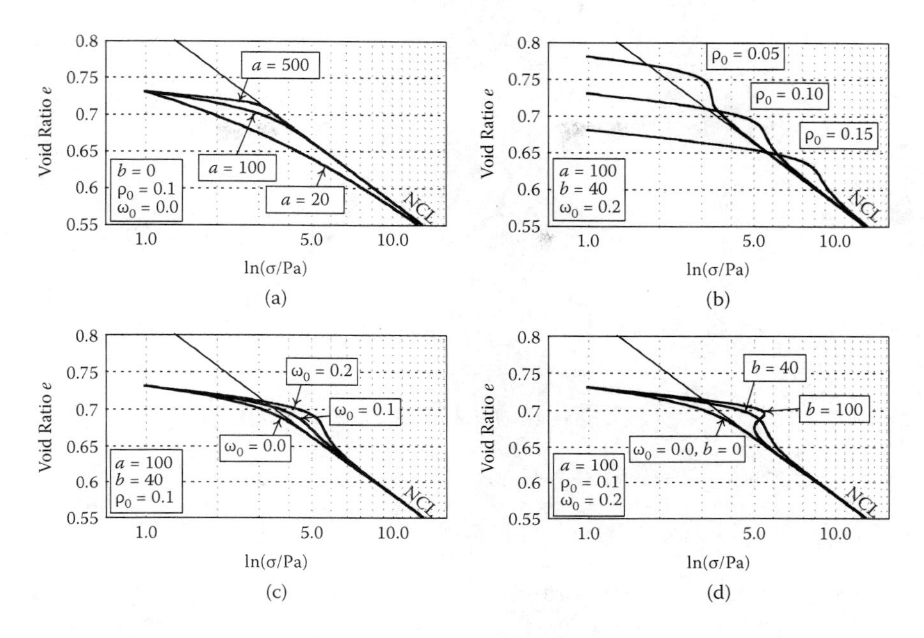

Figure 3.12 Calculated results of clays with different ρ_0, ω_0, *a*, and *b*, where Pa = atmospheric pressure (98 kPa).

different values ($a = 20$, 100, and 500) to check the sensitivity of the stress–strain behavior with this property. Figure 3.12(b) shows the calculated e–$\log\sigma$ relation of the 1D compression of structured clays using the same bonding effect ($\omega_0 = 0.2$) and the different initial void ratios $e_0 = 0.78$, 0.73, and 0.68 ($\rho_0 = 0.05$, 0.10, and 0.15). Figure 3.12(c) shows the calculated results using the same initial void ratio $e_0 = 0.73$ ($\rho_0 = 0.1$) and the different initial values of bonding parameters ($\omega_0 = 0.0$, 0.1, and 0.2). Since the state variable related to density ρ ($= e_N - e$) (represented by the vertical distance between the current void ratio and the void ratio at the NCL) of the clay without bonding ($\omega_0 = 0$) decreases monotonically, its void ratio converges to the NCL from below. The void ratio of the clay with bonding ($\omega_0 > 0$) decreases less than that without bonding ($\omega_0 = 0$) and enters the upper region of the NCL ($\rho < 0$). After that, it converges to the NCL from the negative side of ρ with a sharp reduction of bulk stiffness.

In these figures, the parameters (a and b) that represent the degradation rate of ρ and ω are fixed. It can be seen that it is possible to describe the deformation of structured clays only by considering the effects of density and bonding and their evolution rules. Furthermore, Figure 3.12(d) shows the results in which the initial void ratio and the initial bonding are the same, but the parameter b assumes different values ($b = 0$, 40, and 100). It can be seen that the results with a larger value of b ($= 100$) describe void

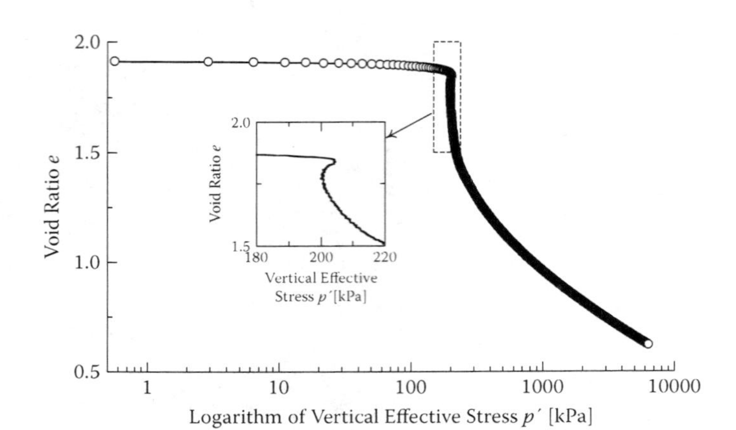

Figure 3.13 Observed result of Louisville clay with strain softening in constant strain rate one-dimensional consolidation test. (Plotted from data in Watabe, Y. et al. 2009. *Proceedings of 64th Annual Meeting on JSCE,* III-011, 21–22.)

ratio–stress relation with strain softening. Figure 3.13 shows the results of a constant strain rate consolidation test on Louisville clay in Canada, which was carried out by Watabe et al. (2009). It can be seen from this figure that this natural clay shows strain softening behavior even in 1D consolidation.

3.5 MODELING OF OTHER FEATURES OF SOILS—ADVANCED ELASTOPLASTIC MODELING AT STAGE III

The model developed so far, up to stage II, can account for the important influences of density and bonding on the behavior of real soils. These effects are dependent on the development of plastic strains as described by conveniently introduced evolution laws. However, there are many other relevant features, such as time-, temperature-, and suction-related phenomena that affect the behavior of soils and that are not dependent on plastic deformation. This section presents a general framework in which these effects or others can be incorporated into an advanced model in a simple and unified manner.

For instance, Figure 3.14 shows schematically the time-dependent behavior of a normally consolidated clay in 1D condition. It is well known that the NCL shifts due to strain rate (rate of void ratio change) and the void ratio (e) change linearly against time in logarithmic scale ($\ln t$) under constant effective vertical stress (creep condition). In addition, experimental

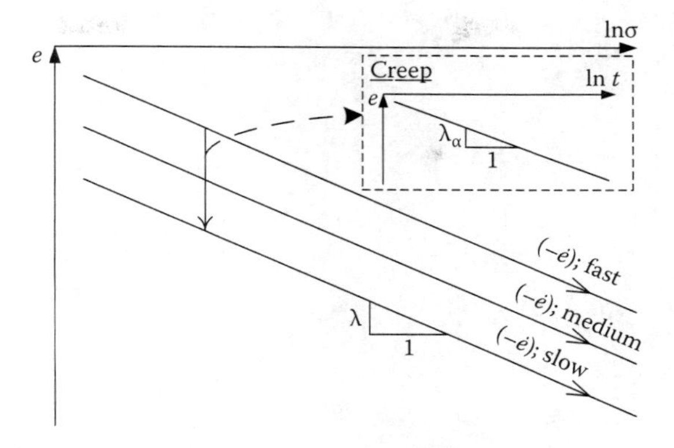

Figure 3.14 Time-dependent behavior of normally consolidated clay.

results show that the NCL (and/or the critical state line in multidimensional state) also change depending on temperature and suction (saturation), among others.

In order to model these features, a state variable, here generically denoted by ψ, is introduced. This variable is independent of the plastic strains and controlled by a function of the strain rate (rate of void ratio change), temperature, suction (degree of saturation), or any other characteristic feature, which shifts the position of NCL as shown in Figure 3.15. Here, ψ_0 is the initial value of ψ, and points I and P indicate the initial ($\sigma = \sigma_0$ and $e = e_0$) and the current ($\sigma = \sigma$ and $e = e$) states, respectively, in the same way as in

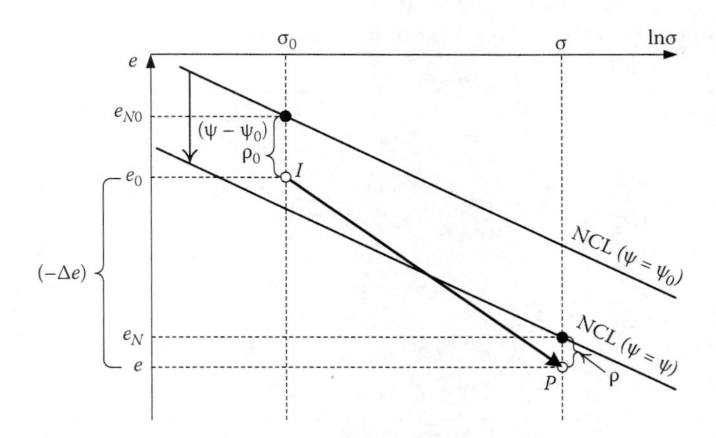

Figure 3.15 Change of void ratio in OC clay and structured clay with some soil features such as time-dependent behavior.

Figures 3.4 and 3.9. By referring to this figure, the plastic change of void ratio $(-\Delta e)^p$ for soil in which the preceding features should be considered is expressed as

$$\begin{aligned}(-\Delta e)^p &= (-\Delta e) - (-\Delta e)^e \\ &= \{(e_{N0} - e_N) - (\rho_0 - \rho)\} - (-\Delta e)^e \\ &= \left\{\lambda \ln\frac{\sigma}{\sigma_0} + (\psi - \psi_0) - (\rho_0 - \rho)\right\} - \kappa \ln\frac{\sigma}{\sigma_0} \\ &= (\lambda - \kappa)\ln\frac{\sigma}{\sigma_0} - (\rho_0 - \rho) - (\psi_0 - \psi)\end{aligned} \tag{3.24}$$

Therefore, the following equation holds between F and H:

$$F + \rho + \psi = H + \rho_0 + \psi_0$$

or

$$f = F - \{H + (\rho_0 - \rho) + (\psi_0 - \psi)\} = 0 \tag{3.25}$$

Figure 3.16 shows eq. (3.25) graphically as the relation between F and $(H + \rho_0 + \psi_0)$, in which the state variable ψ is assumed to be independent of plastic deformation. Although ρ finally becomes zero with increasing plastic strains, ψ does not converge to zero but to some value, depending on the current strain rate, temperature, suction, and/or any other features. Therefore, the solid curve does not necessarily approach the broken

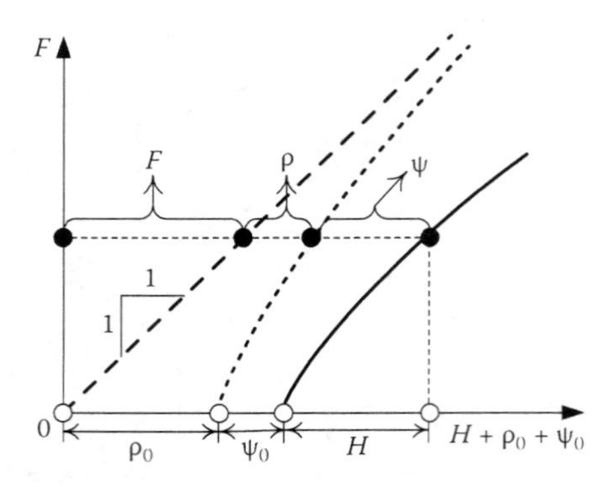

Figure 3.16 Explanation of F and H in clay with some features such as time-dependent behavior.

straight line ($F = H$), but becomes parallel to it. The following equation is obtained from eq. (3.25) and consistency condition ($df = 0$):

$$\begin{aligned} df &= dF - \{dH - d\rho - d\psi\} \\ &= (\lambda - \kappa)\frac{d\sigma}{\sigma} - \{d(-e)^p - d\rho - d\psi\} = 0 \end{aligned} \tag{3.26}$$

Substituting $d\rho$ with plastic deformation in eq. (3.20) into eq. (3.26), the increment of plastic void ratio can be obtained as

$$d(-e)^p = \frac{(\lambda - \kappa)\frac{d\sigma}{\sigma} + d\psi}{1 + G(\rho) + Q(\omega)} \tag{3.27}$$

The increment of total void ratio is expressed as follows from eqs. (3.9) and (3.27):

$$\begin{aligned} d(-e) &= d(-e)^p + d(-e)^e \\ &= \left\{\frac{\lambda - \kappa}{1 + G(\rho) + Q(\omega)} + \kappa\right\}\frac{d\sigma}{\sigma} + \frac{d\psi}{1 + G(\rho) + Q(\omega)} \end{aligned} \tag{3.28}$$

Here, the state variable ψ is a function of the rate of plastic void ratio change $(-\dot{e})^p$ (or time t), temperature T, suction s (or degree of saturation S_r), or others, so its increment $d\psi$ is given in such forms as $d\psi = (\partial\psi/\partial t)dt$, $(\partial\psi/\partial s)ds$, $(\partial\psi/\partial S_r)dS_r$, or others. Furthermore, if the model is formulated considering multiple features, different state variables ω_a, ω_b,... should be considered instead of a single ω. Additionally, both ψ and ψ_0 should be defined as the summation of corresponding factors in such a way that $\psi = \psi_a + \psi_b + \ldots$ and $\psi_0 = \psi_{a0} + \psi_{b0} + \ldots$. Then, the evolution rules of ρ and the increment of ψ ($d\rho$ and $d\psi$ in eq. 3.26) as a superposition of the various effects are given as follows:

$$d\rho = -\{G(\rho) + Q_a(\omega_a) + Q_b(\omega_b) + \cdots\} \cdot d(-e)^p \tag{3.29}$$

$$d\psi = d\psi_a + d\psi_b + \cdots \tag{3.30}$$

Therefore, for considering multiple features, the increment of total void ratio is given by

$$\begin{aligned} d(-e) = &\left\{\frac{\lambda - \kappa}{1 + G(\rho) + Q_a(\omega_a) + Q_b(\omega_b) + \cdots} + \kappa\right\}\frac{d\sigma}{\sigma} \\ &+ \frac{d\psi_a + d\psi_b + \cdots}{1 + G(\rho) + Q_a(\omega_a) + Q_b(\omega_b) + \cdots} \end{aligned} \tag{3.31}$$

The physical meaning of the state variables (ρ, ω, and ψ) in the proposed model is summarized as follows:

ρ: This is a state variable representing the density, which is defined as the difference between the current void ratio and the void ratio on the current NCL at the same stress level.
ω: This is a state variable considering material degradation effects, such as bonding in structured soil, with the development of plastic deformations. These effects are represented by the introduction of an imaginary increase of density.
ψ: This is a state variable describing the soil features such as time-dependent, temperature-dependent, and unsaturated soil behaviors, among others. This variable is not related to the development of plastic deformation. Instead, these features are considered by shifting the NCL as a function of the value of this state variable.

In order to obtain the complete stress–strain relation using the present proposed model, it is necessary to determine the state variable ρ at every calculation step. Since the increment of total void ratio $d(-e)$ is obtained from eq. (3.28), the void ratio $e_{(i)}$ at the current (i)th step can be updated from the void ratio $e_{(i-1)}$ in the previous $(i-1)$th step and the increment $d(-e)$:

$$e_{(i)} = e_{(i-1)} - d(-e) \tag{3.32}$$

Then, the value of the state variable ρ at the current state can be calculated as the difference between the current void ratio $e_{(i)}$ and the void ratio on the NCL at $\psi = \psi$ at the current stress ($\sigma = \sigma$).

As described in Section 3.3, the current stress is considered to be the yield stress regardless of the loading condition in the 1D advanced elastoplastic modeling. It is also shown in Section 2.2 in Chapter 2 that the plastic strain increment should always be positive in 1D conditions. Therefore, the loading conditions of the present advanced models through stages I to III are presented as follows by assuming no occurrence of the plastic volume expansion (negative strain):

$$\begin{cases} d(-e)^p \neq 0 : & \text{if } d(-e)^p > 0 \\ d(-e)^p = 0 : & \text{otherwise} \end{cases} \tag{3.33}$$

3.6 LOADING CONDITION USING INCREMENT OF TOTAL CHANGE OF VOID RATIO AND EXPLICIT EXPRESSION IN ADVANCED ELASTOPLASTIC MODELING

So far, the constitutive models have been expressed in an implicit form in which the increment of void ratio $d(-e)$ (or strain increment $d\varepsilon$) is given by a linear function of the stress increment $d\sigma$. However, it is necessary to describe the constitutive model in an explicit form, such as $d\sigma = D \cdot d\varepsilon$, to be used in finite element analyses. In this section, for applying the proposed constitutive models to finite element analyses, explicit expressions of advanced constitutive modeling in stages I to III with loading conditions as determined following the increment of void ratio (or strain increment) are shown.

In the elastoplastic region, the stress increment $d\sigma$ is expressed as follows using eq. (3.9):

$$d\sigma = \frac{\sigma}{\kappa} d(-e)^e = \frac{\sigma}{\kappa}\left\{d(-e) - d(-e)^p\right\} \tag{3.34}$$

Here, the increment of the plastic void ratio is given by eq. (3.27). Then, substituting eq. (3.27) into eq. (3.34), the following expression is obtained:

$$d(-e)^p = \frac{(\lambda-\kappa)d(-e) + \kappa d\psi}{\lambda + \kappa G(\rho) + \kappa Q(\omega)} \tag{3.35}$$

Usually, the denominator of this equation is positive, so the loading condition is expressed using the increment of void ratio change as

$$\begin{cases} d(-e)^p \neq 0 \ : & \text{if } (\lambda-\kappa)d(-e) + \kappa d\psi > 0 \\ d(-e)^p = 0 \ : & \text{otherwise} \end{cases} \tag{3.36}$$

Substituting eq. (3.35) into eq. (3.34), the following incremental explicit stress–stress relation in the elastoplastic region is obtained:

$$\begin{aligned} d\sigma &= \frac{\left(1 + G(\rho) + Q(\omega)\right)\sigma}{\lambda + \kappa G(\rho) + \kappa Q(\omega)} d(-e) - \frac{\sigma}{\lambda + \kappa G(\rho) + \kappa Q(\omega)} d\psi \\ &= \frac{(1+e_0)\left(1 + G(\rho) + Q(\omega)\right)\sigma}{\lambda + \kappa G(\rho) + \kappa Q(\omega)} d\varepsilon - \frac{\sigma}{\lambda + \kappa G(\rho) + \kappa Q(\omega)} d\psi \end{aligned} \tag{3.37}$$

This equation becomes the stress–strain relation for stage II in the case of $d\psi = 0$ and finally reduces to the expression for stage I in the case of $d\psi = 0$ and $Q(\omega) = 0$.

3.7 APPLICATION OF THE ADVANCED MODEL (STAGE III) TO TIME-DEPENDENT BEHAVIOR OF SOILS

In the previous sections, formulations of the advanced elastoplastic models from stage I to stage III were shown. A method for describing time-dependent behavior of soil using the idea of stage III is described in this section.

Several time-dependent constitutive models for soils are found in the literature. Sekiguchi (1977) proposed a viscoplastic model with a nonstationary flow surface. In this model, the nonstationary flow surface is obtained from the ordinary differential equation in which a unique relation between stress, plastic volumetric strain, and plastic strain rate holds. Then, the viscoplastic strain rate is calculated assuming a flow rule on the flow surface. Nova (1982) also developed a viscoplastic model by extending an inviscid model using nonstationary flow surface theory.

Another type of viscoplastic model is based on the overstress viscoplastic theory by Perzyna (1963) (see Adachi and Oka 1982; Dafalias 1982; Katona 1984), in which the strain rate effects can be described by assuming Bingham-like body and utilizing the difference in sizes of the current static yield surface related to the current plastic strains and the dynamic yield surface related to the real stresses. Hashiguchi and Okayasu (2000) developed a time-dependent subloading surface model introducing a creep potential function. Zhang et al. (2005) worked out a time-dependent model for heavily overconsolidated clays and soft rocks, modifying the subloading t_{ij} model developed by Nakai and Hinokio (2004). A comprehensive review of time-dependent behaviors of soils and their modeling has been written by Sekiguchi (1985).

In this section, a simple method to model time-dependent characteristics of normally consolidated, overconsolidated, and structured soils such as naturally deposited clays is presented—not using the usual viscoplastic theories but rather referring to the previously mentioned formulations of advanced elastoplastic modeling (stage III).

Figure 3.17 shows a well known creep behavior (e–ln$t$ relation) for normally consolidated clays under constant stress. The void ratio at time

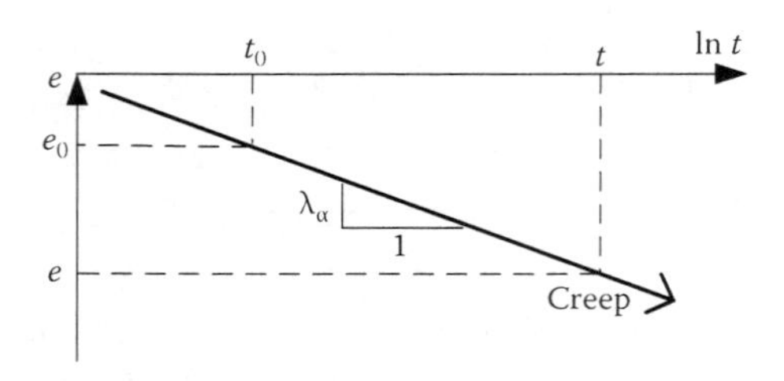

Figure 3.17 Creep characteristics of normally consolidated clay.

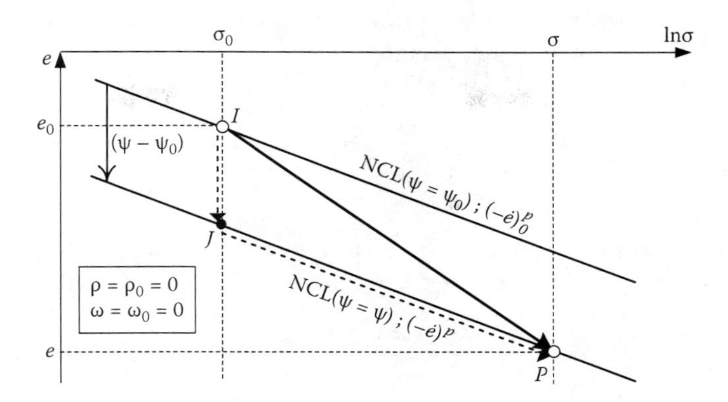

Figure 3.18 Changes in position of NCL and void ratio due to changes in stress and strain rate for normally consolidated clay.

t_0 is e_0, and at time t it is e. Here, λ_α is the coefficient of secondary consolidation. In this time interval, only irreversible plastic change in void ratio can occur because the stress level is fixed.

Figure 3.18 shows the void ratio change induced by the shifting of the NCL caused by a change of plastic strain rate (rate of plastic void ratio change) from $(-\dot{e})_0^p$ to $(-\dot{e})^p$, when the stress condition moves from the initial state I ($\sigma = \sigma_0$) to the current state P ($\sigma = \sigma$) in the normal consolidation condition. This change can be interpreted as a superposition of two effects: (a) the time-dependent effect from point I to J, as depicted in Figure 3.17, and (b) the stress change effect from point J to P, along the NCL, which was shifted due to time (strain rate) effects. Here, e_0 and e are the initial and current void ratios on the NCL at $\psi = \psi_0$ and $\psi = \psi$, respectively, and ψ is a state variable that shifts the position of NCL upward or downward. Thus it is responsible for the time effect as described before. The shift of the NCL $(\psi - \psi_0)$ can be easily obtained, by referring to Figures 3.17 and 3.18, as

$$\psi - \psi_0 = \lambda_\alpha \ln \frac{t}{t_0} = \lambda_\alpha \ln t - \lambda_\alpha \ln t_0 \tag{3.38}$$

Hence, using the plastic void ratio change instead of elapsed time t, it follows readily that

$$\psi - \psi_0 = \lambda_\alpha \ln \frac{(-\dot{e})_0^p}{(-\dot{e})^p} = \left\{-\lambda_\alpha \ln(-\dot{e})^p\right\} - \left\{-\lambda_\alpha \ln(-\dot{e})_0^p\right\} \tag{3.39}$$

To obtain this equation, the following relation during creep deformation in normally consolidated soil was used:

$$(-\dot{e})^p = \frac{d(-\Delta e)^p}{dt} = \frac{d\left(\lambda_\alpha \ln\dfrac{t}{t_0}\right)}{dt} = \lambda_\alpha \frac{1}{t} \tag{3.40}$$

Therefore, ψ and ψ_0, which represent the positions of initial and current NCL, are determined as follows using the rate of plastic void ratio change:

$$\begin{cases} \psi = -\lambda_\alpha \ln(-\dot{e})^p \\ \psi_0 = -\lambda_\alpha \ln(-\dot{e})_0^p \end{cases} \tag{3.41}$$

From eqs. (3.38) and (3.40), the increment $d\psi$ is expressed as

$$d\psi = \frac{\partial \psi}{\partial t} dt = \lambda_\alpha \frac{1}{t} dt = (-\dot{e})^p dt \tag{3.42}$$

Equations (3.41) and (3.42) imply that the position of NCL (ψ) and its increment ($d\psi$) can be expressed by the rate of plastic void ratio change instead of the elapsed time. Now, it is assumed that eqs. (3.41) and (3.42) hold not only for normally consolidated soils, but also for overconsolidated soils and naturally deposited soils. Substituting eq. (3.42) into eq. (3.27), the increment of the plastic void ratio can be obtained as

$$d(-e)^p = \frac{(\lambda - \kappa)\dfrac{1}{\sigma} d\sigma + (-\dot{e})^p \cdot dt}{1 + G(\rho) + Q(\omega)} \cong \frac{(\lambda - \kappa)\dfrac{1}{\sigma} d\sigma + (-\dot{e})^{p^*} \cdot dt}{1 + G(\rho) + Q(\omega)} \tag{3.43}$$

Here, $(-\dot{e})^{p^*}$ denotes the rate of the plastic void ratio change at the previous calculation step. Finally, the total void ratio increment is given by the following equation:

$$d(-e) = d(-e)^p + d(-e)^e = \left(\frac{\lambda - \kappa}{1 + G(\rho) + Q(\omega)} + \kappa\right) \frac{d\sigma}{\sigma} + \frac{(-\dot{e})^{p*}}{1 + G(\rho) + Q(\omega)} dt \tag{3.44}$$

In order to simplify the numerical calculations, the known rate $(-\dot{e})^{p^*}$ in the previous calculation step can be used, instead of the current rate, as described in eqs. (3.43) and (3.44). The error introduced by using the previous known rate is considered to be negligible since small enough steps are adopted in an incremental fashion to solve the nonlinear problem. As can be seen in eq. (3.20), the void ratio approaches the condition satisfying

$G(\rho) + Q(\omega) = 0$, regardless of the sign of ρ. Here, ρ represents the difference between the current real void ratio and the one on the NCL, which shifts depending on the current plastic void ratio change rate $(-\dot{e})^p$, as mentioned before.

Therefore, even if some small error is introduced by using $(-\dot{e})^{p^*}$ in the numerical calculation, updating the rate with the calculated plastic change in void ratio at each time increment and reflecting it in the state variable ψ at the next step, the error caused in the present calculation step is automatically corrected in the next one. The model requires only one additional parameter, which is the coefficient of secondary consolidation λ_α, to account for time effects. An interesting feature of the present model is that it does not include any time variable *t*, but rather is formulated using the rate of void ratio change $(-\dot{e})^p$ alone. If time *t* were used in a model, the results would depend on the choice of the origin for time. Another characteristic of the proposed approach is that by only eliminating the term of the rate effect $(-\dot{e})^{p^*}$ in eq. (3.44), or by setting $\lambda_\alpha = 0$, this model readily reverts to the version of the elastoplastic model without time effects.

The formulation of time effects in the present model is based on experimental evidence that shows that there is a unique relation between stress, strain (void ratio), and strain rate (rate of void ratio change) under loading for normally consolidated clay, in the same way as assumed in Sekiguchi's model (Sekiguchi 1977). This leads to the concept of "isotache" (curves of equal strain rates), which is well known in describing time-dependent behavior of soils (Suklje 1957). Sekiguchi's model was formulated using a nonstationary flow surface obtained by solving the relation (stress–plastic strain–plastic strain rate) in the form of an ordinary differential equation with variables of plastic strain and time. In contrast, the present model is formulated by defining the plastic strain rate (rate of plastic void ratio change) as a state variable in conjunction with the subloading surface concept. The meaning of the present time-dependent model, as well as the common points and differences between the present model and the nonstationary flow surface model proposed by Sekiguchi, are described in the next section.

3.8 MEANING OF PRESENT TIME-DEPENDENT MODEL—COMMON AND DIFFERENT POINTS TO SEKIGUCHI MODEL

Consider initially the 1D behavior of a nonstructured normally consolidated soil without time effects. Upon a stress change, the ensuing increment of plastic void ratio is given by eq. (3.8):

$$d(-e)^p = (\lambda - \kappa) \cdot \frac{d\sigma}{\sigma} \tag{3.8bis}$$

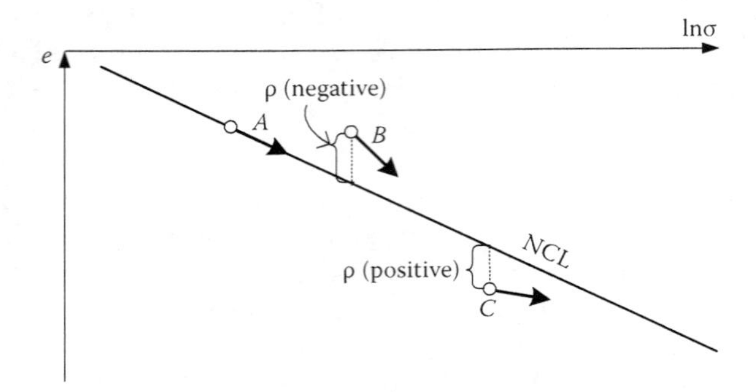

Figure 3.19 Schematic diagram for explaining soil stiffness with positive and negative values of ρ.

In Figure 3.19, the solid straight line (NCL) indicates the ideal e–lnσ relation of a normally consolidated soil, and the slope of each arrow shows schematically the stiffness at some point calculated following the subloading surface concept with ρ as state variable. Here, the increment of plastic void ratio as calculated from eq. (3.17) according to the present model for an overconsolidated soil is given by

$$d(-e)^p = \frac{\lambda - \kappa}{1 + G(\rho)} \cdot \frac{d\sigma}{\sigma} \tag{3.17bis}$$

As described in Section 3.4, it can be seen that positive values of ρ have the effect of increasing the stiffness, while negative values of ρ decrease the stiffness. Even if the initial void ratio is on the NCL ($\rho = 0$) at point A, the void ratios in the subsequent calculation steps may stray from that on the NCL due to numerical errors. For example, the stiffness calculated at point B becomes smaller than that on the NCL because of $\rho < 0$, and the stiffness calculated at point C becomes larger because of $\rho > 0$. Thus the subloading surface concept is very useful not only to consider the influence of density, but also to eliminate the accumulative errors during numerical calculations to converge to the ideal stress–strain curve automatically.

In the following, consider the modeling of the 1D behavior of a nonstructured normally consolidated soil with time effects. As shown in Figure 3.18, it is assumed that the plastic change in void ratio is expressed as the summation of the component from I to J and the plastic component from J to P. This means that there exists a unique relation between void ratio, stress, and rate of plastic void ratio change. When the initial and current values of void ratio, stress, and rate of plastic void ratio change are denoted as

$(e_0, \sigma_0, (-\dot{e})_0^p)$ and $(e, \sigma, (-\dot{e})^p)$, the change in plastic void ratio is given by the following equation (Sekiguchi and Toriihara 1976):

$$
\begin{aligned}
(-\Delta e)^p &= e_0 - e - (-\Delta e)^e = (\lambda - \kappa)\ln\frac{\sigma}{\sigma_0} + \lambda_\alpha \ln\frac{(-\dot{e})_0^p}{(-\dot{e})^p} \\
&= F + \lambda_\alpha \ln\frac{(-\dot{e})_0^p}{(-\dot{e})^p} \qquad \left(\text{where, } \; F = (\lambda - \kappa)\ln\frac{\sigma}{\sigma_0}\right)
\end{aligned}
\tag{3.45}
$$

By solving this ordinary differential equation with variables of time and plastic void ratio change (plastic strain), Sekiguchi (1977) obtained the plastic change of void ratio in the following form:

$$
\begin{aligned}
(-\Delta e)^p &= \lambda_\alpha \ln\left\{\frac{(-\dot{e})_0^p \cdot t}{\lambda_\alpha}\exp\left(\frac{F}{\lambda_\alpha}\right) + 1\right\} \\
&= \lambda_\alpha \ln S \qquad \left(\text{where, } \; S = \frac{(-\dot{e})_0^p \cdot t}{\lambda_\alpha}\exp\left(\frac{F}{\lambda_\alpha}\right) + 1\right)
\end{aligned}
\tag{3.46}
$$

and the increment of plastic void ratio is calculated as

$$
d(-e)^p = \frac{\partial(-\Delta e)^p}{\partial \sigma}d\sigma + \frac{\partial(-\Delta e)^p}{\partial t}dt = (\lambda - \kappa)\left(1 - \frac{1}{S}\right)\frac{1}{\sigma}d\sigma + \lambda_\alpha\left(1 - \frac{1}{S}\right)\frac{1}{t}dt
\tag{3.47}
$$

Sekiguchi (1977) defined the flow surface, which corresponds to the yield surface including the time variable, by replacing F in eq. (3.46) with the Cam clay type yield function in three-dimensional (3D) stresses.

The meaning of the present modeling framework using the subloading surface concept for nonstructured normally consolidated soil ($\rho_0 = 0$ and $\omega_0 = 0$) with time effects is shown below. Although the basic assumption between void ratio, stress, and rate of plastic void ratio change is the same as Sekiguchi's model, the plastic change of void ratio and the increment of plastic void ratio are expressed as follows from eqs. (3.24) and (3.27):

$$
(-\Delta e)^p = (\lambda - \kappa)\ln\frac{\sigma}{\sigma_0} + \rho - (\psi_0 - \psi)
\tag{3.48}
$$

$$
d(-e)^p = \frac{(\lambda - \kappa)\dfrac{1}{\sigma}d\sigma + d\psi}{1 + G(\rho)}
\tag{3.49}
$$

Here, the state variable ρ is introduced and automatically corrects the stiffness of a normally consolidated soil. The state variable ψ, which determines the position of the NCL, and its increment $d\psi$ are given by eqs. (3.41) and (3.42) using the coefficient of secondary consolidation λ_α and the rate of plastic void ratio change $(-\dot{e})^p$. Then, this rate of plastic void ratio change $(-\dot{e})^p$ is determined from the increment of plastic void ratio changes, stress increment, and time increment from the previous calculation step to the current calculation step. This is used only to determine the position of the NCL and the increment $d\psi$ for the next incremental calculation.

Figure 3.20 shows schematically the calculation process based on the present time-dependent model in the e–lnσ plane. The current condition (ith step) is represented with void ratio $e = e_i$, stress $\sigma = \sigma$, time $t = t$, rate of plastic void ratio change $(-\dot{e})^p = (-\dot{e})_i^p$, state variable $\rho = \rho_i$, and the NCL with ψ_i. These variables in subsequent condition ((i + 1)th step) are represented as e_{i+1}, $\sigma + d\sigma$, $t + dt$, $(-\dot{e})_{i+1}^p$, ρ_{i+1}, and the NCL with ψ_{i+1}. Here, the increment of void ratio from the ith step (point A) to the (i + 1)th step (point C) is expressed as a summation of the component due to creep (A to B) and the component due to stress increment (B to C). Looking at eqs. (3.9), (3.42), and (3.49), these components are given by

$$d(-e)_{creep} = d(-e)_{creep}^p = \frac{(-\dot{e})^p\, dt}{1+G(\rho_i)} \tag{3.50}$$

$$d(-e)_{stress} = d(-e)_{stress}^p + d(-e)^e = \left\{\frac{\lambda-\kappa}{1+G(\rho_i)} + \kappa\right\}\frac{1}{\sigma}d\sigma \tag{3.51}$$

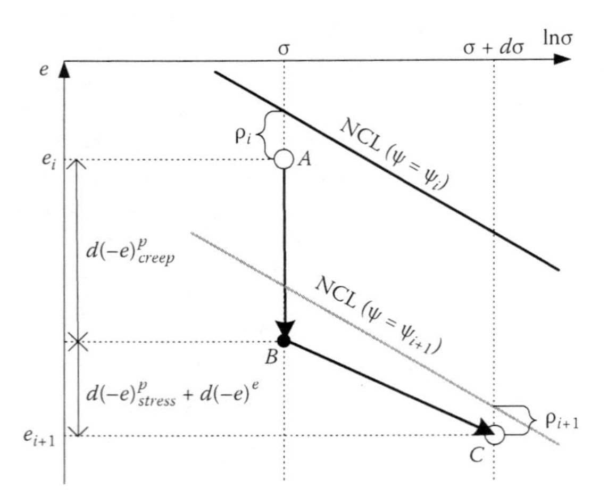

Figure 3.20 Schematic diagram for explanation of incremental calculation in time-dependent model.

Then, the rate of plastic void ratio change $(-\dot{e})^p_{i+1}$ used in the following calculation step is determined as

$$(-\dot{e})^p_{i+1} = \left\{ e_i - e_{i+1} - d(-e)^e \right\} / dt \quad \text{or} \quad (-\dot{e})^p_{i+1} = \left\{ d(-e)^p_{creep} + d(-e)^p_{stress} \right\} / dt \tag{3.52}$$

From eq. (3.41), the position of the NCL with ψ_{i+1} can be determined using the rate obtained by eq. (3.52), and the value of ρ_{i+1} can be defined as the difference between the void ratio e_{i+1} and that on the NCL with ψ_{i+1} at the same stress level. These are used in the subsequent calculation step. The errors that arise when using the previous values of the rate and not solving the differential equation are self-corrected by the state variable ρ in the same way as in the case without time effect, as mentioned before. Although it is possible to carry out iterative calculations so that $(-\dot{e})^p_{i+1} \cong (-\dot{e})^p$, there is not much difference between the results obtained with and without iterations, unless large increments of stress and time are used.

For the case of overconsolidated soil and/or structured soil, $G(\rho)$ and $Q(\omega)$ converge to zero with the development of plastic deformation and the void ratio finally becomes the same as that in normally consolidated soil. Therefore, determining the position of the NCL (ψ) by the current rate of plastic void ratio change alone, even in overconsolidated soils and structured soils, in the same way as that in normally consolidated soils, the model described here can be applied to overconsolidated soils and structured soils as well.

The validity of the present model has been verified by the simulations of various kinds of time-dependent tests on soils in the subsequent section. It should be noted that although nonstationary flow surface models such as Sekiguchi's model are useful for normally consolidated soils alone, the model includes a time variable that is nonobjective. The present models (eqs. 3.41–3.44) are applicable not only to normally consolidated soils but also to overconsolidated soils and structured soils. They are also objective in character as the time variable is not used.

3.9 SIMULATION OF TIME-DEPENDENT BEHAVIOR OF SOIL IN ONE-DIMENSIONAL CONDITIONS

The validity of the proposed time-dependent model is checked by simulating 1D behavior under constant strain rate for an infinitesimal soil element, as well as by simulations of conventional oedometer tests with instantaneous loading of constant stress as a boundary value problem using 1D soil water coupled finite element analyses using Tamura's method (Akai and Tamura 1978). The adopted parameters for the

Table 3.2 Values of material parameters for simulations of time-dependent behavior

λ	0.104
κ	0.010
N (e_N at σ = 98 kPa)	0.83
a	100
b	40 unless otherwise stated
ω_0	0.0 (without bonding)
	0.2 (with bonding)
λ_α	0.003 unless otherwise stated
$(-\dot{e})^p_{ref}$	1×10^{-7}/min

Note: N is the void ratio of NCL $(-\dot{e})^p$ at = $(-\dot{e})^p_{ref}$.

simulations are the same as those used in Sections 3.3 and 3.4 (i.e., compression index $\lambda = 0.104$, swelling index $\kappa = 0.010$, void ratio on the NCL at $\sigma = 98$ kPa, $N = 0.83$).

The evolution rule for ρ is considered to be a linear function, $G(\rho) = a\rho$. Similarly, the evolution rule for ω is also considered to be linear such that $Q(\omega) = b\omega$, as indicated in Figure 3.11. The parameter for density and confining pressure is chosen as $a = 100$, whereas the bonding degradation parameter is given by $b = 40$, unless otherwise stated. The initial value of ω for the soil with bonding is $\omega_0 = 0.20$. The rate of the plastic void ratio change at reference state is $(-\dot{e})^p_{ref} = 1.0 \times 10^{-7}$/min. Here, the value of the coefficient of secondary consolidation, λ_α, is 0.003, unless otherwise stated. The values of material parameters of the clay for the time-dependent model are summarized in Table 3.2.

Let us show some methods to determine the material parameters of the evolution rules of ρ, ω, and the position of NCL (ψ) for a naturally deposited clay. The compression index λ, the swelling index κ, and the coefficient of secondary consolidation λ_α are derived from the loading and unloading e–lnσ relation and the e–ln$t$ relation obtained from the usual oedometer tests on remolded and normally reconsolidated samples. The evolution rule of ρ can be determined by fitting the calculated e–lnσ relation of remolded and overconsolidated sample to the observed one, referring to the sensitivity of the parameter a (e.g., Figure 3.12a) because the bonding effect is considered to be zero in remolded reconsolidated samples.

After that, the evolution rule of bonding (b and ω_0) is determined by fitting the calculated e–lnσ relation of the undisturbed naturally deposited clay to the observed one with reference to the sensitivity of these parameters (e.g., Figures 3.12b–3.12d). The position of the NCL (ψ) in the time-dependent model is determined from the rate $(-\dot{e})^p$ and the coefficient of secondary consolidation λ_α (see eq. 3.41). Here, it can be considered that

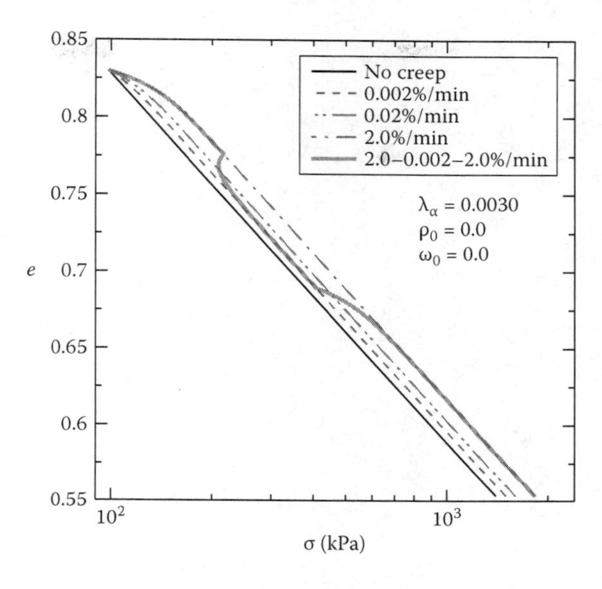

Figure 3.21 Calculated e: lnσ relations under constant strain rates for normally consolidated clays.

although the material parameters, except for ω_0, are independent of the initial conditions (void ratio, stress level, and others), ω_0 depends on the degree of the initial natural cementation (bonding).

Figure 3.21 shows the simulated results of 1D compression behavior of normally consolidated clay for different strain rates, plotted in e–logσ axes. The initial rate of plastic void ratio change is the same as that at reference state ($(-\dot{e})_0^p = (-\dot{e})_{ref}^p = 1.0 \times 10^{-7}$/min). In the figure, the solid straight line (No creep) shows the simulated relation without time effect. The simulation results for the case of constant rate of $(-\dot{e})^p = (-\dot{e})_0^p = (-\dot{e})_{ref}^p$ also coincide with the solid straight line.

It is seen from this figure that, with the increase in strain rate, the resistance to compression increases and the lines of constant strain rate are parallel to each other, which is in agreement with published experimental results (Bjerrum 1967). It is also seen that when the strain rate is changed at a certain point, the curve follows exactly the same path, which is supposed to follow for the new rate. This is valid for both increasing and decreasing the strain rates, and in the case where the strain rate is increased, the numerical simulations gradually latch onto the target curve following the phenomenon of isotache. Therefore, it can be said that the present model can describe properly the strain rate effects of a nonstructured normally consolidated clay ($\omega_0 = 0$ and $\rho_0 = 0$) under constant strain rate consolidation tests.

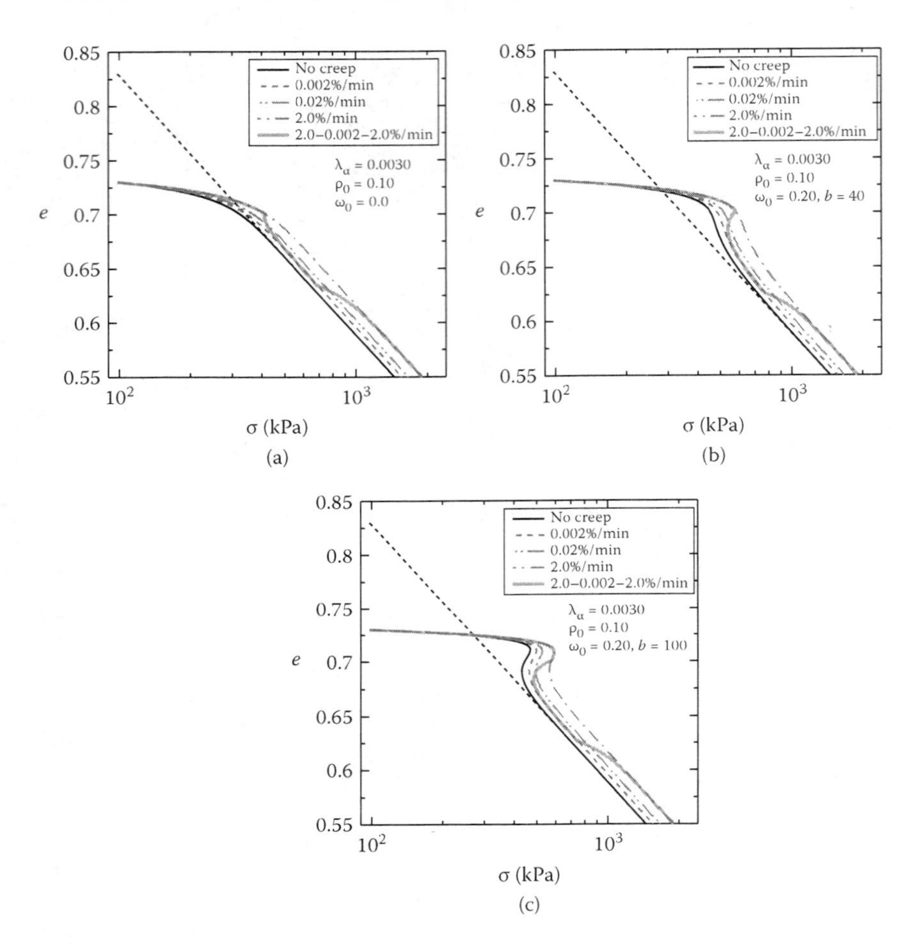

Figure 3.22 Calculated e: lnσ relations under constant strain rates for OC and OC-structured clays. (a) OC clay; (b) OC-structured clay; (c) OC-structured clay with softening.

Figure 3.22(a) represents the calculated e–logσ relations for an overconsolidated nonstructured clay ($\omega_0 = 0$ and $\rho_0 \neq 0$) subjected to different strain rates. Here, at the initial stress condition (σ_0 = 98 kPa), the void ratio is $e_0 = 0.73$, the rate of plastic void ratio change is the same as that at reference state $(-\dot{e})_0^p = (-\dot{e})_{ref}^p = 1.0 \times 10^{-7}$/min, and the void ratio on the NCL at σ = 98 kPa is $e_{N0} = 0.83$. Therefore, the initial value of state variable is $\rho_0 = 0.10$. Figure 3.22(b) illustrates the results for an overconsolidated-structured soil. Here, the initial void ratio and the initial rate of strain are the same as those in Figure 3.22(a). The initial value of the state variable reflecting the bonding effect is $\omega_0 = 0.20$, and the bonding degradation parameter is $b = 40$. In the figures, the dotted straight line denotes the NCL for $(-\dot{e})^p = (-\dot{e})^p_{ref} = 1.0 \times 10^{-7}$/min. It can be observed that the strain rate dependency of the soil for overconsolidated states is less significant compared to normally consolidated states.

However, each e–logσ relation for overconsolidated soils and structured soils finally approaches the line simulated for normally consolidated soil under corresponding strain rate. Figure 3.22(c) shows the results of the overconsolidated-structured soil for the same material parameters as illustrated in Figure 3.22(b), but using a different bonding degradation parameter value ($b = 100$). The results show both hardening and softening behaviors of soil for different strain rates. In these figures, the thick curves represent the results in which the strain rate decreases and increases again during loading in the same way as that in Figure 3.21. The simulated results of the overconsolidated clay and the structured clays also tend to the curve that corresponds to the new strain rates.

The phenomenon of isotache is simulated for overconsolidated and structured soils, in the same way as observed for the normally consolidated soil. This phenomenon was observed by Leroueil et al. (1985) during 1D compression tests with a natural clay (see Figure 3.23). In diagrams (a) to (c) in Figure 3.22, each solid curve designated as "No creep" is the same as the corresponding curve without time effect in Figure 3.12(d). It is also seen that the "preconsolidation" stress (effective consolidation yield stress) p_c increases with increasing strain rates, particularly in structured clays. These simulations in Figure 3.22 describe well the strain rate effects in 1D compression for overconsolidated clay and structured clay reported in the literature (Leroueil et al. 1985; Tanaka, Udaka, and Nosaka 2006; Watabe, Udaka, and Morikawa 2008).

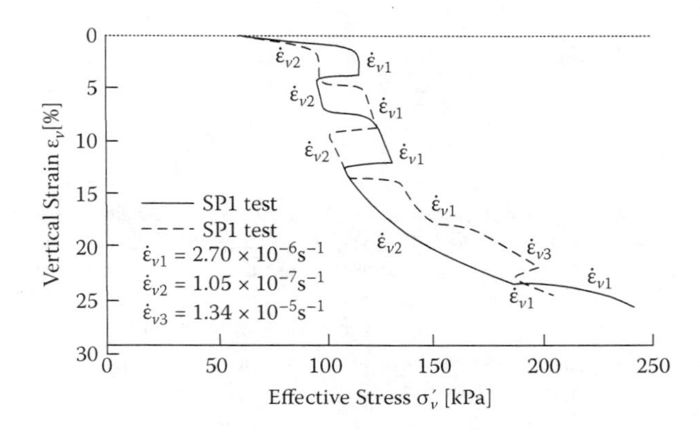

Figure 3.23 Observed result of constant strain rate oedometer test on Batiscan natural clay. (Replotted from data in Leroueil, S. et al. 1985. *Geotechnique* 35 (2): 159–180.)

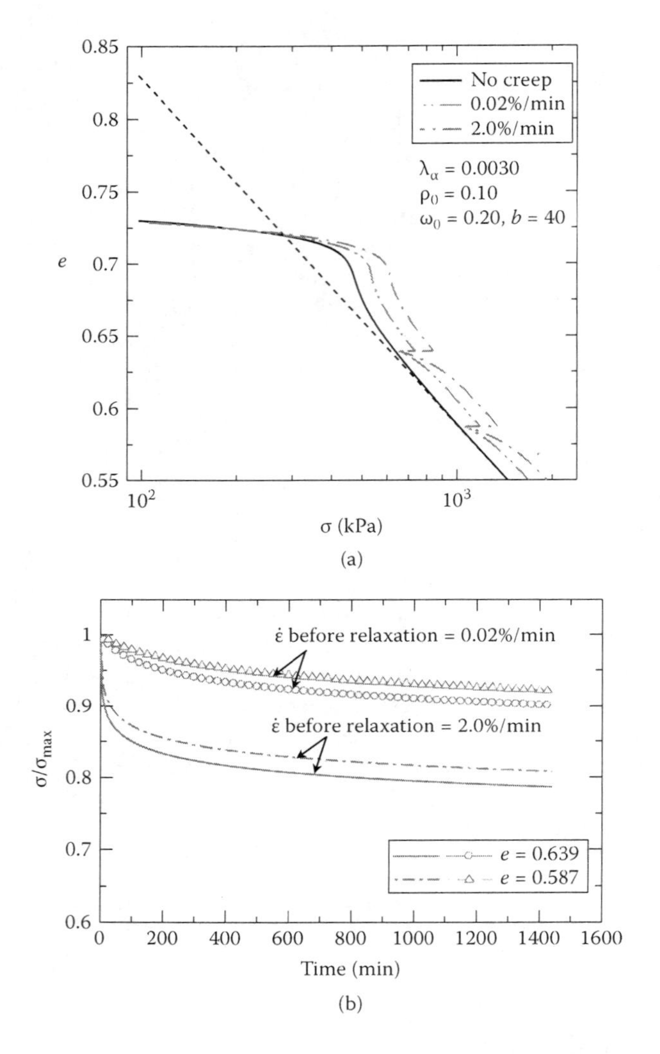

Figure 3.24 Calculated consolidation behavior under constant strain rate including some stress relaxation periods for structured clay. (a) Vertical stress in log scale versus void ratio; (b) reduction of stress during stress relaxation period.

Figure 3.24 shows the simulated results of the 1D compression tests under constant strain rate tests including some stress relaxation periods on a structured soil. Here, the initial condition is the same as that in Figure 3.22(b); that is, $\rho_0 = 0.10$, $\omega_0 = 0.20$, and $b = 40$. Figure 3.24(a) shows the results arranged in terms of the relation between void ratio and stress in log scale for different strain rates, and Figure 3.24(b) shows the reduction of stress σ during the stress relaxation period. The stress in Figure 3.24(b)

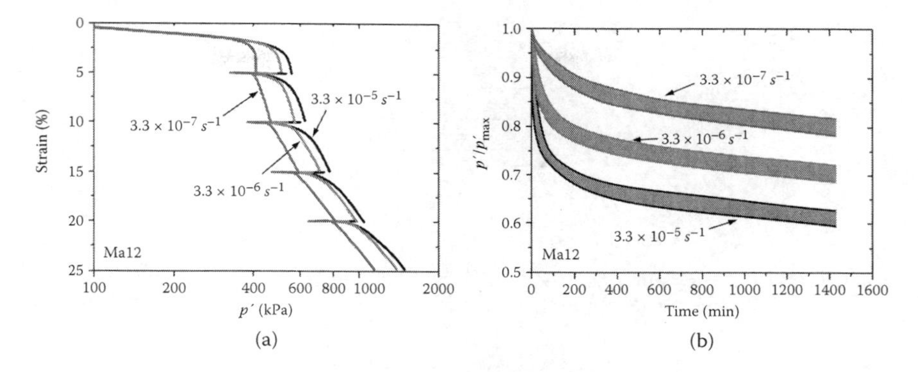

Figure 3.25 Observed results of constant strain rate consolidation tests including some stress relaxation periods on Osaka Pleistocene clay. (a) Vertical stress in log scale versus strain; (b) reduction of stress during stress relaxation period. (Replotted from data in Tanaka, H. et al. 2006. *Soils and Foundations* 46 (3): 315–322.)

is normalized by the maximum stress value σ_{max} imposed at the start of the corresponding relaxation.

Tanaka et al. (2006) carried out 1D constant strain rate consolidation tests including some stress relaxation periods for different strain rates on undisturbed Osaka Pleistocene clay (Ma12 layer). Figure 3.25 shows the observed results of their tests, arranged in the same form as those in Figures 3.24(a) and (b). It can be seen from these figures that the present model simulates well the observed features of the naturally deposited clay (i.e., the yield stress becomes larger with the increase of strain rate). The shape of e–logσ relation (or ε–logσ relation) is almost independent of the strain rate, and the normalized relaxation stress σ/σ_{max} shows more or less the same tendency if the strain rates at the start of the relaxation are the same.

One-dimensional soil water coupled finite element analyses of oedometer tests with instantaneous loading of constant vertical stress were carried out to investigate the consolidation characteristics of clays. In the simulations of oedometer tests, the height H of the sample was divided into elements with the thickness of 0.1 cm, as shown in Figure 3.26. Drainage was allowed at the top boundary of the sample, while the bottom boundary was considered undrained. In order to make the coefficient of consolidation c_v constant during normal consolidation, regardless of the stiffness of the soil, the following relationship between the coefficient of permeability, k, and the current void ratio, e, was used in these simulations:

$$k = k_0 \cdot \exp\left(\frac{e - e_{N0}}{\lambda_k}\right) \tag{3.53}$$

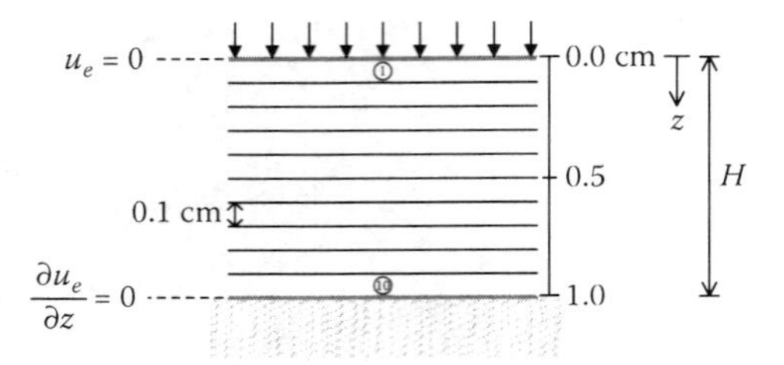

Figure 3.26 Finite element mesh and boundary condition in one-dimensional consolidation analysis (H = 1 cm).

where $e_{N0} = 0.83$, $k_0 = 1.0 \times 10^{-5}$ cm/min, and $\lambda_k = 0.104$, which is the same as the compression index λ.

Figure 3.27 shows the computed e–log$t$ simulations of conventional oedometer tests for normally consolidated clay, where the initial stress was σ_0 = 98 kPa, the initial rate of void ratio change was $(-\dot{e})_0^p = 1.0 \times 10^{-7}$/min, and the instantaneous increment of stress $\Delta\sigma$ = 98 kPa was applied. After applying the stress increment, the consolidation behavior of soil was investigated for different values of the coefficient of secondary consolidation (λ_α). The vertical axis ($\underline{e}$) represents the average void ratio of the soil mass.

Here, H is the sample height, which represents the maximum drainage distance in the sample. The solid curve (No creep) represents the results where the effect of secondary consolidation is not considered. It is seen that delayed consolidation occurs when the time effect is considered, which shows the creep behavior of the soil. With the increase of the value of λ_α, the delay in consolidation becomes more remarkable. In the cases where the

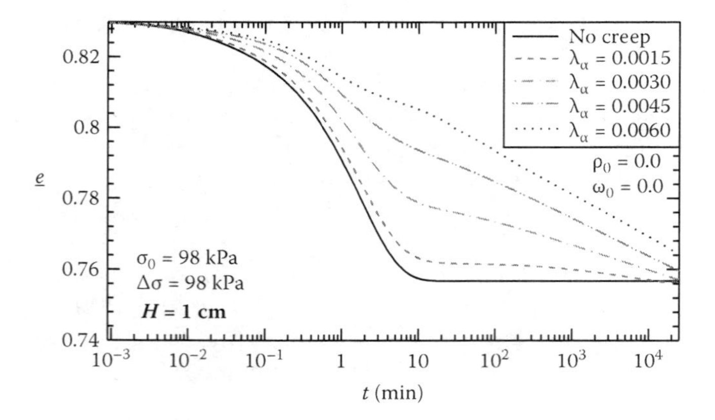

Figure 3.27 Simulation of oedometer tests on NC clay for different coefficients of secondary consolidation.

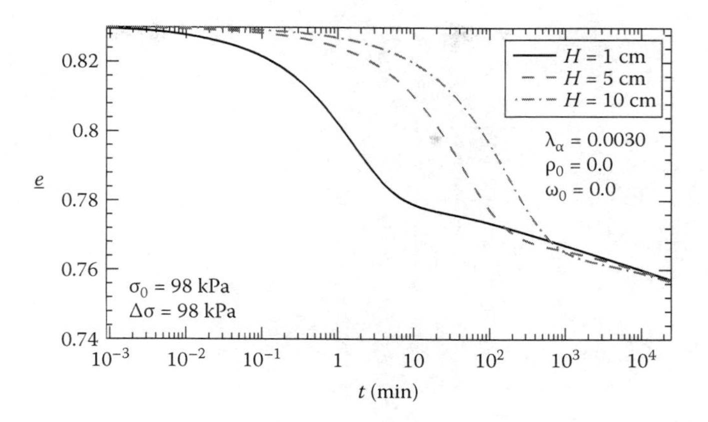

Figure 3.28 Simulation of oedometer tests on NC clays for different heights of sample.

time effect is not as prominent, the curves of void ratio (settlement) versus logarithm of time have the shape of reverse "s" during the dissipation process of pore water pressure, as is commonly seen in the literature. During secondary consolidation, the slopes of the curves are the same as the coefficients of secondary consolidation (λ_α) employed in the simulations.

Figure 3.28 represents the computed e–ln$t$ relation for the normally consolidated clay with different heights of the sample (H = 1, 5, and 10 cm). The initial condition of each sample was $\sigma_0 = 98$ kPa, $(-\dot{e})_0^p = 1.0 \times 10^{-7}$/min, and the increment of the stress was $\Delta\sigma = 98$ kPa. This figure describes the well known effects of sample height (Aboshi 1973; Ladd et al. 1977). It can be observed that after some time from the application of the load, the consolidation curves for different sample heights converge to a single curve. This tendency of e–log$t$ curves under different sample heights corresponds to the curve of type B named by Ladd et al. (1977).

Figure 3.29 illustrates the computed e–log$t$ relation for the normally consolidated clay with different stress increments ($\Delta\sigma$). Here, the height of the sample was 1 cm and the coefficient of secondary consolidation (λ_α) was 0.003 in the same way as those in Figure 3.28. It is seen from the figure that final slope of each curve after the dissipation of the excess pore water pressure is independent of the increment of stress ($\Delta\sigma$) and is the same as the coefficient of secondary consolidation. The tendency of the computed results in Figure 3.29 has good qualitative correspondence with published experimental results (Leonards and Girault 1961; Oshima, Ikeda, and Masuda 2002). Figure 3.30 shows the observed ε–log$t$ relation of oedometer tests on remolded normally consolidated Osaka Nanko clay under different stress increments.

Figure 3.31 shows the computed e–log$t$ response for an overconsolidated clay with different values of the coefficient of secondary consolidation (λ_α).

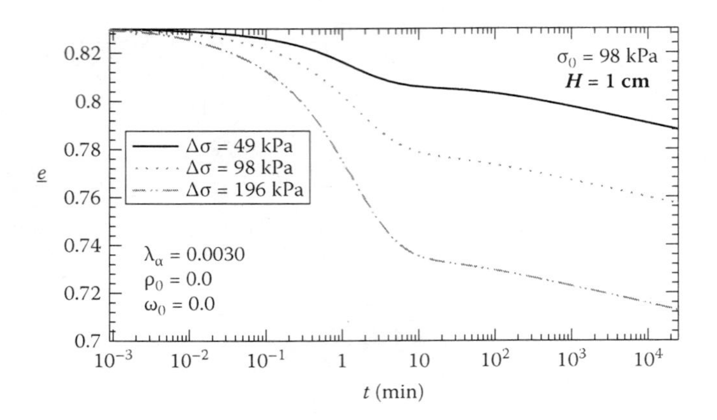

Figure 3.29 Simulation of oedometer tests on NC clays for different stress increments.

Here, the overconsolidation ratio was 1.70 ($\rho_0 = 0.05$). The other conditions were the same as those for the normally consolidated clay in Figure 3.27. It is seen that after the excess pore water pressure dissipates, the slope of the curve for the overconsolidated clay is much flatter than that of the normally consolidated clay in Figure 3.27 as a whole, but becomes steeper again over time.

Figure 3.32 shows the observed result of oedometer tests on overconsolidated Hiroshima clay performed by Yoshikuni et al. (1990). Here, the overconsolidated clay, which was obtained by unloading the vertical stress from $\sigma = 314$ kPa to $\sigma = 157$ kPa on a remolded normally consolidated clay,

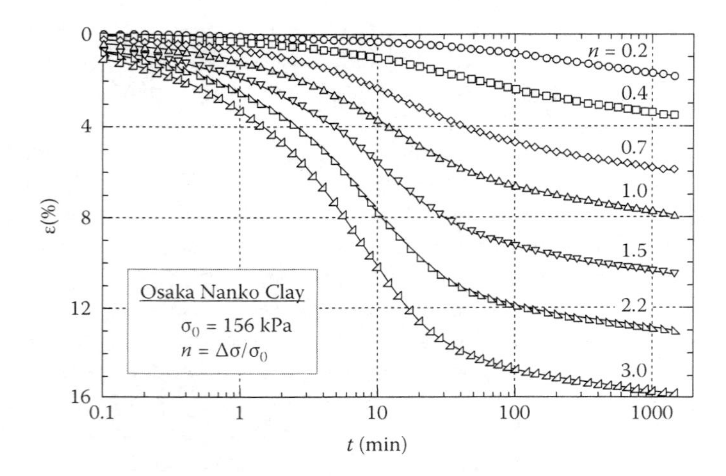

Figure 3.30 Observed results of oedometer tests on remolded NC Osaka Nanko clay for different stress increments. (Ohshima, A. et al. 2002. *Proceedings of 37th Annual Meeting on JGS* 1:289–290.)

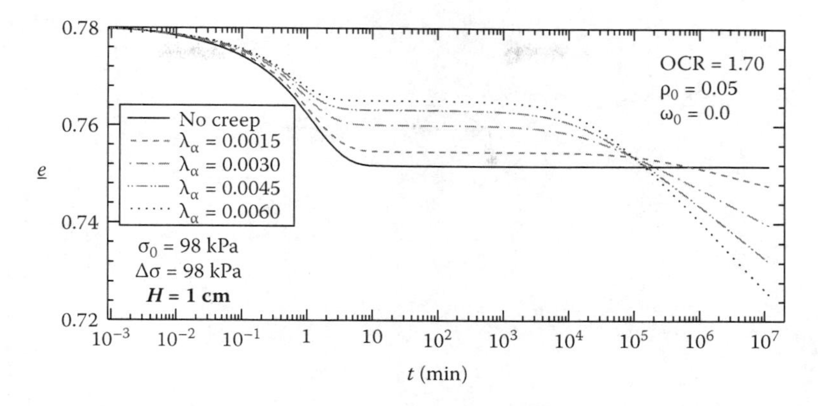

Figure 3.31 Simulation of oedometer tests on OC clay for different coefficients of secondary consolidation.

is reloaded instantaneously to σ = 294 kPa. "Time in unloading" in the legend denotes the time during unloaded state at σ = 157 kPa. The results of the simulation in Figure 3.31 describe well the observed feature of the overconsolidated clay in Figure 3.32.

A sudden increase and delay in settlement after excess pore water pressure is almost dissipated have been observed in oedometer tests on natural clay (Leroueil et al. 1985). Asaoka et al. (2000b) simulated such behavior with a soil water coupled finite element analysis using an inviscid model for structured soils; they stated that the strain-softening characteristics with volume contraction for highly structured clay and the

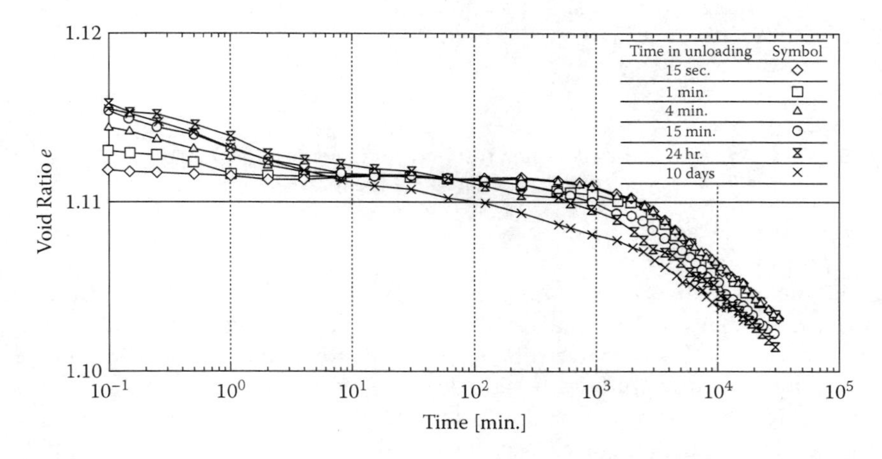

Figure 3.32 Experimental results of oedometer tests on OC clays. (Yoshikuni, H. et al. 1990. *Proceedings of 25th Annual Meeting on JGS* 1:381–382.)

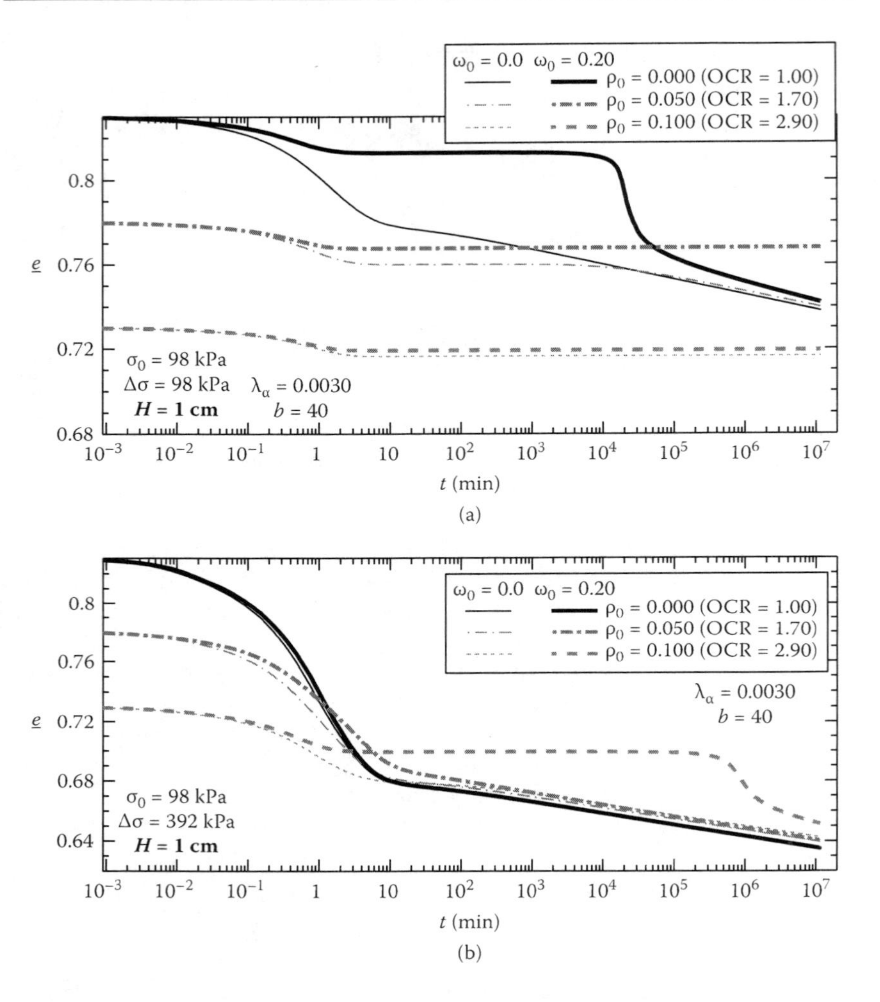

Figure 3.33 Simulation of oedometer tests on nonstructured and structured clay for different initial void ratios. (a) $\Delta\sigma/\sigma_0 = 1$; (b) $\Delta\sigma/\sigma_0 = 4$.

migration of pore water in the clay due to Darcy's law caused such delay of compression.

Figure 3.33 shows the computed e–log$t$ response of nonstructured and structured clays in normally consolidated and overconsolidated states. Figure 3.33(a) shows the results under small stress increment (the ratio of stress increment to initial stress: $\Delta\sigma/\sigma_0 = 1$), and 3.33(b) shows the results under large stress increment ($\Delta\sigma/\sigma_0 = 4$). Here, the "thin" curves indicate the results of nonstructured clay ($\omega_0 = 0.0$), and the "thick" curves show the results of structured clay ($\omega_0 = 0.2$). Although the behavior of the normally consolidated structured clay (OCR = 1.0) is different from that of the normally consolidated nonstructured clay under small stress increment, there is not much difference between them under large stress increment.

On the other hand, the behavior of overconsolidated clays (OCR = 2.9) is highly influenced by the effect of structure (bonding)—not under small stress increment, but rather under large stress increment. Figure 3.34

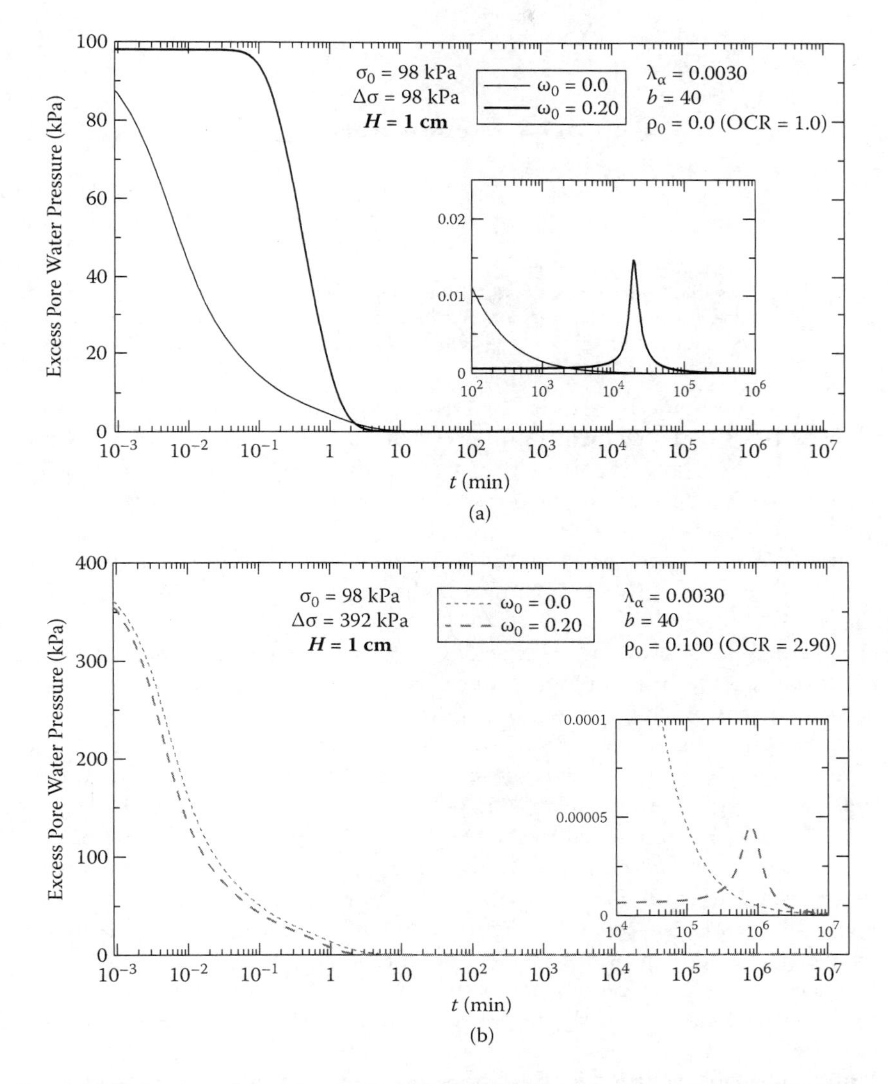

Figure 3.34 Computed variations of excess pore water pressure at element ⑩ in Figure 3.26 with time in oedometer tests on nonstructured and structured clay for different initial void ratios. (a) $\Delta\sigma/\sigma_0 = 1$ and $\rho_0 = 0.00$ (OCR = 1.0); (b) $\Delta\sigma/\sigma_0 = 4$ and $\rho_0 = 0.10$ (OCR = 2.9).

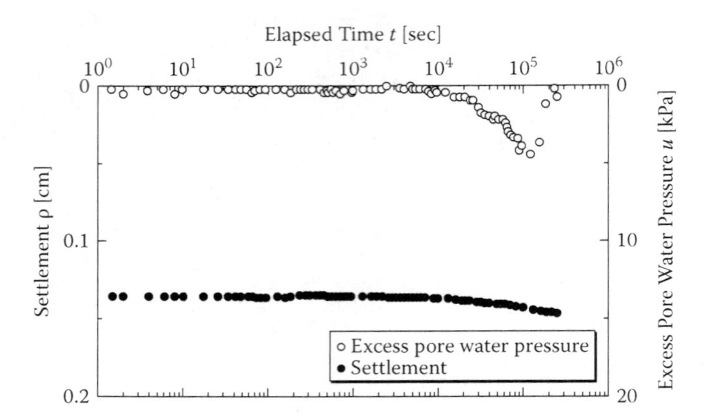

Figure 3.35 Observed variations of settlement and excess pore water pressure with time in oedometer test on crushed mudstone pebbles. (Kaneda, K. 1999. Dissertation for PhD in engineering, Nagoya University, 125–126.)

shows the computed variations of the excess pore water pressure with elapsed time at the element number 10 in Figure 3.26 for structured and nonstructured clays. It can be seen from Figures 3.33 and 3.34 that when delayed settlements occur for the structured clay, the pore water pressure, which is almost dissipated but is not exactly zero, increases again and then decreases to zero. As described before, an overconsolidated structured clay ($\rho_0 = 0.1$, $\omega_0 = 0.2$, $b = 40$) may not exhibit strain softening behavior (see Figure 3.22c).

Figure 3.35 shows the observed variations of the settlement and the excess pore water pressure with time in an oedometer test ($\sigma = 78$ kPa) on a saturated crushed mudstone; this is described in Kaneda (1999). It has been reported that the crushed mudstone shows a similar behavior with structured clays (Nakano, Asaoka, and Constantinescu 1998). In the experiment, drainage is allowed only at the top of the specimen, and the pore water pressure is measured at the bottom of the specimen in the same way as the condition of the simulation in Figure 3.26. The computed results in Figures 3.33 and 3.34 describe well the features observed in oedometer tests on structured soil in Figure 3.35—that is, delayed settlement and delayed development of pore water pressure.

Therefore, it can be considered that the delay of compression on saturated natural clay is mostly due to the bonding effect and the time-dependent behavior of clay. According to the present results, it is presumed that when a large overburden load is applied on an overconsolidated structured clayey ground, large delayed creep settlement may occur after excess pore water pressure is almost dissipated.

3.10 APPLICATION OF ADVANCED METHODS TO MODELING OF SOME OTHER FEATURES

As described in Section 3.5, other features such as temperature effect and behavior of unsaturated soils can be described using the present advanced modeling framework. In this section, only the outline and the key concepts for applying the present advanced methods (stages I to III) to the modeling of these features are described. The concrete derivation processes and the details of each modeling are described in the papers introduced in the following sections.

3.10.1 Temperature-dependent behavior

Figure 3.36 shows the observed results of oedometer tests of an undisturbed silty clay under constant temperatures in the range of 5°C to 45°C reported in the literature by Eriksson (1989). It is seen from this figure that the apparent yield stress decreases with an increase of temperature, and the consolidation curves shift almost parallel with temperature. Interestingly, these observed results are similar to the features displayed by strain rate effects described in Section 3.9.

Kikumoto et al. (2011) applied the present advanced method (stage III) to the modeling of temperature effects on soil behavior. In order to account for temperature effects, the difference of the current state variable ψ from its initial one ψ_0, which determines the position of the NCL on e–lnσ, is assumed to take the following form:

$$\psi - \psi_0 = \lambda_T (T - T_0) \tag{3.54}$$

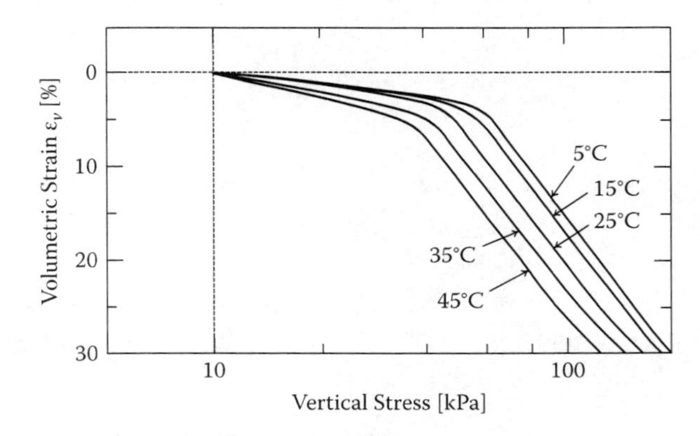

Figure 3.36 Observed results of oedometer tests on a silty clay with different temperatures. (Replotted from data in Eriksson 1989.)

In calculating the recoverable (elastic) change in void ratio, it is also necessary to consider the recoverable (elastic) change in void ratio $(-\Delta e)^{eT}$ due to a temperature change in addition to the elastic change of void ratio in eq. (3.2). Thus,

$$(-\Delta e)^e = \kappa \ln \frac{\sigma}{\sigma_0} + (-\Delta e)^{eT} = \kappa \ln \frac{\sigma}{\sigma_0} + \kappa_T (T - T_0) \tag{3.55}$$

Here, κ_T is related to the coefficient of thermal expansion α_T as follows:

$$\kappa_T = -3\alpha_T (1 + e_0) \tag{3.56}$$

The difference in void ratio due to the shift of the NCL given by eq. (3.54) also involves the thermoelastic component $(-\Delta e)^{eT}$, so eq. (3.54) can be rewritten as the sum of the plastic and elastic components:

$$\begin{aligned} \psi - \psi_0 &= \left(\psi^p - \psi_0^p\right) + \left(\psi^e - \psi_0^e\right) \\ &= (\lambda_T - \kappa_T)(T - T_0) + \kappa_T (T - T_0) \end{aligned} \tag{3.57}$$

Then, referring to Figure 3.15, the plastic change in void ratio $(-\Delta e)^p$ for soil due to temperature effects is expressed as

$$\begin{aligned} (-\Delta e)^p &= (-\Delta e) - (-\Delta e)^e \\ &= \left\{(e_{N0} - e_N) - (\rho_0 - \rho)\right\} - (-\Delta e)^e \\ &= \left\{\lambda \ln \frac{\sigma}{\sigma_0} + (\psi - \psi_0) - (\rho_0 - \rho)\right\} - \kappa \ln \frac{\sigma}{\sigma_0} - \left(\psi^e - \psi_0^e\right) \\ &= (\lambda - \kappa) \ln \frac{\sigma}{\sigma_0} - (\rho_0 - \rho) - \left(\psi_0^p - \psi^p\right) \end{aligned} \tag{3.58}$$

The yield function then follows in the same way as in eq. (3.25):

$$f = F - \left\{H + (\rho_0 - \rho) + \left(\psi_0^p - \psi^p\right)\right\} = 0 \tag{3.59}$$

The consistency condition ($df = 0$) gives the increment of plastic void ratio change following the same way as in eq. (3.27):

$$d(-e)^p = \frac{(\lambda - \kappa)\dfrac{d\sigma}{\sigma} + d\psi^p}{1 + G(\rho) + Q(\omega)} = \frac{(\lambda - \kappa)\dfrac{d\sigma}{\sigma} + (\lambda_T - \kappa_T)dT}{1 + G(\rho) + Q(\omega)} \tag{3.60}$$

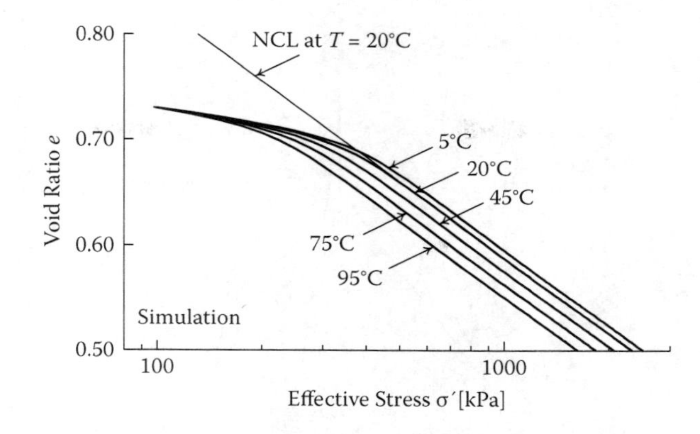

Figure 3.37 Simulations of oedometer tests on OC clay at different constant temperatures.

From eq. (3.55), the elastic component is expressed as

$$d(-e)^e = \kappa \frac{d\sigma}{\sigma} + \kappa_T dT \tag{3.61}$$

whereas the loading condition is described just as in eq. (3.33).

Figure 3.37 shows the computed results of 1D compression tests on an overconsolidated clay under different temperatures. The material parameters used for the simulations are the same as those in Table 3.1. The values for the additional material parameters for the simulation are as follows: $\lambda_T = 0.003$, $\kappa_T = -0.0001$, and $T_0 = 20°C$. It can be seen that the numerical results describe well the experimentally observed temperature effects shown in Figure 3.36.

Figure 3.38 shows the calculated relation between temperature and volumetric strain for clays under isotropic confining pressure ($p = 98$ kPa) with different OCRs as the temperature is increased from 20°C to 95°C and then cooled back to 20°C. It is interesting to note that volume contraction occurs during both increasing and decreasing temperatures for normally or lightly overconsolidated clays. On the other hand, the clay expands upon an increase in temperature for heavily overconsolidated clays. These tendencies agree well with the experimental results observed by Baldi et al. (1991) on Boom clay (see Figure 3.39).

We next examine lab experimental results (Boudali, Leroueil, and Srinivasa Murthy 1994) as shown in Figure 3.40, which pertains to constant strain rate oedometer tests conducted at various temperatures on naturally deposited clay. It is observed that changes in temperature and strain rate affect similarly the volumetric strain–stress behavior of the clay by shifting the compressibility curves accordingly. Thus it is possible to

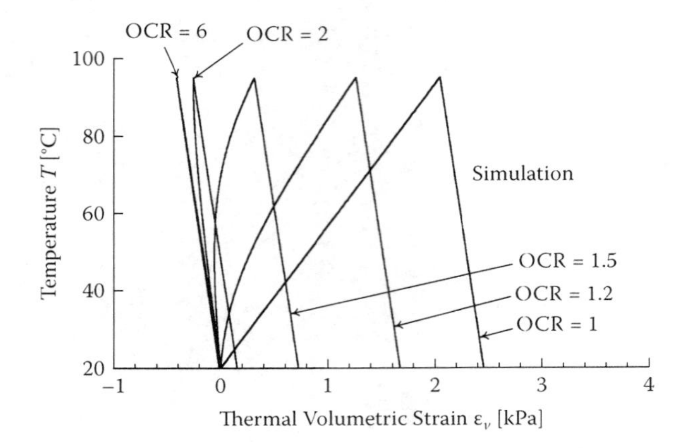

Figure 3.38 Calculated variations of volumetric strains in clays at different OC ratios with change in temperature under isotropic stress conditions.

model both of these effects by using the proposed constitutive modeling framework in stage III.

Figure 3.41 shows the numerical simulation results of both temperature and strain rate effects using the stage III constitutive model with the material parameters listed in Table 3.2. Figure 3.41(a) illustrates the relation

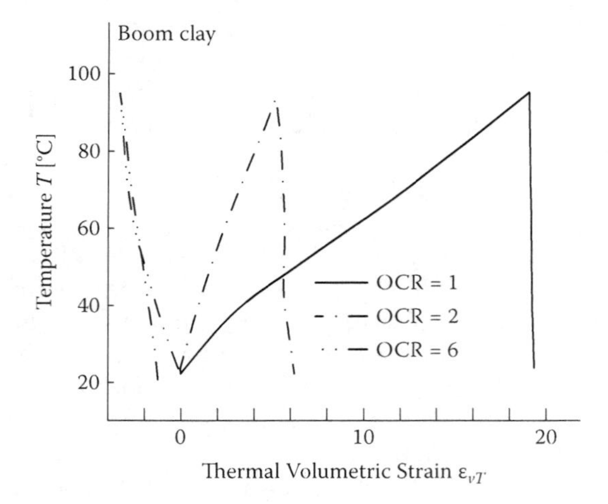

Figure 3.39 Observed variations of volumetric strains in Boom clay at different OC ratios with change of temperature under isotopic stress conditions. (Replotted from data in Baldi, G. et al. 1991. Report ER 13365, Commission of the European Communities, Nuclear Science and Technology.)

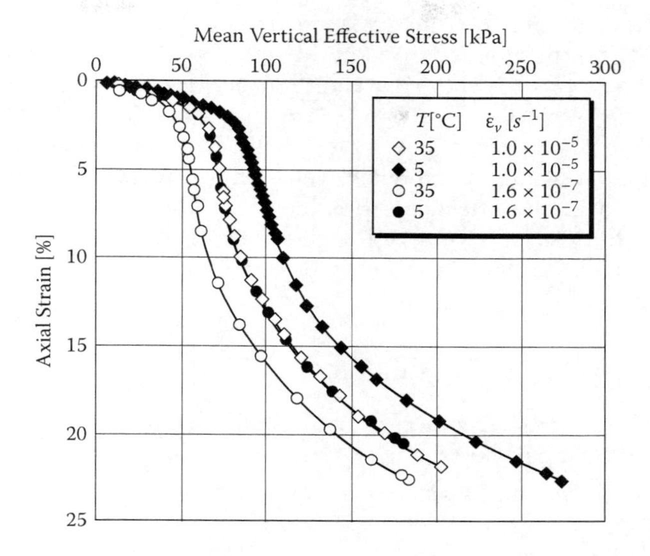

Figure 3.40 Observed constant strain rate oedometer tests on naturally deposited clay under different strain rates and temperature. (Replotted from data in Boudali, M. et al. 1994. *Proceedings of 13th International Conference on Soil Mechanics and Foundation Engineering,* New Delhi, 1:411–416.)

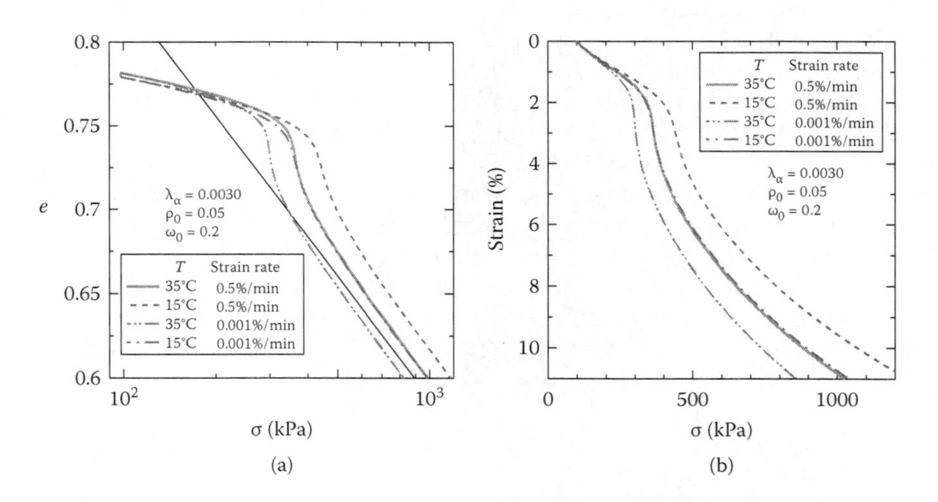

Figure 3.41 Calculated results of one-dimensional compression tests on structured clay under different strain rates and temperatures. (a) Void ratio: lnσ relation; (b) axial strain: σ relation.

between the void ratio and the stress in log scale, whereas in Figure 3.41(b), the computed results are replotted in the same way as those in Figure 3.40. Note that in Figure 3.41(a), the straight line denotes the NCL at a reference state ($(-\dot{e})^p = (-\dot{e})^p_{ref} = 1\times 10^{-7}$/min, $T = T_{ref} = 20°\text{C}$). It can be seen that the simulated results describe well the coupling features of temperature and strain rate effects quite comprehensively. Although the modeling of both temperature and time effects has been developed by Modaressi and Laloui (1997), Yashima et al. (1998), Zhang and Zhang (2009), and others, the present modeling is thought to be simpler and more effective in describing the same physical phenomenon.

3.10.2 Unsaturated soil behavior

As an effective stress common to saturated, partially saturated, and dry soils, the following Bishop's effective stress σ'' has been most widely known (Bishop 1959):

$$\sigma'' = \sigma - u_a + \chi s \cong \sigma - u_a + S_r s \qquad (\text{where} \quad s = u_a - u_w \geq 0) \tag{3.62}$$

Here, u_a and u_w are the pore air pressure and the pore water pressure, respectively. The difference $s(= u_a - u_w)$ is named as suction. The variable χ varies from zero to unity depending on the saturation S_r—that is, $\chi = 0$ at $S_r = 0$ and $\chi = 1$ at $S_r = 1$, so the second approximate equality sign in eq. (3.62) holds.

Now it is experimentally known that when a partially saturated soil ($0 < S_r < 1$)—loose soil in particular—is soaked so that suction is lost, large volume contraction may occur, leading to the so-called collapse phenomenon. On the other hand, according to eq. (3.62), the effective stress decreases during this soaking process. Hence, it is impossible to describe the collapse phenomenon by replacing the well known effective stress $\sigma' = \sigma - u_w$ used in models for saturated soils with Bishop's effective stress σ'' alone, because elastic volume expansion occurs with a decrease of effective stress, according to the usual concept for saturated soils.

Some models have been developed to describe this typical behavior of unsaturated soils by introducing other variables related to the suction (or the saturation) in addition to the effective stress σ'' or combining the net stress $\sigma^{net}(= \sigma - u_a)$ and the suction s—for example, Karube and Kato (1989), Alonso, Gens, and Josa (1990), Fredlund and Morgenstern (1977), Wheeler and Sivakumar (1995), Kohgo, Nakano, and Miyazaki (1993), Khalilia and Khabbaz (1998), and others.

Figure 3.42 shows the observed void ratio (e) versus Bishop's vertical stress (σ'') relation obtained from oedometer tests carried out by Honda

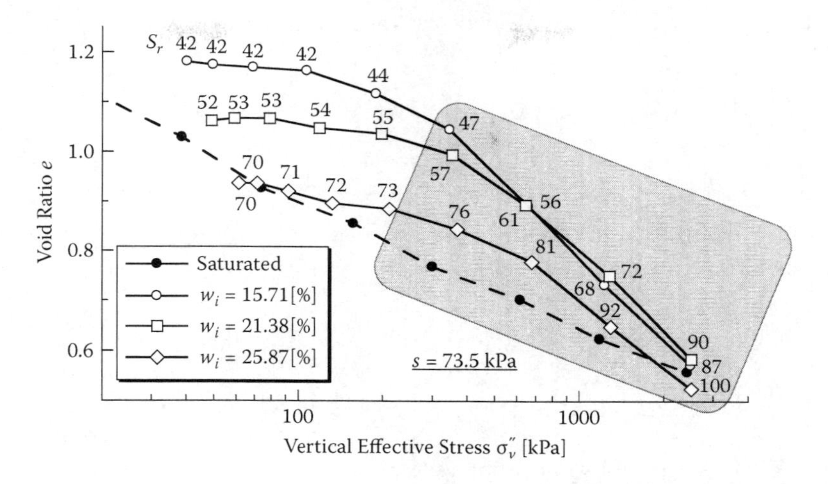

Figure 3.42 Observed results of oedometer tests on unsaturated clays under constant suction. (Replotted from data in Honda, M. 2000. Dissertation for PhD in engineering, Kobe University, 125–126.)

(2000) on unsaturated Catalpo clay with different initial water contents (w_i) under constant suction ($s = 73.5$ kPa). The solid symbol curve refers to the saturated case, whereas the other curves correspond to unsaturated samples with the evolution of saturation (percentage) shown at each stress level during consolidation. In these tests, although the region where σ'' is relatively small may be under overconsolidation state, the region at relatively large σ'' can be considered to be at normally consolidated state.

It is seen from this figure that the void ratio of partially saturated soils at the normally consolidated state decreases with the increase in degree of saturation. Although there is some scattering of the data, the NCL seems to be subject to an approximately parallel shift with degree of saturation. Based on these observations, Kikumoto et al. (2010) and Kyokawa et al. (2010) applied the present advanced method (stage III) to model unsaturated soil behavior. Referring to Figure 3.15, the plastic change of void ratio $(-\Delta e)^p$ for unsaturated soil is expressed in the same form as eq. (3.24):

$$\begin{aligned}(-\Delta e)^p &= (-\Delta e) - (-\Delta e)^e \\ &= (\lambda - \kappa)\ln\frac{\sigma''}{\sigma_0''} - (\rho_0 - \rho) - (\psi_0 - \psi)\end{aligned} \tag{3.63}$$

Then, the yield function is also obtained in the same form as eq. (3.25):

$$f = F - \{H + (\rho_0 - \rho) + (\psi_0 - \psi)\} = 0 \quad \left(\text{where } F = (\lambda - \kappa)\ln\frac{\sigma''}{\sigma''_0},\ H = (-\Delta e)^p \right) \tag{3.64}$$

The current and initial state variables ψ and ψ_0 are simply related to degree of saturation S_r as follows:

$$\psi = lS_r, \qquad \psi_0 = lS_{r0} \qquad (\text{where} \quad l : \text{material parameter}) \tag{3.65}$$

Then, when the initial state (ψ_0) is considered to be under fully saturated condition ($S_{r0} = 1$), the difference of the void ratio between the current NCL and that of saturated soil is expressed as

$$\psi - \psi_0 = -l(1 - S_r) \tag{3.66}$$

From the consistency condition ($df = 0$) and eq. (3.20), the increment of plastic void ratio change is given by

$$d(-e)^p = \frac{(\lambda - \kappa)\dfrac{d\sigma''}{\sigma''} + d\psi}{1 + G(\rho) + Q(\omega)} = \frac{(\lambda - \kappa)\dfrac{d\sigma''}{\sigma''} + \dfrac{\partial \psi}{\partial S_r} dS_r}{1 + G(\rho) + Q(\omega)} \tag{3.67}$$

The increment of elastic void ratio change is expressed in the same way as eq. (3.9):

$$d(-e)^e = \kappa \frac{d\sigma''}{\sigma''} \tag{3.68}$$

In formulating the constitutive model for unsaturated soils, the relation between the suction s and the saturation S_r, which is called "soil water characteristic curve (SWCC)," is also necessary. The degree of saturation S_r is related not only to the suction s, but also to the history of wetting and drying and others. Thus several soil water characteristic curves have been presented—for example, in van Genuchten (1980). Figure 3.43 replots the results of Figure 3.42 in terms of the relation between the saturation S_r and vertical effective stress σ'' (Honda 2000). It is seen that the saturation S_r increases with the increasing of the effective stress σ'' (i.e., the decreasing

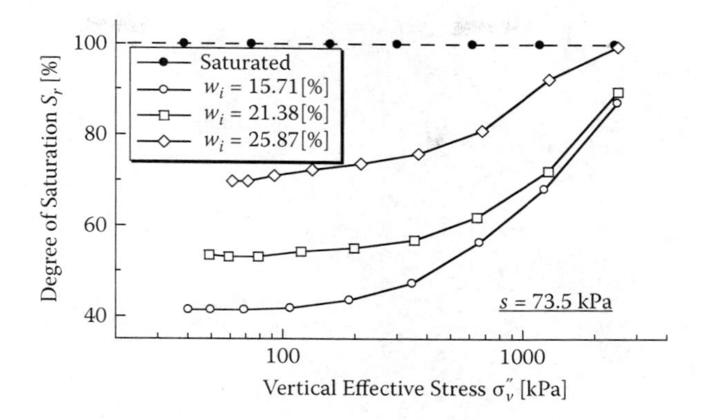

Figure 3.43 Observed variations of saturation in oedometer tests on unsaturated clays under constant suction. (Replotted from data in Honda, M. 2000. Dissertation for PhD in engineering, Kobe University, 125–126.)

of the void ratio e), even if the suction s is constant. Kikumoto et al. (2010) presented the relation between the saturation and the suction in which the influence of the void ratio as well as the influence of the history of wetting and drying are taken into consideration. To consider the influence of void ratio, the following modified suction s^* is introduced:

$$s^* = s\left(\frac{e}{N_{sat}}\right)^{\xi_e} \quad \text{(where } N_{sat} = \text{void ratio of NCL for saturated soil at 98 kPa)} \tag{3.69}$$

Then, they obtained the SWCC in the form of

$$f_{SW}(S_r, s^*, I_w) = 0 \tag{3.70}$$

As shown in Figure 3.44, I_w is the variable that represents the current relative position of S_r, which is determined by the history of drying and wetting and then is defined as

$$I_w = \frac{S_r - S_r^w}{S_r^d - S_r^w} \tag{3.71}$$

Here, S_r^d and S_r^w represent the saturation on the main drying and wetting curves. From the preceding equations and the consistency condition on

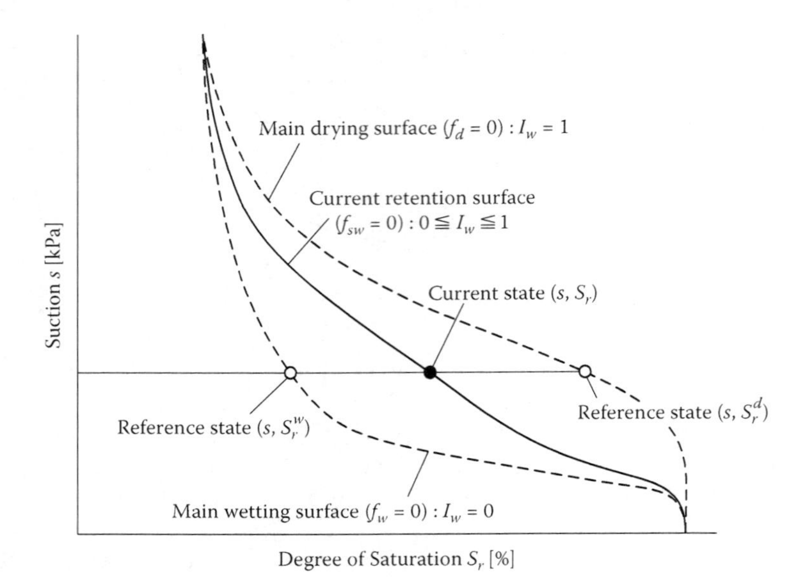

Figure 3.44 Schematic diagram of soil water characteristic curve (SWCC) using main wetting and drying curves.

SWCC ($df_{SW} = 0$), the increment of the saturation is expressed as

$$dS_r = \frac{\dfrac{\partial f_{SW}}{\partial s^*} ds^*}{\dfrac{\partial f_{SW}}{\partial S_r} + \dfrac{\partial f_{SW}}{\partial I_w}\dfrac{\partial I_w}{\partial S_r}} \tag{3.72}$$

The details of the formulation and the material parameters used for the present SWCC are described in papers by Kikumoto, Kyokawa, et al. (2009) and Kikumoto et al. (2010).

Figure 3.45 shows the calculated results of the oedometer tests of unsaturated clay with different initial water contents under constant suctions, corresponding to the observed results in Figures 3.42 and 3.43. In the simulations, the material parameters in Table 3.1 are used, and the value of the added material parameter l for considering the behavior of unsaturated soils is 0.5. It can be seen that the present model describes the observed features of unsaturated soil in oedometer tests.

Figure 3.46 shows the calculated results of the oedometer tests in which the unsaturated clays are consolidated to some stresses under constant suction of 294 kPa and then are soaked to fully saturated condition. Each number in the figure denotes the saturation at the corresponding condition. Observed

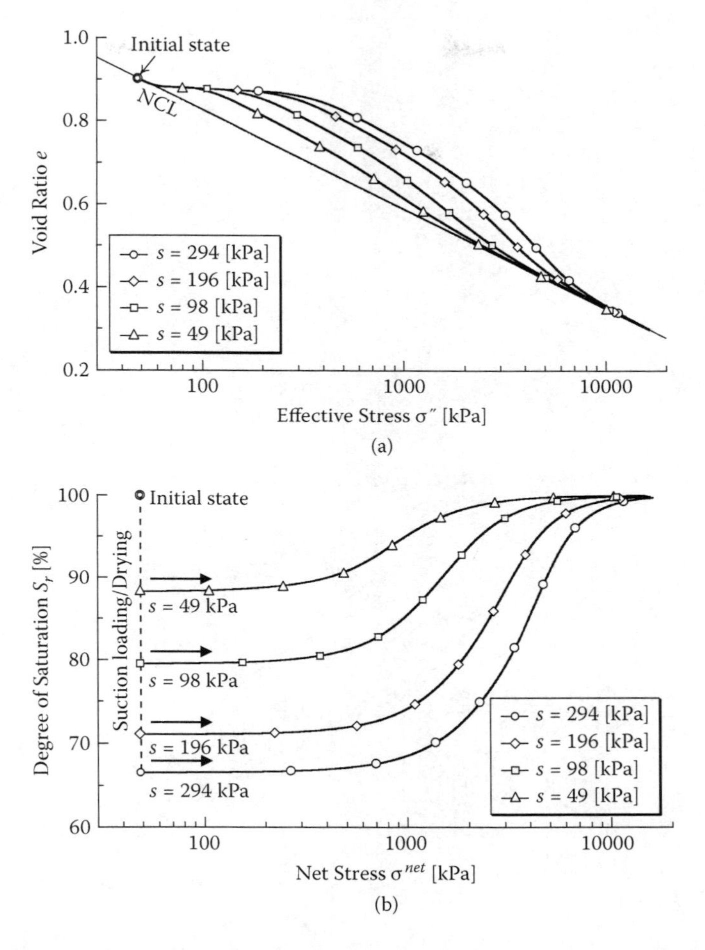

Figure 3.45 Calculated results of oedometer tests of unsaturated clays under constant suctions. (a) σ'' versus e relation; (b) σ^{net} versus S_r relation.

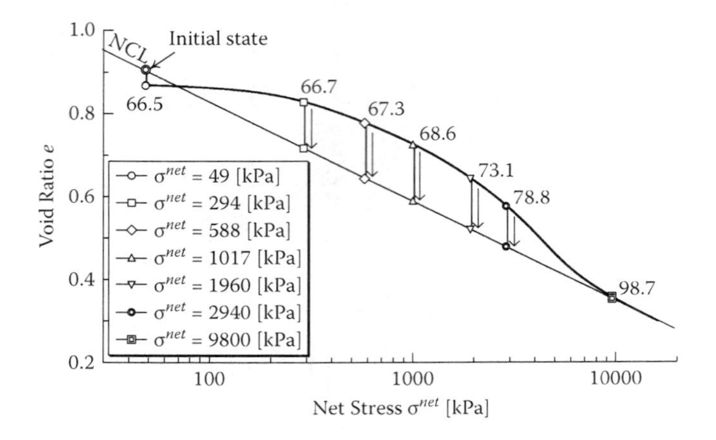

Figure 3.46 Calculated results of consolidation and soaking tests of unsaturated soils.

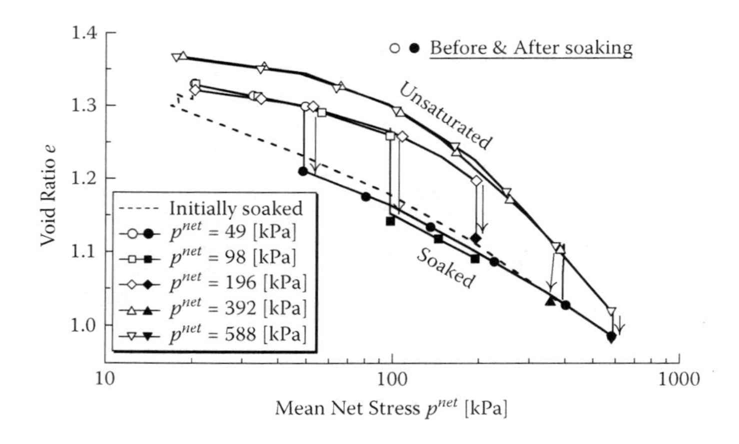

Figure 3.47 Observed results of consolidation and soaking tests on unsaturated clays. (Replotted from data in Sun, D. A. et al. 2007. *Canadian Geotechnical Journal* 44:673–686.)

results of such oedometer tests with soaking were reported by Jennings and Burland (1962), Sun, Sheng, and Xu (2007), and others. Figure 3.47 shows the observed results of an unsaturated soil upon which consolidation under constant suction and subsequent soaking are carried out. The simulated results in Figure 3.46 describe the observed features of unsaturated soil in Figure 3.47 (e.g., the collapse phenomenon and the agreement of the void ratios after soaking with that of NCL for saturated soil).

Chapter 4

Ordinary modeling of three-dimensional soil behavior

4.1 INTRODUCTION

In this chapter, the formulations of elastoplastic models for soils using the stress invariants (p and q) used in ordinary soil models are presented. As a three-dimensional (3D) extension of the well known one-dimensional (1D) consolidation behavior of normally consolidated clay (linear relation between e and $\ln\sigma$), the original and modified Cam clay models are explained. The validity of these models in general 3D stress conditions is also discussed based on the experimental results.

4.2 OUTLINE OF ORDINARY ELASTOPLASTIC MODELS SUCH AS THE CAM CLAY MODEL

As described in Section 2.3 in Chapter 2, the yield function of 3D elastoplastic models is given in the form of eq. (2.10). In a precise term, the yield surface is formulated using the stress tensor σ_{ij} and plastic strain tensor ε_{ij}^{p}. However, the stress tensor and the plastic strain tensor each consists of six independent variables, so it is rather difficult to formulate the yield function using these tensors directly. Even assuming isotropic materials, the yield function depends on three stress invariants (e.g., three principal stresses) and three plastic strain invariants (e.g., three principal plastic strains).

Moreover, most elastoplastic models are formulated using one or two stress invariants, such as the mean stress (p) and the deviatoric stress (q), and the corresponding plastic strain increment invariants like the plastic volumetric strain (ε_{v}^{p}) and the plastic deviatoric strains (ε_{d}^{p}). For instance, one of the most well known constitutive models for soils such as the Cam clay model (Schofield and Wroth 1968; Roscoe and Burland 1968) has been formulated using stress invariants such as the mean stress p and the deviatoric stress q and the corresponding plastic strain and strain increment invariants.

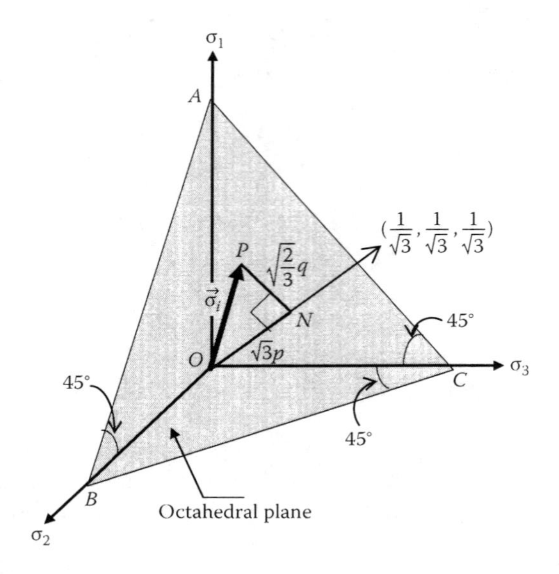

Figure 4.1 Definitions of p and q.

Figure 4.1 shows the octahedral plane in principal stress space, where the current principal stress vector is represented by $\overrightarrow{\mathrm{OP}}$. Then, the mean stress p and the deviatoric stress q are defined by the normal and in-plane components of the stress with respect to the octahedral plane and by eqs. (4.1) and (4.2) using three principal stresses or a stress tensor:

$$p = \sqrt{\frac{1}{3}}\,\overline{\mathrm{ON}} = \frac{1}{3}(\sigma_1 + \sigma_2 + \sigma_3) = \frac{1}{3}\sigma_{ij}\delta_{ij} \tag{4.1}$$

$$q = \sqrt{\frac{3}{2}}\,\overline{\mathrm{NP}} = \frac{1}{\sqrt{2}}\sqrt{(\sigma_1 - \sigma_2)^2 + (\sigma_2 - \sigma_3)^2 + (\sigma_3 - \sigma_1)^2}$$

$$= \sqrt{\frac{3}{2}(\sigma_{ij} - p\delta_{ij})(\sigma_{ij} - p\delta_{ij})} \tag{4.2}$$

The volumetric strain increment $d\varepsilon_v$ and the deviatoric strain increment $d\varepsilon_d$ are also defined by the normal and in-plane components of the principal strain increment vector $\overrightarrow{\mathrm{O'P'}}$ with respect to the octahedral plane as shown in Figure 4.2. They are expressed as

$$d\varepsilon_v = \sqrt{3}\,\overline{\mathrm{O'N'}} = d\varepsilon_1 + d\varepsilon_2 + d\varepsilon_3 = d\varepsilon_{ij}\delta_{ij} \tag{4.3}$$

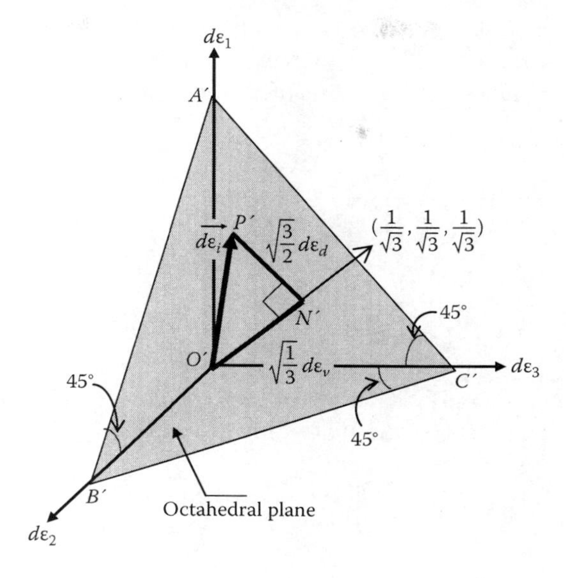

Figure 4.2 Definitions of $d\varepsilon_v$ and $d\varepsilon_d$.

$$d\varepsilon_d = \sqrt{\frac{2}{3}}\,\overline{\mathrm{N'P'}} = \frac{\sqrt{2}}{3}\sqrt{(d\varepsilon_1 - d\varepsilon_2)^2 + (d\varepsilon_2 - d\varepsilon_3)^2 + (d\varepsilon_3 - d\varepsilon_1)^2}$$

$$= \sqrt{\frac{2}{3}\left(d\varepsilon_{ij} - \frac{d\varepsilon_v}{3}\delta_{ij}\right)\left(d\varepsilon_{ij} - \frac{d\varepsilon_v}{3}\delta_{ij}\right)} \qquad (4.4)$$

Under triaxial compression ($\sigma_1 > \sigma_2 = \sigma_3$) and extension ($\sigma_1 = \sigma_2 > \sigma_3$) conditions, these stress and strain increment invariants are expressed as

$$\begin{cases} p = \dfrac{\sigma_1 + 2\sigma_3}{3}, \qquad q = (\sigma_1 - \sigma_3) \\ \\ d\varepsilon_v = d\varepsilon_1 + 2d\varepsilon_2, \quad d\varepsilon_d = \dfrac{2}{3}(d\varepsilon_1 - d\varepsilon_3) \end{cases} \quad \text{(under triaxial compression)} \qquad (4.5)$$

$$\begin{cases} p = \dfrac{2\sigma_1 + \sigma_3}{3}, \qquad q = \sigma_1 - \sigma_3 \\ \\ d\varepsilon_v = 2d\varepsilon_1 + d\varepsilon_2, \quad d\varepsilon_d = \dfrac{2}{3}(d\varepsilon_1 - d\varepsilon_3) \end{cases} \quad \text{(under triaxial extension)} \qquad (4.6)$$

The plastic volumetric strain (ε_v^p) and plastic deviatoric strain (ε_d^p) are also defined as the normal and in-plane components of the plastic strain with respect to the octahedral plane in the same way as eqs. (4.3) and (4.4):

$$\varepsilon_v^p = \varepsilon_1^p + \varepsilon_2^p + \varepsilon_3^p = \varepsilon_{ij}^p \delta_{ij} \tag{4.7}$$

$$\begin{aligned} \varepsilon_d^p &= \frac{\sqrt{2}}{3}\sqrt{\left(\varepsilon_1^p - \varepsilon_2^p\right)^2 + \left(\varepsilon_2^p - \varepsilon_3^p\right)^2 + \left(\varepsilon_3^p - \varepsilon_1^p\right)^2} \\ &= \sqrt{\frac{2}{3}\left(\varepsilon_{ij}^p - \frac{\varepsilon_v^p}{3}\delta_{ij}\right)\left(\varepsilon_{ij}^p - \frac{\varepsilon_v^p}{3}\delta_{ij}\right)} \end{aligned} \tag{4.8}$$

Under triaxial compression ($\sigma_1 > \sigma_2 = \sigma_3$) and extension ($\sigma_1 = \sigma_2 > \sigma_3$) conditions, these plastic strain invariants are expressed as

$$\varepsilon_v^p = \varepsilon_1^p + 2\varepsilon_3^p, \qquad \varepsilon_d^p = \frac{2}{3}\left(\varepsilon_1^p - \varepsilon_3^p\right) \qquad \text{(under triaxial compression)} \tag{4.9}$$

$$\varepsilon_v^p = 2\varepsilon_1^p + \varepsilon_3^p, \qquad \varepsilon_d^p = \frac{2}{3}\left(\varepsilon_1^p - \varepsilon_3^p\right) \qquad \text{(under triaxial extension)} \tag{4.10}$$

In metal plasticity, it is usually assumed that there is no plastic volumetric strain (ε_v^p) and that both plastic deformation and strength are independent of mean stress (p), but rather depend on deviatoric stress (q) alone. Then, the yield function for such a material can be written using solely deviatoric stress (q) and plastic deviatoric strains (ε_d^p) and following eq. (2.10) in Chapter 2:

$$f = F(q) - H\left(\varepsilon_d^p\right) = 0 \tag{4.11}$$

On the other hand, it is well known that plastic deformations in soils and their strength are greatly influenced not only by the deviatoric stress but also by the mean stress. Typically, plastic volumetric strains (or plastic void ratio change) occur either as irrecoverable compression in consolidation tests or as dilatancy in shear tests. Therefore, the yield function for soils is fundamentally formulated using the mean and deviatoric stresses (p and q) and the plastic volumetric and/or deviatoric strains ($\varepsilon_v^p, \varepsilon_d^p$):

$$\begin{cases} f = F(p, q) - H(\varepsilon_v^p, \varepsilon_d^p) = 0 \\ \text{or} \\ f = F(p, \eta = q/p) - H(\varepsilon_v^p, \varepsilon_d^p) = 0 \end{cases} \tag{4.12}$$

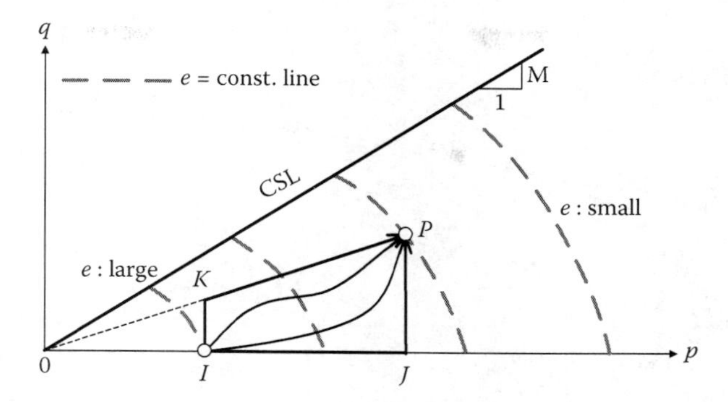

Figure 4.3 Explanation of equivoid ratio lines by D. J. Henkel (1960. *Geotechnique* 10 (2): 41–45.)

In order to construct a proper yield function for soils, the following two experimental findings are taken into consideration. One is the existence of lines of equal void ratio (water content) in the q versus p space for normally consolidated clays as found by Henkel (1960) and shown by broken lines in Figure 4.3. Therefore, the change in void ratio ($-\Delta e$) of a normally consolidated clay is determined only by the initial and current stress states alone and does not depend on the stress paths between the initial and current states (e.g., from point I to point P in Figure 4.3).

The other experimental finding pertains to shear tests at constant mean principal stress on normally consolidated clays, where the stress ratio versus strain relations are found to be independent of the magnitude of mean principal stress. Figure 4.4 shows

(a) the observed relation between stress ratio $\eta = q/p$, deviatoric strain ε_d, and volumetric strain ε_v for triaxial tests on normally consolidated Fujinomori clay under different mean principal stresses
(b) schematic changes of void ratio in the same tests in the e–ln$p$ plane

Figure 4.5 illustrates the stress ratio–strain volume change curves in Figure 4.4 schematically. With an increase of the stress ratio, positive volumetric strain (volume contraction) occurs, and clays finally reach a perfectly plastic state without volume change at the stress ratio $\eta = \mathrm{M}$, which is called the "critical state." The solid lines in diagram (b) of Figure 4.4 indicate both the normal consolidation line (NCL) and the critical state line (CSL). Therefore, the volumetric strain under constant mean stress can be expressed as an increasing function $\xi(\eta)$ in spite of the magnitude of the mean stresses, as shown in Figure 4.6.

$$\varepsilon_v = \xi(\eta) \qquad (\text{where } \xi(0) = 0) \tag{4.13}$$

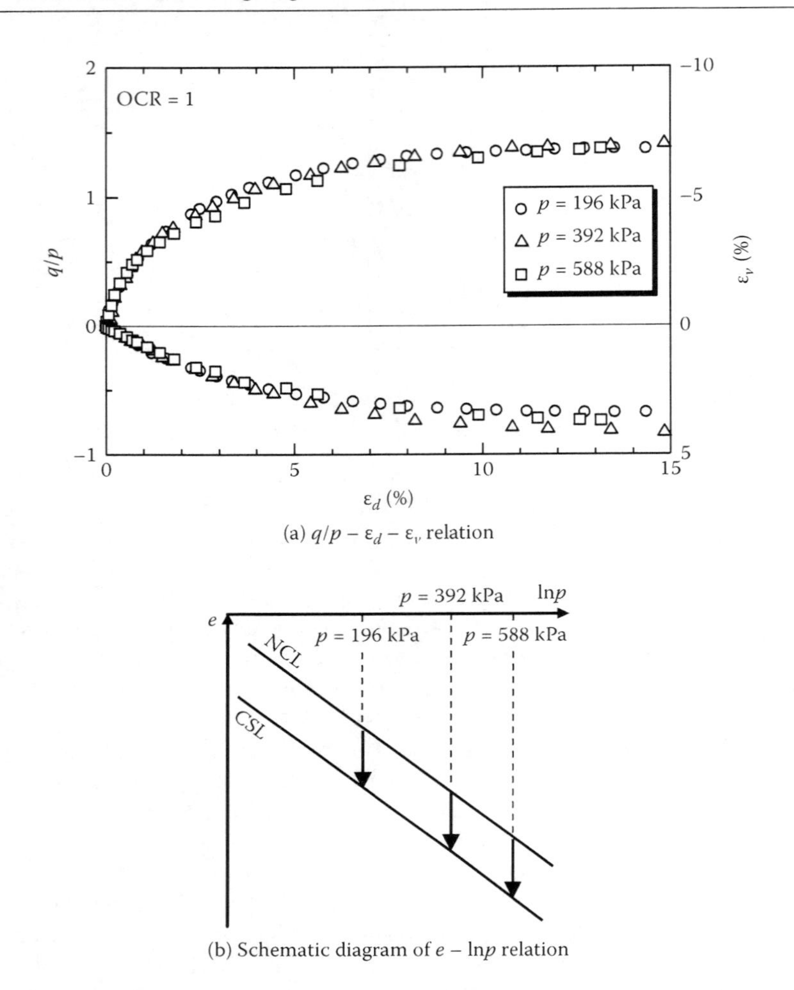

Figure 4.4 Test results of triaxial compression tests on normally consolidated clays (OCR = 1) under constant mean principal stress.

In the case that the volumetric strain is expressed by a linear function of the stress ratio in the form of

$$\varepsilon_v = \xi(\eta) = D\eta, \tag{4.14}$$

the coefficient D is called Shibata's dilatancy coefficient (Shibata 1963).

Now, let us turn to Figure 4.3, where the mean stress, stress ratio, and void ratio at the initial state (point I) and the current state (point P) are represented by ($p = p_0$, $\eta = 0$, $e = e_0$) and ($p = p$, $\eta = \eta$, $e = e$), respectively. Furthermore, consider two different stress paths that both connect

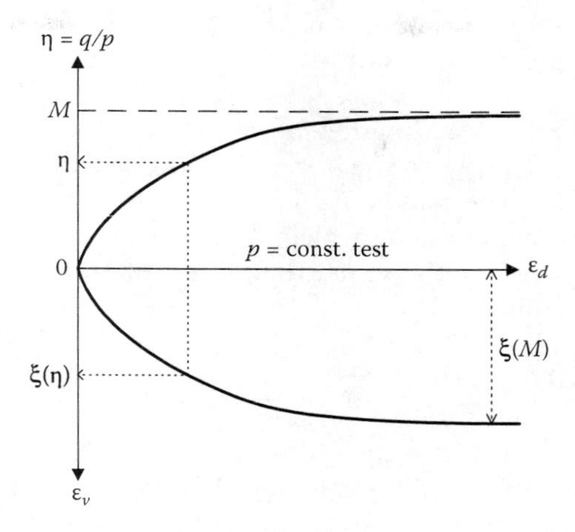

Figure 4.5 Stress ratio (η) versus deviatoric strain (ε_d) versus volumetric strain (ε_v) curves of normally consolidated soils under constant mean principal stresses.

point I to point P: path I–J–P (isotropic compression at $\eta = 0$ and subsequent pure shear loading with $p = p$) and path I–K–P (pure shear loading at $p = p_0$ and subsequent anisotropic compression with $\eta = \eta$). As shown in Figure 4.3, the deviatoric strain (ε_d) and the volumetric strain (ε_v) developed along paths I–K and J–P must be the same because the stress ratio–strain–dilatancy relation under shear loading is independent of the mean stresses. As a result, the volumetric strain (ε_v) along paths K–P and I–J is also the same because the lines of equal void ratio shown in Figure 4.3 are unique.

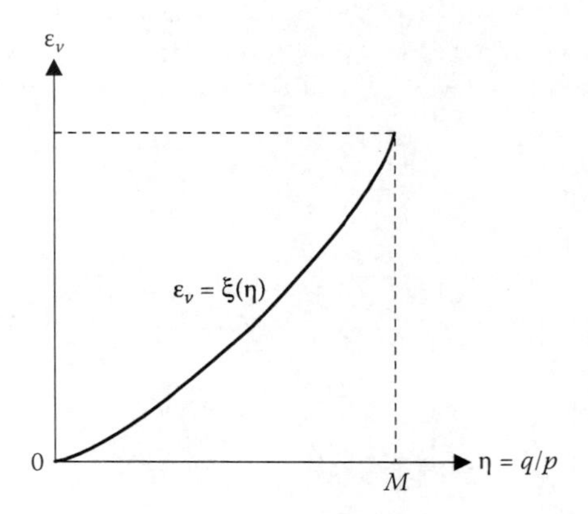

Figure 4.6 Relation between volumetric strain (ε_v) and stress ratio (η) for normally consolidated soils under constant mean principal stresses.

On the other hand, although no deviatoric strain (ε_d) occurs along path I–J (isotropic compression of η = 0), deviatoric strain inevitably develops along path K–P under anisotropic compression with η = η. Therefore, we conclude that deviatoric strain depends on the stress paths followed, even if both the initial and current stresses are the same. It can be seen that the plastic deviatoric strain (ε_d^p) is not suitable as a hardening parameter for the yield function because elastic strain is independent of stress path. Thus, the yield function for soils proposed in eq. (4.12) is reduced to

$$f = F(p, \eta = q/p) - H\left(\varepsilon_v^p\right) = 0 \tag{4.15}$$

The yield function used in the Cam clay model will be derived based on the preceding discussions. For example, when the stress state evolves from point I to point P along any stress paths in Figure 4.3, the ensuing change in void ratio is obtained as the summation of void ratio changes under isotropic consolidation (I–J) and pure shear loading (J–P). When the stress condition moves from the point I to point P in Figure 4.3, total change of void ratio ($-\Delta e$) can be expressed as a summation of the component due to isotropic compression between I and J and the component due to shear loading between J and K:

$$(-\Delta e) = e_0 - e = \lambda \ln \frac{p}{p_0} + (1 + e_0)\xi(\eta) \tag{4.16}$$

Here, the component due to isotropic compression is expressed using the compression index λ, and the component due to shear loading is given by eq. (4.13).

On the other hand, assuming the volume change due to shear (volumetric strain expressed by eq. 4.13) is irrecoverable (plastic), the elastic change of void ratio that occurs between I and P is expressed as follows using swelling index κ:

$$(-\Delta e)^e = \kappa \ln \frac{p}{p_0} \tag{4.17}$$

From eqs. (4.16) and (4.17), the plastic change in void ratio is expressed as

$$(-\Delta e)^p = (-\Delta e) - (-\Delta e)^e = (\lambda - \kappa) \ln \frac{p}{p_0} + (1 + e_0)\xi(\eta)$$

$$= (\lambda - \kappa) \left\{ \ln \frac{p}{p_0} + \zeta(\eta) \right\} \tag{4.18}$$

Here, because of $\zeta(\eta)=\{(1+e_0)/(\lambda-\kappa)\}\xi(\eta)$, $\zeta(\eta)$ is also an increasing function of η that satisfies $\zeta(\eta)=0$. Rewriting eq. (4.18), the yield function of normally consolidated soils in multidimensional conditions is obtained in the same form as that in 1D conditions (see eqs. 3.4–3.6):

$$f = F(p, \eta = q/p) - H((-\Delta e)^p) = 0 \tag{4.19}$$

where

$$F = (\lambda - \kappa)\left\{\ln\frac{p}{p_0} + \zeta(\eta)\right\} = (\lambda - \kappa)\ln\frac{p_1}{p_0} \tag{4.20}$$

$$H = (-\Delta e)^p = (1+e_0)\varepsilon_v^p \tag{4.21}$$

The solid curve in Figure 4.7 shows the yield surface on p–q plane represented by eq. (4.20). The broken curve shows the initial yield surface when $p = p_0$ and $\eta = 0$. The yield function (yield surface) is convex in the stress space, as illustrated in the figure. Here, p_0 and p_1 are the values of the mean stress on the p-axis for the initial yield surface (broken curve) and the current yield surface (solid curve), respectively, which determine the size of these surfaces. Comparing eqs. (4.19) to (4.21) with the corresponding eqs. (3.4) to (3.6) in the 1D model for normally consolidated soils, it is seen that the yield function of an ordinary elastoplastic model for normally consolidated soils such as the Cam clay model can be easily obtained by replacing σ_0 and σ in the function F of the 1D model with the mean quantities p_0 and p_1 in 3D conditions. Now, the stress ratio function $\zeta(\eta)$ in eq. (4.20) is given as follows for the original Cam clay

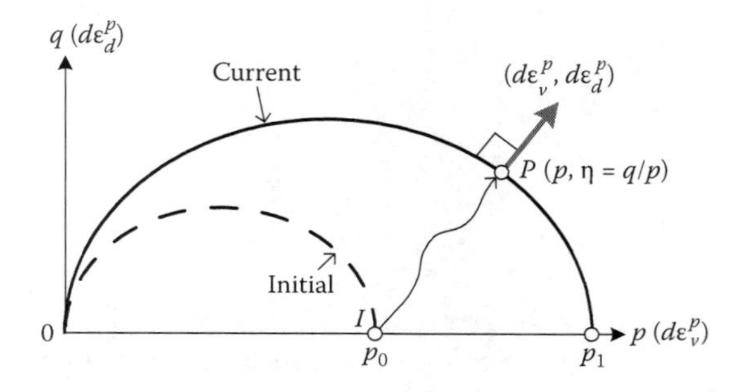

Figure 4.7 Initial and current yield surfaces in the p–q plane and direction of plastic flow in an ordinary model such as the Cam clay model.

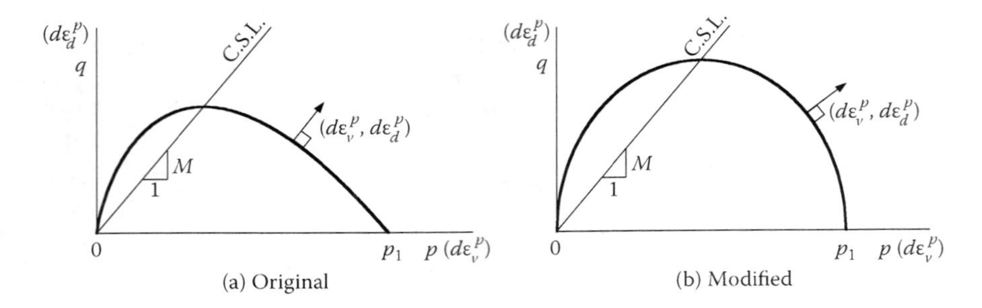

Figure 4.8 Yield surface of Cam clay model on p–q plane.

model (Schofield and Wroth 1968) and the modified Cam clay model (Roscoe and Burland 1968):

$$\varsigma(\eta) = \frac{1}{\mathrm{M}}\eta \qquad \text{(original)} \tag{4.22}$$

$$\varsigma(\eta) = \ln\frac{\mathrm{M}^2 + \eta^2}{\mathrm{M}^2} \qquad \text{(modified)} \tag{4.23}$$

Here, M is the stress ratio η at critical state, as shown in Figure 4.5. The shape of yield surfaces in the original Cam clay model and the modified Cam clay model are illustrated in Figure 4.8.

As described in Chapter 2, since the direction of the plastic strain increment is normal to the yield surface according to the associated flow rule in eq. (2.11), the plastic strain increment can be obtained as follows:

$$d\varepsilon_{ij}^p = \Lambda\frac{\partial F}{\partial \sigma_{ij}} = \Lambda\left(\frac{\partial F}{\partial p}\frac{\partial p}{\partial \sigma_{ij}} + \frac{\partial F}{\partial \eta}\frac{\partial \eta}{\partial \sigma_{ij}}\right) \tag{4.24}$$

From the consistency condition ($df = 0$), the proportionality constant Λ in eq. (4.24) is given by

$$\Lambda = \frac{dF}{(1+e_0)\frac{\partial F}{\partial \sigma_{kk}}} = \frac{dF}{h^p} \qquad \left(\text{where } dF = \frac{\partial F}{\partial \sigma_{ij}}d\sigma_{ij}\right) \tag{4.25}$$

Here, h^p represents the plastic modulus. As the directions of the principal plastic strain increments coincide with those of the principal stresses in an isotropic elastoplastic model, the axes of the plastic volumetric strain increment $d\varepsilon_v^p$ and the plastic deviatoric strain increment $d\varepsilon_d^p$ coincide with those of the mean principal stress p and the deviatoric stress q, respectively, Then, the direction of plastic strain increment ($d\varepsilon_v^p, d\varepsilon_d^p$) is also normal to the yield surface, as shown in Figures 4.7 and 4.8. From this normality condition, the ratio of the plastic volumetric strain increments $d\varepsilon_v^p/d\varepsilon_d^p$ is calculated as

$$\frac{d\varepsilon_v^p}{d\varepsilon_d^p} = \frac{\frac{\partial F}{\partial p} + \frac{\partial F}{\partial \eta}\frac{\partial \eta}{\partial p}}{\frac{\partial F}{\partial \eta}\frac{\partial \eta}{\partial q}} = \frac{1 - \zeta'(\eta)\cdot\eta}{\zeta'(\eta)} \tag{4.26}$$

where $\zeta'(\eta)$ means $d\zeta(\eta)/d\eta$. Therefore, it can be seen that there holds a unique relation between the stress ratio and the plastic strain increment ratio, which is called the "stress–dilatancy relation." For the cases of the original and modified Cam clay models, this relation is expressed as follows from eqs. (4.22), (4.23), and (4.26):

$$\frac{d\varepsilon_v^p}{d\varepsilon_d^p} = \mathrm{M} - \eta \qquad \text{(original)} \tag{4.27}$$

$$\frac{d\varepsilon_v^p}{d\varepsilon_d^p} = \frac{\mathrm{M}^2 - \eta^2}{2\eta} \qquad \text{(modified)} \tag{4.28}$$

Figure 4.9 shows the stress–dilatancy relation for both the original and modified Cam clay models. Now, as can be seen from eqs. (4.1) and (4.2), the stress invariants (p and q) used for formulating ordinary models such as the Cam clay model are denoted by ON and NP, respectively, in Figure 4.1. The axis of p coincides with the direction of the space diagonal (normal to the octahedral plane) in 3D stress space, whereas q represents the deviation of the stress state from the space diagonal as measured in the octahedral plane containing it. Therefore, the shape (trace) of the yield surface, which is formulated using p and q or p and η, is inevitably a circle

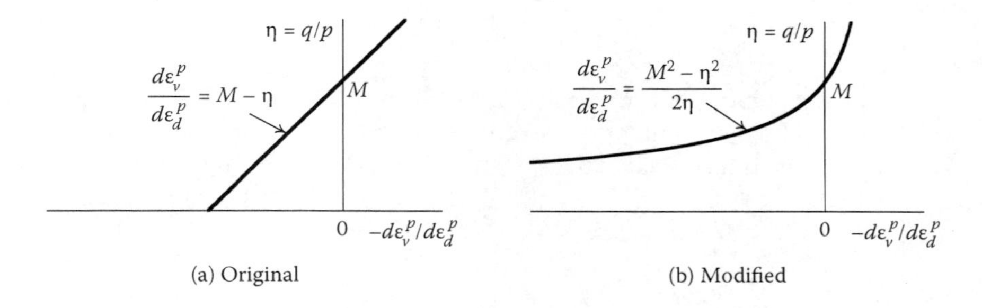

Figure 4.9 Stress–dilatancy relation used in Cam clay model.

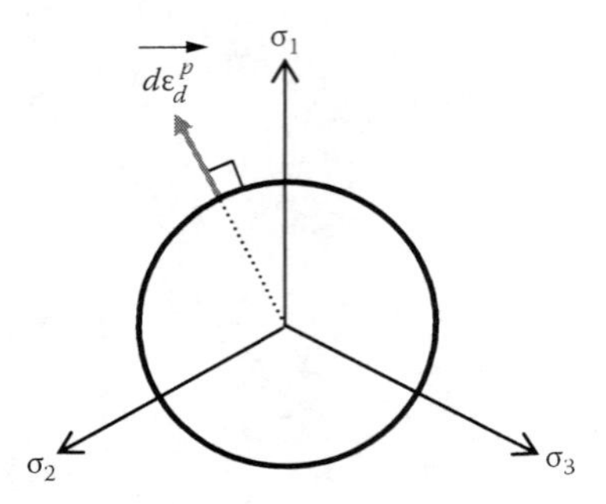

Figure 4.10 Shape of yield surface of ordinary model such as the Cam clay model on the octahedral plane.

on the octahedral plane; the direction of plastic strain increments on the octahedral plane (deviatoric strain increment $d\varepsilon_d^p$) is always in the radial direction as shown in Figure 4.10.

The elastic strain increment is given by the generalized Hooke's law:

$$d\varepsilon_{ij}^e = \frac{1+\nu_e}{E_e} d\sigma_{ij} - \frac{\nu_e}{E_e} d\sigma_{mm}\delta_{ij} = C_{ijkl}^e d\sigma_{kl} = \left[D_{ijkl}^e\right]^{-1} d\sigma_{kl} \tag{4.29}$$

The fourth order tensors C_{ijkl}^e and D_{ijkl}^e are usually called the elastic compliance tensor and the elastic stiffness tensor, respectively. The Young's modulus E_e is expressed in terms of the swelling index κ and Poisson's ratio ν_e as

$$E_e = \frac{3(1-2\nu_e)(1+e_0)p}{\kappa} \tag{4.30}$$

Therefore, the total strain increment is given by

$$d\varepsilon_{ij} = d\varepsilon_{ij}^e + d\varepsilon_{ij}^p \tag{4.31}$$

The following loading condition for usual elastoplastic models is adapted in the Cam clay to simulate the strain-hardening behavior of soils such as normally consolidated clays in the same way as eq. (2.14) in Chapter 2:

$$\begin{cases} d\varepsilon_{ij}^p \neq 0 & \text{if } f = 0 \;\&\; dF \geq 0 \\ d\varepsilon_{ij}^p = 0 & \text{otherwise} \end{cases} \tag{4.32}$$

The yield surface is fixed when no plastic strain occurs.

Historically speaking, the Cam clay models were not originally formulated based on this framework. Instead of eq. (4.13), the following energy dissipation equations were assumed for the original and modified Cam clay models (Schofield and Wroth 1968; Roscoe and Burland 1968):

$$dW^p = p \cdot d\varepsilon_v^p + q \cdot d\varepsilon_d^p = Mpd\varepsilon_d^p \qquad \text{(original)} \qquad (4.33)$$

$$dW^p = p \cdot d\varepsilon_v^p + q \cdot d\varepsilon_d^p = \sqrt{\left(Mpd\varepsilon_d^p\right)^2 + \left(pd\varepsilon_v^p\right)^2} \qquad \text{(modified)} \qquad (4.34)$$

Here, dW^p denotes an increment of the plastic work, and the right-hand side is an assumed dissipated energy increment at critical state. These equations lead to the stress–dilatancy equations (eqs. 4.27 and 4.28). Then, solving the ordinary differential equation, which is derived from the stress–dilatancy equation and the normality condition ($dp \cdot d\varepsilon_v^p + dq \cdot d\varepsilon_d^p = 0$), the function $\zeta(\eta)$ for the original and modified Cam clay models (eqs. 4.22 and 4.23) has been obtained. However, in order to satisfy the expression of the incremental plastic work,

$$dW^p = \sigma_1 \cdot d\varepsilon_1^p + \sigma_2 \cdot d\varepsilon_2^p + \sigma_3 \cdot d\varepsilon_3^p = p \cdot d\varepsilon_v^p + q \cdot d\varepsilon_d^p, \qquad (4.35)$$

it is necessary that the direction of $d\varepsilon_d^p$ coincides with that of q (see Figures 4.1 and 4.2).

However, this condition is only true under asymmetric conditions (triaxial compression and extension), as described in the following section. Also, the physical meaning of the energy dissipation equations given in eqs. (4.33) and (4.34) is not clear. Alternatively, these equations should be derived from plausible forms of the stress–dilatancy relationship and shape of yield surface chosen as starting points. In fact, Ohta (1971) obtained the same yield function as the one in the original Cam clay model, based on the framework described in this chapter, which seems more logical.

Figures 4.11–4.13 show (a) the observed results of triaxial compression tests on overconsolidated (overconsolidation ratio [OCR] = 2, 4, and 8) Fujinomori clay under different mean principal stresses, and (b) the schematic diagram of void ratio change, arranged with respect to the same relations as those of the normally consolidated Fujinomori clay in Figure 4.4.

Figure 4.14 illustrates schematically the idealized relation between stress ratio, deviatoric strain, and volumetric strain for normally consolidated (NC) clay and overconsolidated (OC) clay under shear loading. Solid and broken curves are for NC clay and OC clay, respectively. Here, M denotes the stress ratio at the critical state, and η_f denotes the peak stress ratio for OC clay. Very loose sand and dense sand also show qualitatively the same stress–strain–dilatancy characteristics as those of NC and OC clays.

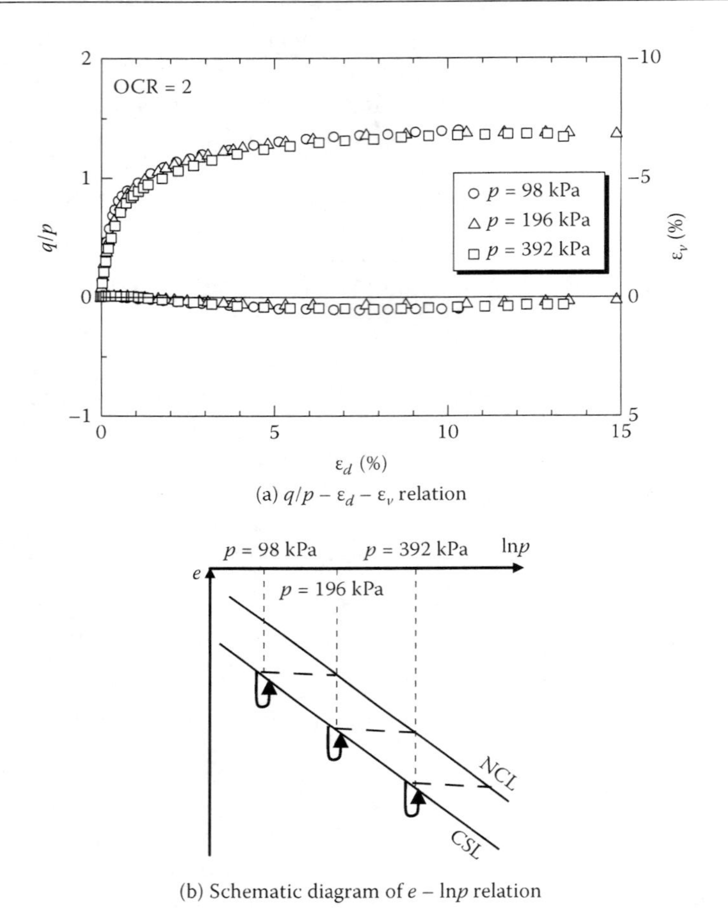

Figure 4.11 Test results of triaxial compression on overconsolidated clays (OCR = 2) under constant mean principal stress.

As explained before, NC clay exhibits only strain hardening with volume contraction and finally reaches the critical state (perfect plastic) condition of $\eta = M$ and $d\varepsilon_v = 0$.

On the other hand, OC clay shows strain hardening with volume contraction in the region of $0 < \eta < M$, strain hardening with volume expansion in the region of $M < \eta < \eta_f$, and strain softening with volume expansion in the region of $\eta_f > \eta > M$. It then approaches the critical state condition of $\eta = M$ and $d\varepsilon_v = 0$, with sustained development of shear strains. It is also understood that the strain increment ratio $(-d\varepsilon_v/d\varepsilon_d)$, which corresponds to the tangential slope of the dilatancy curves in the figure,

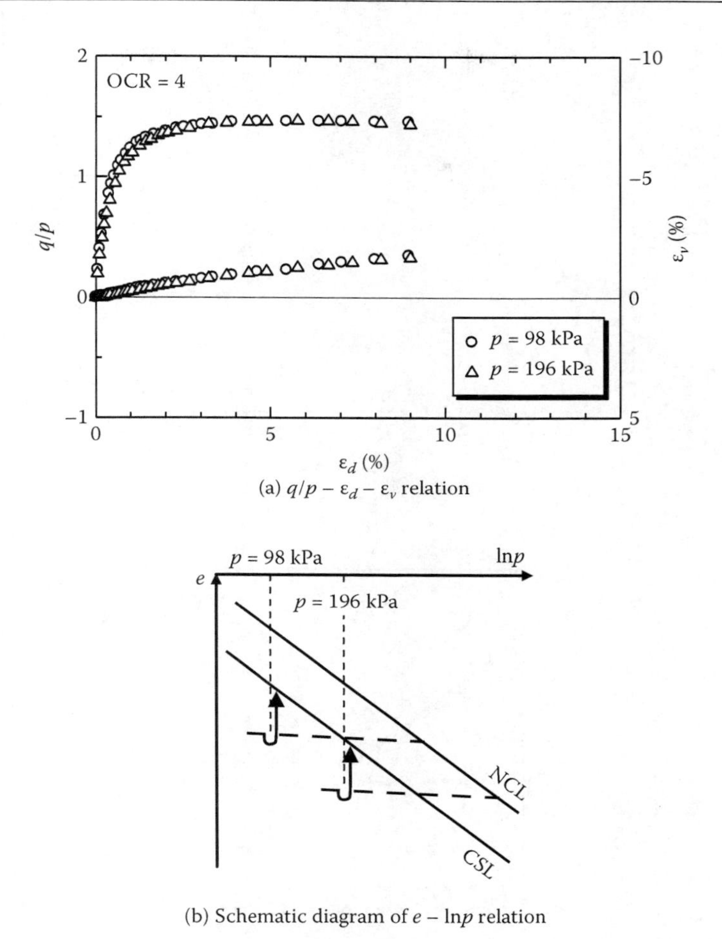

Figure 4.12 Test results of triaxial compression on overconsolidated clays (OCR = 4) under constant mean principal stress.

increases monotonically with increasing stress ratio and decreases with decreasing stress ratio. Additionally, the ratio ($-d\varepsilon_v/d\varepsilon_d$) becomes zero (no volume change) at η = M not only under the critical state but also under the strain-hardening region in OC clay. These tendencies of soil dilatancy point to the conclusion that the strain increment ratio ($-d\varepsilon_v/d\varepsilon_d$) may be determined only by the stress ratio and is rather independent of the density (initial void ratio).

Therefore, the stress–dilatancy relation expressed by eq. (4.26) and the shape of yield function given in the form of eq. (4.20) are applicable not

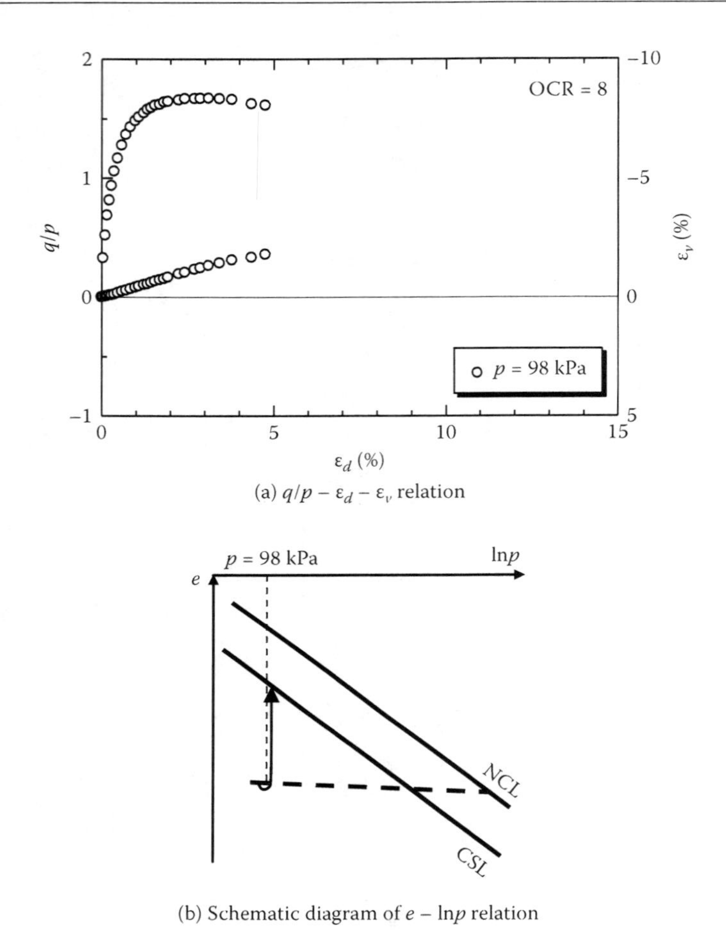

Figure 4.13 Test results of triaxial compression on overconsolidated clay (OCR = 8) under constant mean principal stress.

only to the normally consolidated soils with negative dilatancy but also to the overconsolidated soils with negative and positive dilatancy. Strictly investigating, the stress–dilatancy relation is influenced a little by the density of soil and whether it is in a hardening region or softening region (Wan and Guo 1999; Hinokio et al. 2001), but the difference is not as much. Here, the elastic strains are much smaller than the plastic strains under shear loadings, so the plastic strain increment ratio $(-d\varepsilon_v^p/d\varepsilon_d^p)$ can be considered to have almost the same tendencies as the strain increment ratio $(-d\varepsilon_v/d\varepsilon_d)$. As mentioned in Chapter 2, to adapt the Cam clay model to the strain softening region as well, the following loading condition is employed

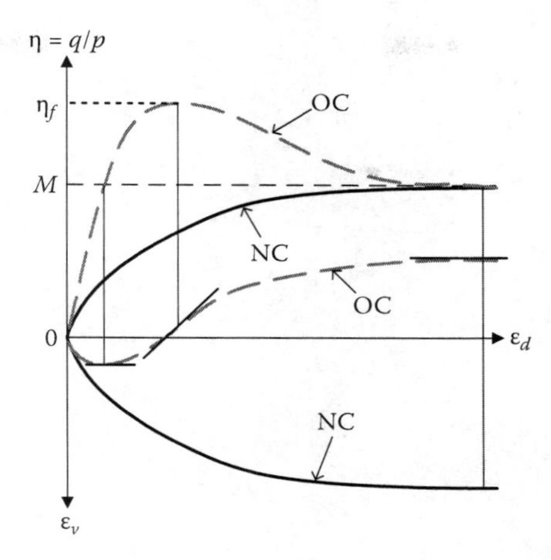

Figure 4.14 Schematic illustration of stress–strain–dilatancy curves for normally consolidated and overconsolidated clays.

instead of eq. (4.32) (Zienkiewicz and Taylor 1991; Asaoka, Nakano, and Noda 1994):

$$\begin{cases} d\varepsilon_{ij}^{p} \neq 0 & \text{if } f = 0 \ \& \ \Lambda = \dfrac{dF}{h^{p}} > 0 \\ d\varepsilon_{ij}^{p} = 0 & \text{otherwise} \end{cases} \tag{4.36}$$

The simulations of drained tests (constant mean principal tests) and undrained tests using the original Cam clay model are shown in Figures 4.15 and 4.16 for normally consolidated clay, lightly overconsolidated clay (OCR = 1.5), and heavily overconsolidated clay (OCR = 4). The following material parameters are employed in the simulations:

compression index $\lambda = 0.104$
swelling index $\kappa = 0.010$
void ratio on the NCL at $\sigma = 98$ kPa (atmospheric pressure) $N = 0.83$
stress ratio q/p at critical state M = 1.36
Poisson's ratio of elastic component $\nu_e = 0.2$

Here, the parameters, except for M and ν_e, are the same as those used in the simulations of 1D models in Chapter 3. The added material parameter $M = (q/p)_{CS} = 1.36$ corresponds to the major minor principal stress ratio $R_{cs} = (\sigma_1/\sigma_3)_{CS} = 3.5$ at the critical state under triaxial compression condition, and $\nu_e = 0.2$ is assumed.

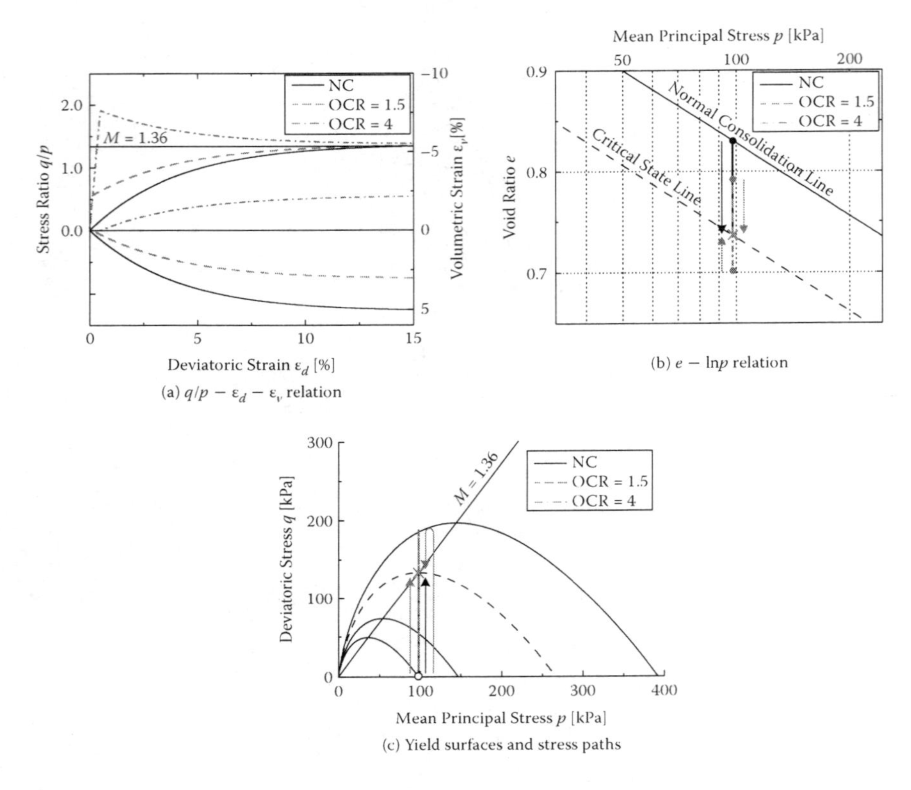

Figure 4.15 Simulated results of constant mean principal tests by original Cam clay model.

Figure 4.15 shows the simulated results of constant mean principal stress tests. Diagram (a) shows the stress–strain–dilatancy curves, arranged with respect to the relation between q/p, ε_d, and ε_v. Diagram (b) shows the variations of the void ratio of these constant mean principal tests. The solid curves in diagram (c) refer to the yield surfaces for the NC and OC clays (OCR = 1.5 and OCR = 4) at the initial condition (p = 98 kPa, η = 0.0) before the shear loadings.

The following characteristics of the Cam clay model can be seen from Figure 4.15(a). The Cam clay model can describe well the strain-hardening elastoplastic behaviors with negative dilatancy (volume contraction) observed on normally consolidated and lightly overconsolidated clays. It can also capture strain-softening behaviors with positive dilatancy (volume expansion) observed in heavily overconsolidated clays after peak strength. However, positive dilatancy, which is usually observed during strain hardening in overconsolidated clays and medium and dense sands, cannot be described as elastic behavior is assumed in the Cam clay model.

It is also seen from Figures 4.15(b) and (c) that even if the soil is under overconsolidated state before shear loadings, it reaches the same critical state as that of normally consolidated soil. In diagram (b), the solid straight line and the broken straight line are the normal consolidation line and the critical state line, respectively. The solid dots and the cross mark represent the initial void ratios and the void ratio at critical state. In diagram (c), the dotted curve represents the yield surface when the clays reach the critical state, whereas the open dot and the cross mark represent the initial state and the critical state on the p–q plane.

Diagrams (a) to (c) in Figure 4.16 also illustrate simulated results in undrained shear tests, arranged with respect to the same relations as those in Figure 4.15. In diagrams (b) and (c), the solid and open dots and the

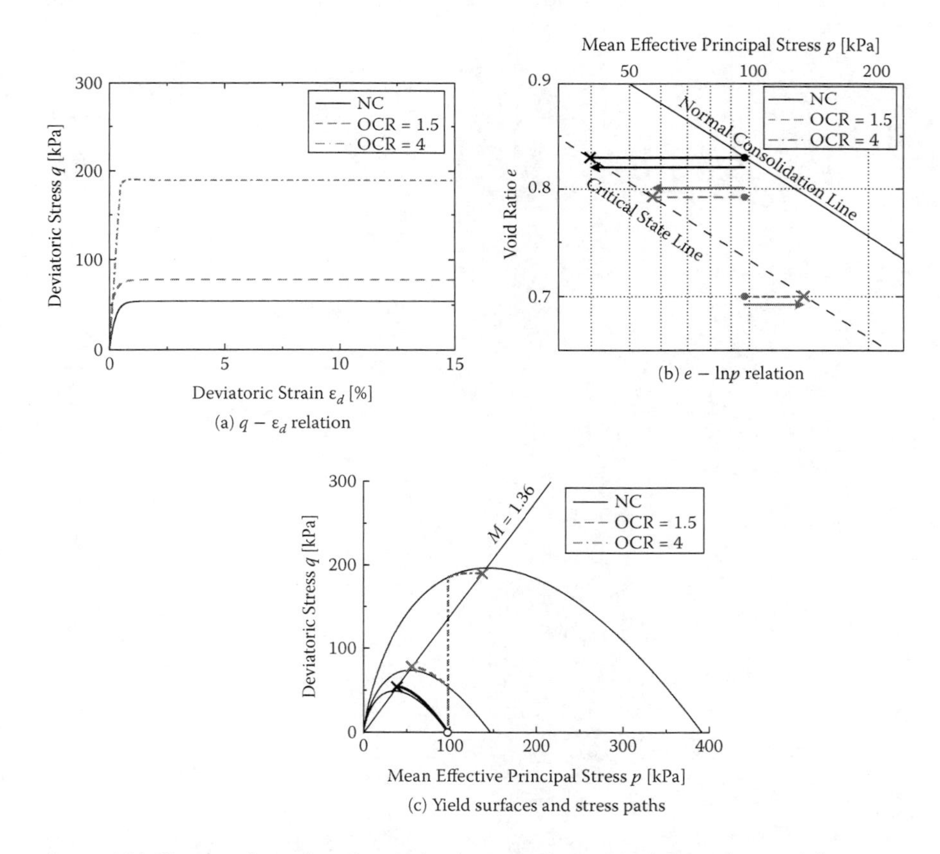

Figure 4.16 Simulated results of undrained tests using original Cam clay model.

cross marks denote the stresses and the void ratios at the initial states and the critical state. Although the normally consolidated clay behaves as a strain-hardening elastoplastic material with the decrease of mean principal stress under undrained conditions, the overconsolidated clays under undrained condition show only elastic behavior with constant mean principal stress inside the initial yield surfaces. After these stresses reach the respective yield surfaces, the lightly overconsolidated clay (OCR = 1.5) shows the strain-hardening elastoplastic behavior with the decrease of mean principal stress; the heavily overconsolidated clay (OCR = 4) shows the strain-softening (shrinkage of the yield surface) behavior with the increase of mean principal stress.

Figures 4.17 and 4.18 show the simulated results of the constant mean principal tests and undrained tests based on the modified Cam clay model. The material parameters and the conditions for the simulations are the same as those in Figures 4.15 and 4.16. The overall tendencies of the results in both cases (drained and undrained conditions) are the same as those in

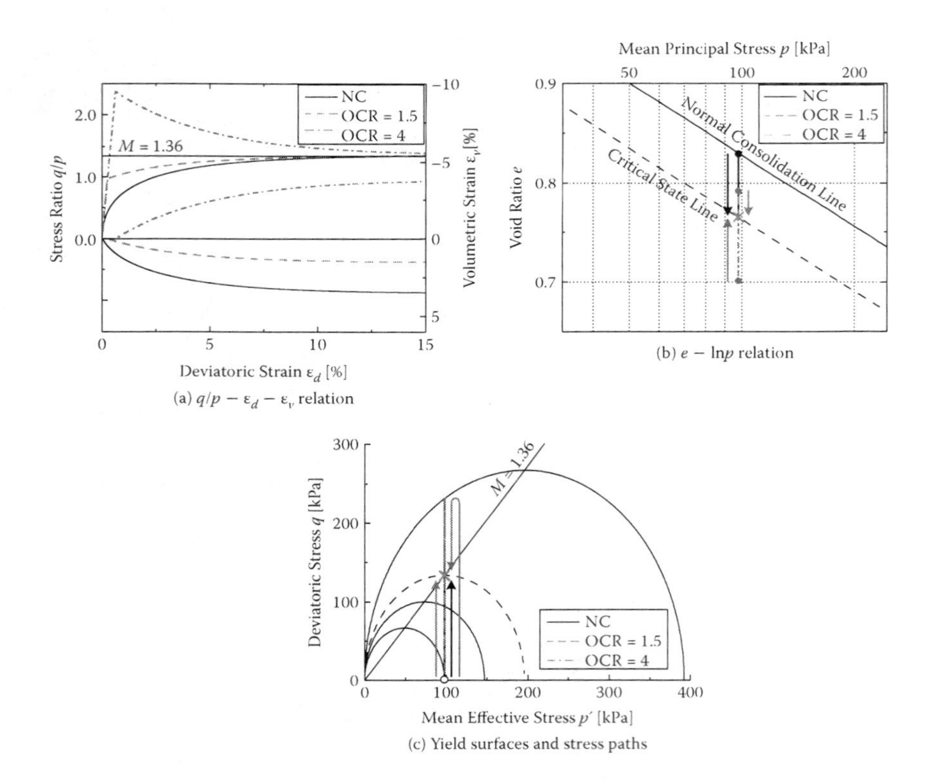

Figure 4.17 Simulated results of constant mean principal tests using modified Cam clay model.

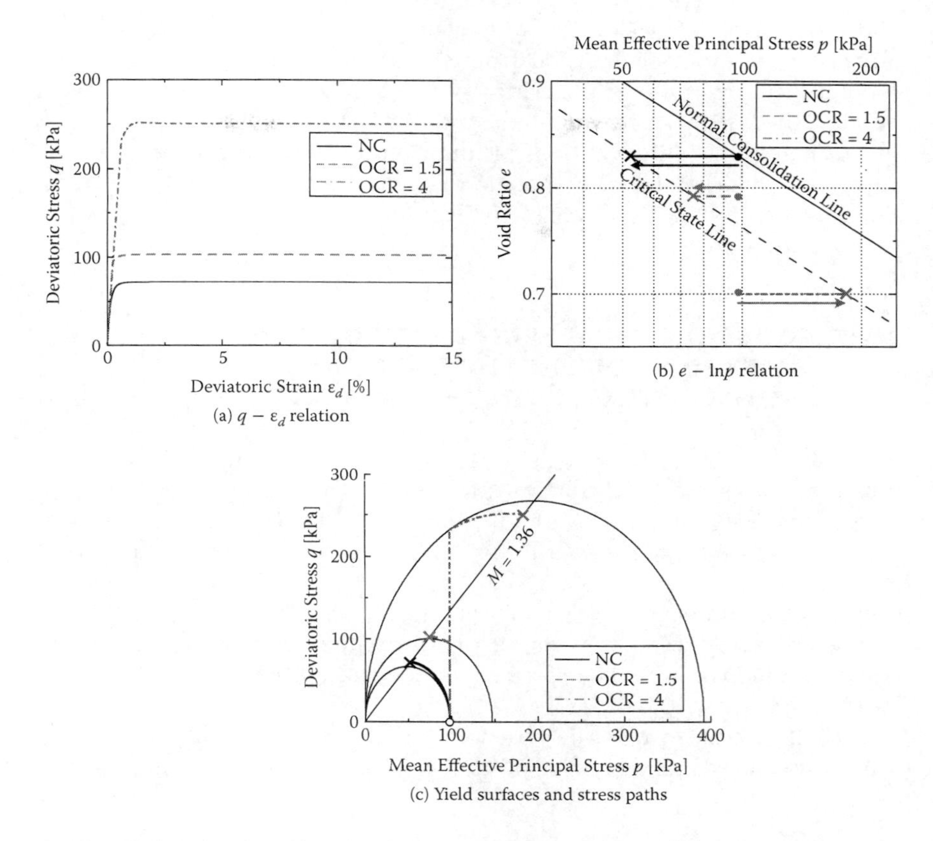

Figure 4.18 Simulated results of undrained tests using modified Cam clay model.

Figures 4.15 and 4.16. However, in the drained tests, the results based on the modified Cam clay model show higher stiffness and less compression because the difference of void ratio between the normal consolidation line and the critical state line in the modified Cam clay model is smaller than that in the original Cam clay model (see diagram (b) for both models). Here, the differences in void ratio between the normal consolidation line and the critical state line for both models are expressed as follows from eqs. (4.20), (4.22), and (4.23):

$$(\lambda - \kappa)\varsigma(M) = \lambda - \kappa \qquad \text{(original)} \tag{4.37}$$

$$(\lambda - \kappa)\varsigma(M) = (\lambda - \kappa)\ln 2 \qquad \text{(modified)} \tag{4.38}$$

In the undrained tests, it can be seen from diagrams (a) and (c) for both models that the simulated undrained strengths of normally consolidated and overconsolidated clays using the modified Cam clay model become larger than the ones calculated from the original Cam clay model, even though the stress ratio at the critical state (M = 1.36) is the same. This is because the critical state line by the modified Cam clay model is closer to the normally consolidated line as mentioned previously, so the mean principal stresses at critical state by the modified Cam clay model are larger than those by the original Cam clay model in undrained conditions (see diagram (b) for both models).

4.3 DISCUSSION ON APPLICABILITY OF CAM CLAY TYPE MODEL IN THREE-DIMENSIONAL CONDITIONS BASED ON TEST RESULTS

Figures 4.19 and 4.20 show the observed results of drained triaxial compression ($\sigma_1 > \sigma_2 = \sigma_3$) and triaxial extension ($\sigma_1 = \sigma_2 > \sigma_3$) tests on normally consolidated Fujinomori clay and medium dense Toyoura sand under constant mean principal stress, in terms of stress ratio (q/p) versus deviatoric strain (ε_d) and volumetric strain (ε_v). These are stress and strain invariants used in the Cam clay model mentioned in the previous section. It can be seen from these figures that the deformation and strength of soils in 3D stress conditions cannot be described uniquely using these stress and strain invariants.

It is interesting to plot the observed stress–dilatancy relations for the same previously mentioned tests in the same fashion as in Figure 4.9.

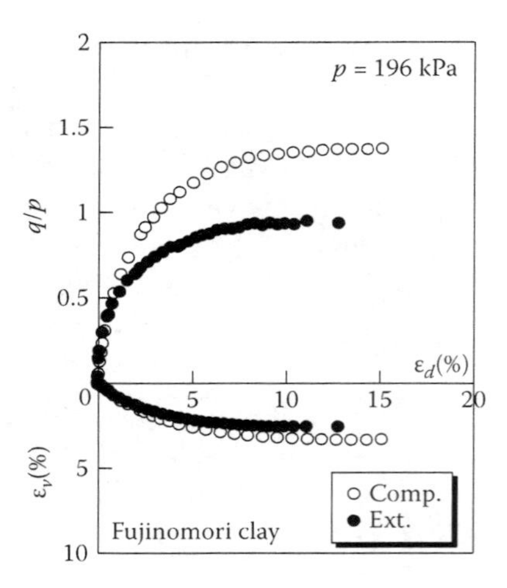

Figure 4.19 Observed stress–strain relation in triaxial compression and extension tests on normally consolidated clay.

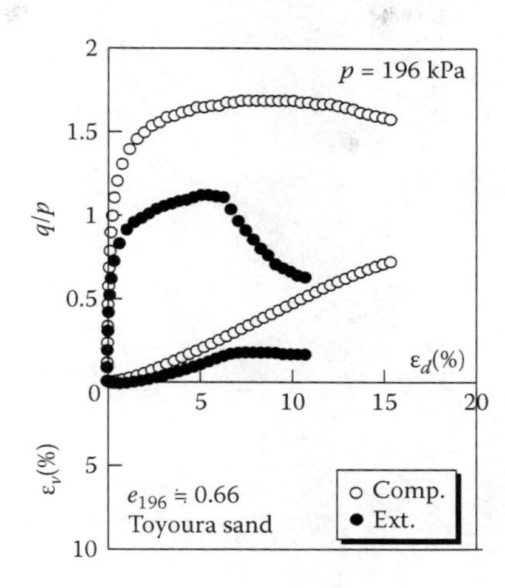

Figure 4.20 Observed stress–strain relation in triaxial compression and extension tests on medium dense sand.

Figures 4.21 and 4.22 show the stress–dilatancy plots for Fujinomori clay and medium dense Toyoura sand, respectively, where the strain increment ratio includes the elastic component. However, under shear loading, developed plastic strains are much larger than the elastic ones, and hence the total plastic strain increment ratio is almost equal to the plastic strain increment ratio. It is found that there is no unique relation between $d\varepsilon_v/d\varepsilon_d$ and q/p in Figures 4.21 and 4.22 as the shape of yield surface on the p–q plane is dependent on the relative magnitude of the intermediate principal stress.

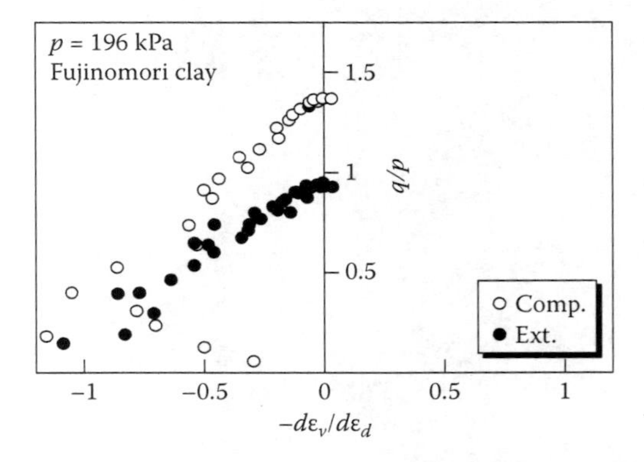

Figure 4.21 Observed q/p–$d\varepsilon_v/d\varepsilon_d$ relation in triaxial compression and extension tests on normally consolidated clay.

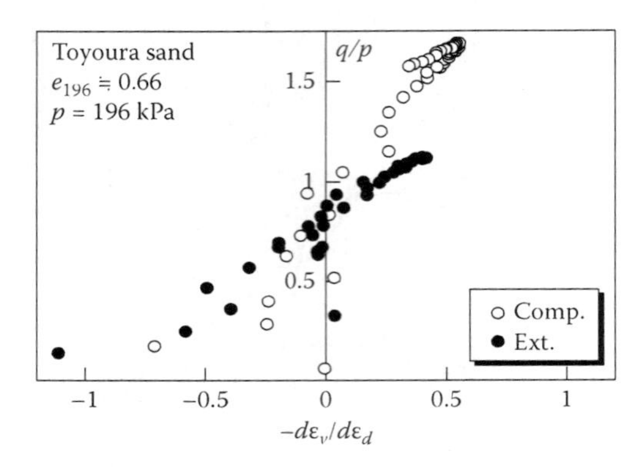

Figure 4.22 Observed q/p–$d\varepsilon_v/d\varepsilon_d$ relation in triaxial compression and extension tests on medium dense sand.

Figures 4.23 and 4.24 show the directions and the magnitudes of the observed shear strain increments represented by bold lines on the octahedral plane for true triaxial ($\sigma_1 > \sigma_2 > \sigma_3$) tests ($\theta = 15°$, 30°, and 45°). Here, the length of each bold line is proportional to the value of shear strain increment divided by the shear–normal stress ratio increment on the octahedral plane. In the figures, θ denotes the angle between the σ_1-axis and the corresponding radial stress path on the octahedral plane, where $\theta = 0°$ and 60° represent the stress path under triaxial compression and triaxial extension conditions, respectively.

It can be seen that the direction of the observed shear strain increments deviates leftward from the direction of shear stress (radial direction) with the increase in stress ratio under three different principal stresses. On the

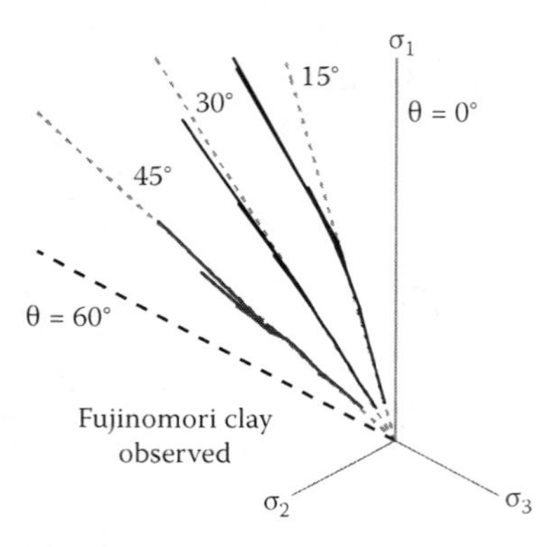

Figure 4.23 Observed direction of $d\varepsilon_d$ on octahedral plane under three different principal stresses ($\sigma_1 > \sigma_2 > \sigma_3$) on normally consolidated clay.

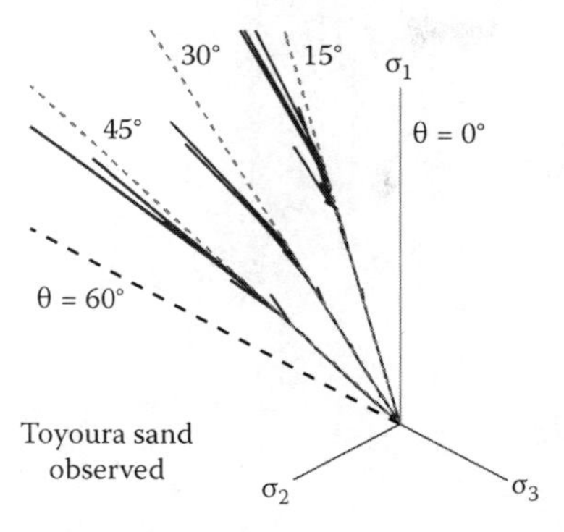

Figure 4.24 Observed direction of $d\varepsilon_d$ on octahedral plane under three different principal stresses ($\sigma_1 > \sigma_2 > \sigma_3$) on medium dense sand.

other hand, since the plastic potential (yield surface) formulated using the stress invariants p and q is a circle on the octahedral plane, as shown in Figure 4.10, the calculated plastic strain increments are also radial in that plane. It goes without saying that the elastic strain increments as calculated by Hooke's law are radial, just as the stress increment vector is in the octahedral plane.

The deviation of the strain increment vector from the radial direction as observed in Figures 4.23 and 4.24 cannot be simulated by the Cam clay model. It is noted that the direction of the strain increments in the octahedral plane has been traditionally less studied in the literature than the stress–dilatancy relation. For example, turning to Figures 4.23 and 4.24, the directions of the strain increments at the 0° and 60° branches in the octahedral plane imply that the intermediate principal strain increment $d\varepsilon_2$ coincides with the minor principal strain increment $d\varepsilon_3$ and the major principal strain increment $d\varepsilon_1$, respectively. Hence, a small deviation of the directions of the strain increment on the octahedral plane has much influence on the 3D behaviors of soils. It is understood that the elastoplastic constitutive models whose yield functions are formulated using the invariants p and q or p and η are not capable of describing the influence of the intermediate principal stress on the deformation and strength characteristics of soils, even if the initial stress condition and the density of soils are the same.

Chapter 5

Unified modeling of three-dimensional soil behavior based on the t_{ij} concept

5.1 INTRODUCTION

As mentioned in the previous chapter, the ordinary models formulated using the stress invariants (p and q) cannot describe uniquely the deformation and strength of soils under three different principal stresses. To describe such soil behavior, some material parameters in the models are given by a function of the magnitude of the intermediate principal stress. However, such a method is not essential in rational modeling.

In this chapter, a simple and unified method to describe stress–strain behavior in general three-dimensional (3D) stress conditions, which is called the t_{ij} concept, is presented. Since this concept is found from the idea that the frictional law essentially governs soil behavior, its meaning is very clear. Using this concept, any kinds of one-dimensional (1D) models and/or 3D models formulated by using p and q can be easily extended to general 3D models. As an example, the Cam clay model (corresponding to the linear e–lnσ relation of normally consolidated soil in 1D condition) is extended to the model valid in the general 3D condition. Detailed methods for formulating the model using t_{ij} and the meanings of the t_{ij} concept are also presented.

5.2 CONCEPT OF MODIFIED STRESS t_{ij}

5.2.1 Definition of t_{ij} and stress and strain increment invariants based on the t_{ij} concept

The deformation and strength characteristics of soils in 3D stress conditions can be uniquely described by the extended concept of spatially mobilized plane (SMP*) proposed by Nakai and Matsuoka (1983). Based on the generalized concept of the SMP*, Nakai and Mihara (1984) developed a method to formulate an elastoplastic model in which the influence of the intermediate principal stress can be automatically taken into consideration,

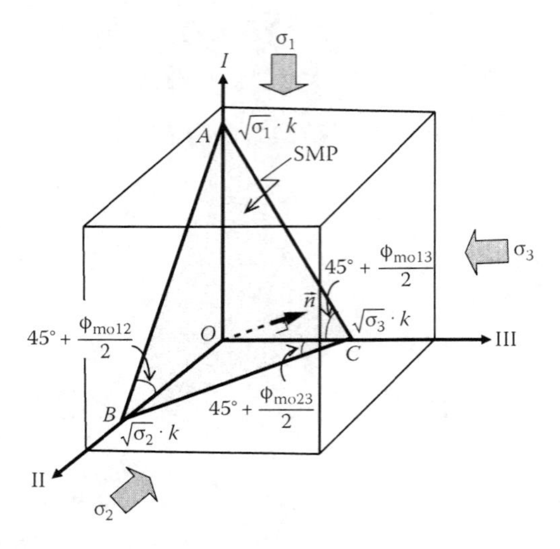

Figure 5.1 Spatially mobilized plane in three-dimensional space.

by introducing the modified stress tensor t_{ij}. In the t_{ij} concept, attention is focused on the so-called spatially mobilized plane (SMP; Matsuoka and Nakai 1974) instead of the octahedral plane used in classical models, such as the Cam clay.

The plane ABC in Figure 5.1 is the SMP in the 3D stress space, where axes I, II, and III imply the directions of three principal stresses. At each of the three sides AB, AC, and BC of plane ABC, the shear–normal stress ratio is maximized between two principal stresses as shown in Figure 5.2. Here, the plane AB, where the shear normal stress ratio (τ/σ) is maximized, has been called the "mobilized plane" or "plane of maximum mobilization" by Murayama (1964), and the three planes AB, AC, and BC have been collectively called "compounded mobilized planes" by Matsuoka (1974). It can be seen from Figures 5.1 and 5.2 that the values of the coordinate axes intersected by the

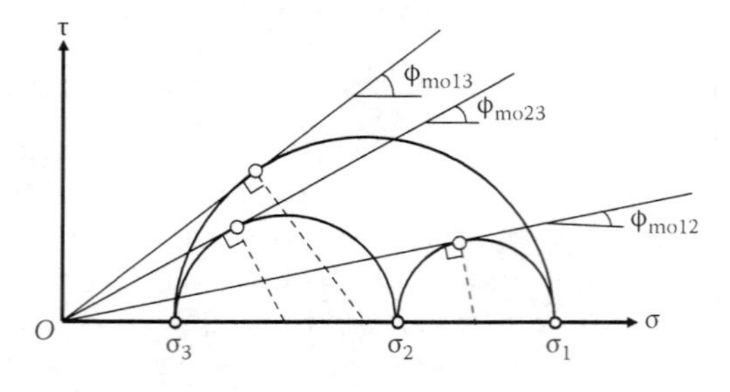

Figure 5.2 Three Mohr's stress circles under three different principal stresses.

plane ABC (SMP) are proportional to the square root of the ratio between the corresponding principal stresses because the following equation holds:

$$\tan\left(45^\circ + \frac{\phi_{moij}}{2}\right) = \sqrt{\frac{1+\sin\phi_{moij}}{1-\sin\phi_{moij}}} = \sqrt{\frac{\sigma_i}{\sigma_j}} \quad (i,\ j = 1,\ 2,\ 3;\ i < j) \tag{5.1}$$

Therefore, the SMP coincides with the octahedral plane only under isotropic stress conditions and varies with possible changes of stress ratio. The direction cosines (a_1, a_2, and a_3) of the normal to the SMP and the unit tensor whose principal values are determined by these direction cosines are expressed as follows (Nakai 1989):

$$a_1 = \sqrt{\frac{I_3}{I_2\sigma_1}},\ a_2 = \sqrt{\frac{I_3}{I_2\sigma_2}},\ a_3 = \sqrt{\frac{I_3}{I_2\sigma_3}} \quad \left(\text{where} \quad a_1^2 + a_2^2 + a_3^2 = 1\right) \tag{5.2}$$

$$a_{ij} = \sqrt{\frac{I_3}{I_2}} \cdot r_{ij}^{-1} = \sqrt{\frac{I_3}{I_2}} \cdot (\sigma_{ik} + I_{r2}\delta_{ik})(I_{r1}\sigma_{kj} + I_{r3}\delta_{kj})^{-1} \tag{5.3}$$

where
δ_{ij} is the unit tensor
σ_i ($i = 1,2,3$) are the three principal stresses
I_1, I_2, and I_3 are the first, second, and third invariants of σ_{ij}
I_{r1}, I_{r2}, and I_{r3} are the first, second, and third invariants of r_{ij}, which is the square root of the stress tensor or $r_{ik}r_{kj} = \sigma_{ij}$

These invariants are expressed using principal stresses and stress tensors as

$$\left.\begin{aligned} I_1 &= \sigma_1 + \sigma_2 + \sigma_3 = \sigma_{ii} \\ I_2 &= \sigma_1\sigma_2 + \sigma_2\sigma_3 + \sigma_3\sigma_1 = \frac{1}{2}\left\{(\sigma_{ii})^2 - \sigma_{ij}\sigma_{ji}\right\} \\ I_3 &= \sigma_1\sigma_2\sigma_3 = \frac{1}{6}e_{ijk}e_{lmn}\sigma_{il}\sigma_{jm}\sigma_{kn} \end{aligned}\right\} \tag{5.4}$$

$$\left.\begin{aligned} I_{r1} &= \sqrt{\sigma_1} + \sqrt{\sigma_2} + \sqrt{\sigma_3} = r_{ii} \\ I_{r2} &= \sqrt{\sigma_1\sigma_2} + \sqrt{\sigma_2\sigma_3} + \sqrt{\sigma_3\sigma_1} = \frac{1}{2}\left\{(r_{ii})^2 - r_{ij}r_{ji}\right\} \\ I_{r3} &= \sqrt{\sigma_1\sigma_2\sigma_3} = \frac{1}{6}e_{ijk}e_{lmn}r_{il}r_{jm}r_{kn} \end{aligned}\right\} \tag{5.5}$$

where e_{ijk} is the permutation tensor. As can be seen from this equation, a_{ij} is a function of the stress ratio and its principal axes coincide with those of σ_{ij}. The detailed expression of a_{ij} is described in Section 5.2.2.

The modified stress tensor t_{ij} is then defined by the product of a_{ik} and σ_{kj} as follows:

$$t_{ij} = a_{ik}\sigma_{kj} \tag{5.6}$$

Its principal values are given by

$$t_1 = a_1\sigma_1, \; t_2 = a_2\sigma_2, \; t_3 = a_3\sigma_3 \tag{5.7}$$

In conventional models, the stress invariants (p and q) and strain increment invariants ($d\varepsilon_v$ and $d\varepsilon_d$) are given by the normal and in-plane components of the ordinary stress and strain increment with respect to the octahedral plane (see Figures 4.1 and 4.2 in Chapter 4). On the other hand, the stress invariants (t_N and t_S) and strain increment invariants ($d\varepsilon_N^*$, $d\varepsilon_S^*$) in the t_{ij} concept are defined as the normal and in-plane components of the modified stress tensor t_{ij} and the strain increment with respect to the SMP (see Figures 5.3 and 5.4). Hence, these invariants are given by

$$t_N = \overline{\mathrm{ON}} = t_1a_1 + t_2a_2 + t_3a_3 = t_{ij}a_{ij} \tag{5.8}$$

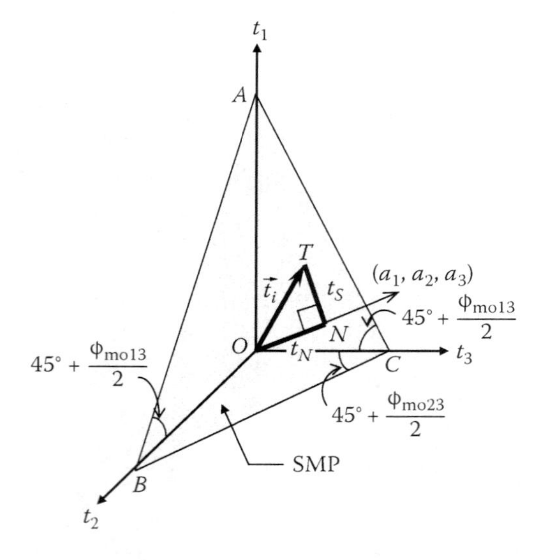

Figure 5.3 Definitions of t_N and t_S.

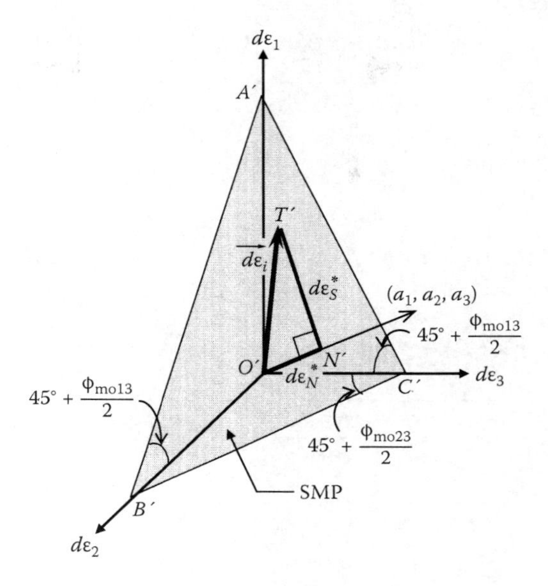

Figure 5.4 Definitions of $d\varepsilon_N^*$ and $d\varepsilon_S^*$.

$$t_S = \overline{\text{NT}} = \sqrt{(t_1a_2 - t_2a_1)^2 + (t_2a_3 - t_3a_2)^2 + (t_3a_1 - t_1a_3)^2}$$

$$= \sqrt{t_1^2 + t_2^2 + t_3^2 - (t_1a_1 + t_2a_2 + t_3a_3)^2} = \sqrt{t_{ij}t_{ij} - (t_{ij}a_{ij})^2} \tag{5.9}$$

$$d\varepsilon_N^* = \overline{\text{O'N'}} = d\varepsilon_1a_1 + d\varepsilon_2a_2 + d\varepsilon_3a_3 = d\varepsilon_{ij}a_{ij} \tag{5.10}$$

$$d\varepsilon_S^* = \overline{\text{N'T'}} = \sqrt{(d\varepsilon_1a_2 - d\varepsilon_2a_1)^2 + (d\varepsilon_2a_3 - d\varepsilon_3a_2)^2 + (d\varepsilon_3a_1 - d\varepsilon_1a_3)^2}$$

$$= \sqrt{d\varepsilon_1^2 + d\varepsilon_2^2 + d\varepsilon_3^2 - (d\varepsilon_1a_1 + d\varepsilon_2a_2 + d\varepsilon_3a_3)^2}$$

$$= \sqrt{d\varepsilon_{ij}d\varepsilon_{ij} - (d\varepsilon_{ij}a_{ij})^2} \tag{5.11}$$

Here, the stress invariants t_N and t_S are also equivalent to the normal and shear stresses σ_{SMP} and τ_{SMP} on the spatially mobilized plane (ABC) in Figure 5.1, so these invariants are expressed using the three ordinary principal stresses (σ_1, σ_2, and σ_3) or the first, second, and third stress invariants (I_1, I_2, and I_3), as well (Matsuoka and Nakai 1974).

$$t_N = t_1a_1 + t_2a_2 + t_3a_3$$

$$= \sigma_1a_1^2 + \sigma_2a_2^2 + \sigma_3a_3^2 = 3\frac{I_3}{I_2} = \sigma_{SMP} \tag{5.12}$$

$$\begin{aligned} t_S &= \sqrt{(t_1 a_2 - t_2 a_1)^2 + (t_2 a_3 - t_3 a_2)^2 + (t_3 a_1 - t_1 a_3)^2} \\ &= \sqrt{(\sigma_1 - \sigma_2)^2 a_1^2 a_2^2 + (\sigma_2 - \sigma_3)^2 a_2^2 a_3^2 + (\sigma_3 - \sigma_1)^2 a_3^2 a_1^2} \\ &= \frac{\sqrt{I_1 I_2 I_3 - 9 I_3^2}}{I_2} = \tau_{SMP} \end{aligned} \tag{5.13}$$

Therefore, the stress ratio X based on the t_{ij} concept is also expressed using the three principal stresses (σ_1, σ_2, and σ_3), the mobilized friction angles between two respective principal stresses (ϕ_{mo12}, ϕ_{mo23}, and ϕ_{mo13}; see eq. 5.1 and Figure 5.2), or the three stress invariants (I_1, I_2, and I_3) as

$$\begin{aligned} X &= \frac{t_S}{t_N} = \frac{2}{3}\sqrt{\frac{(\sigma_1 - \sigma_2)^2}{4\sigma_1\sigma_2} + \frac{(\sigma_2 - \sigma_3)^2}{4\sigma_2\sigma_3} + \frac{(\sigma_3 - \sigma_1)^2}{4\sigma_3\sigma_1}} \\ &= \frac{2}{3}\sqrt{\tan\phi_{mo12}^2 + \tan\phi_{mo23}^2 + \tan\phi_{mo13}^2} \\ &= \sqrt{\frac{I_1 I_2}{9 I_3} - 1} \end{aligned} \tag{5.14}$$

The condition of X = constant gives the following criterion using the three stress invariants:

$$\frac{I_1 I_2}{I_3} = \text{const.} \tag{5.15}$$

This is called the Matsuoka–Nakai criterion or the SMP criterion (Matsuoka and Nakai 1974). The intersection between this criterion and the octahedral plane in principal stress space is described in Figure 5.5 together with the (extended) von Mises criterion and the Mohr–Coulomb criterion. It is seen from this figure that the shape of the SMP criterion on the octahedral plane is a rounded triangle in the σ_{ij} space, and it is circumscribed with the Mohr–Coulomb criterion at triaxial compression and extension conditions. Lade and Duncan (1975) also proposed the following 3D criterion for soils using the first and third stress invariants alone:

$$\frac{I_1^3}{I_3} = \text{const.} \tag{5.16}$$

The shape of the Lade–Duncan criterion is also a rounded triangle, but extends a little more outward than the Mohr–Coulomb criterion at triaxial extension condition.

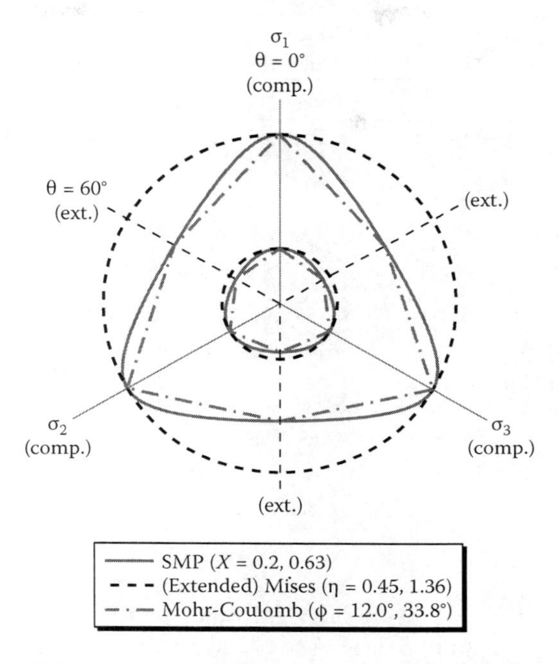

Figure 5.5 Shape of Matsuoka–Nakai SMP criterion, (extended) von Mises criterion, and Mohr–Coulomb criterion on the octahedral plane in three principal stresses space. (a) Stress quantities used in the ordinary concept; (b) stress quantities used in the t_{ij} concept.

As can be seen from eqs. (5.8), (5.9), (5.12), and (5.13), it is important in the modeling based on the t_{ij} concept that the stress invariants t_N and t_S can be expressed by the modified stress t_{ij} as well as the Cauchy stress σ_{ij}. The details of modeling based on the t_{ij} concept are described later.

A comparison between the stress and strain increment tensors and their invariants used classically and those based on the t_{ij} concept is shown in Table 5.1. The definitions of some quantities related to stress in (a) the classical approach and (b) the t_{ij} concept are illustrated in Figure 5.6 as well. It can be observed from Table 5.1 and Figure 5.6 that by merely substituting the SMP for the octahedral plane as the reference plane, the modified stress t_{ij}, together with various scalars and tensors related to stress and strain increments, can be analogically calculated (Nakai, Fujii, and Taki 1989). For reference, tensor η_{ij} in Table 5.1 is the stress ratio tensor defined by Sekiguchi and Ohta (1977) to extend the ordinary isotropic models such as the Cam clay model to anisotropic ones. The corresponding stress ratio tensor based on the t_{ij} concept is given by x_{ij}.

As mentioned in Chapter 4, in a classical model such as Cam clay, the yield function is given in the form of eq. (4.15), and the flow rule is assumed

Table 5.1 Comparison between tensors and scalars related to stress and strain increments in the ordinary concept and the t_{ij} concept

	Ordinary Concept	t_{ij} *Concept*
Tensor normal to reference plane	δ_{ij} (unit tensor)	a_{ij} (tensor normal to SMP)
Stress tensor	σ_{ij}	t_{ij}
Mean stress	$p = \sigma_{ij}\,\delta_{ij}/3$	$t_N = t_{ij}a_{ij}$
Deviatoric stress tensor	$s_{ij} = \sigma_{ij} - p\delta_{ij}$	$t'_{ij} = t_{ij} - t_N a_{ij}$
Deviatoric stress	$q = \sqrt{(3/2)s_{ij}s_{ij}}$	$t_s = \sqrt{t'_{ij}t'_{ij}}$
Stress ratio tensor	$\eta_{ij} = s_{ij}/p$	$x_{ij} = t'_{ij}/t_N$
Stress ratio	$\eta = q/p$	$X = t_s/t_N$
Strain increment normal to reference plane	$d\varepsilon_v = d\varepsilon_{ij}\delta_{ij}$	$d\varepsilon^*_N = d\varepsilon_{ij}a_{ij}$
Deviatoric strain increment tensor	$de_{ij} = d\varepsilon_{ij} - d\varepsilon_v\delta_{ij}/3$	$d\varepsilon'_{ij} = d\varepsilon_{ij} - d\varepsilon^*_N a_{ij}$
Strain increment parallel to reference plane	$d\varepsilon_d = \sqrt{(2/3)de_{ij}de_{ij}}$	$d\varepsilon^*_s = \sqrt{d\varepsilon'_{ij}d\varepsilon'_{ij}}$

in the Cauchy stress space as in eq. (4.24). On the other hand, in the modeling based on the t_{ij} concept, the yield function f is formulated using the stress invariants (t_N and t_S) instead of p and q in such a form as

$$f = F(t_N, X = t_S/t_N) - H = 0 \tag{5.17}$$

The flow rule is defined in the t_{ij} space and not in the σ_{ij} space:

$$d\varepsilon^p_{ij} = \Lambda\frac{\partial F}{\partial t_{ij}} \quad (\Lambda: \text{positive proportional constant}) \tag{5.18}$$

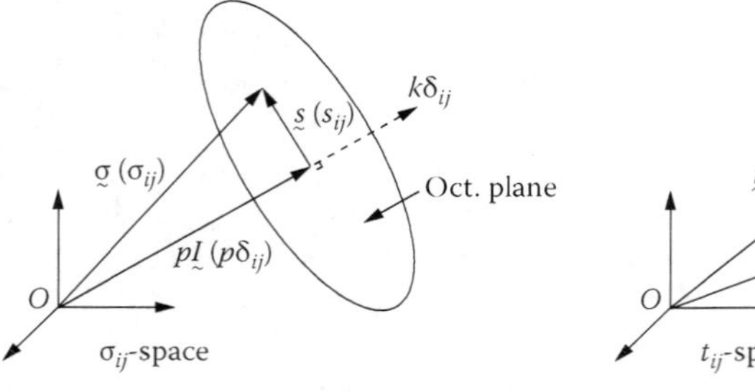

(a) Stress quantities used in the ordinary concept

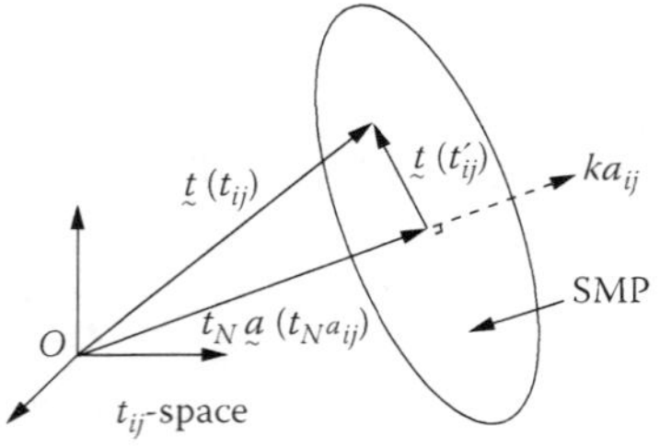

(b) Stress quantities used in the t_{ij} concept

Figure 5.6 Illustration of stress quantities.

5.2.2 Explicit expression of a_{ij}

The tensor a_{ij} is a dimensionless symmetric tensor whose principal axes coincide with those of the stress tensor, σ_{ij}, so that it is obtained by transformation from its principal values $\hat{a}_{ij}$ as follows:

$$a_{ij} = Q_{im}Q_{jn}\hat{a}_{mn} \tag{5.19}$$

Here, Q_{ij} is an orthogonal transformation that converts the stress tensor (σ_{ij}) with respect to the current global system of coordinates to a new stress tensor ($\hat{\sigma}_{ij}$), with reference to the principal axes, so that each component of Q_{ij} can be determined using the eigenvectors of σ_{ij}:

$$\hat{\sigma}_{ij} = Q_{mi}Q_{nj}\sigma_{mn} \tag{5.20}$$

where

$$\hat{\sigma}_{ij} = \begin{pmatrix} \sigma_1 & 0 & 0 \\ 0 & \sigma_2 & 0 \\ 0 & 0 & \sigma_3 \end{pmatrix} \tag{5.21}$$

From eq. (5.2), $\hat{a}_{ij}$ is expressed as follows:

$$\begin{cases} \hat{a}_{ij} = \sqrt{\dfrac{I_3}{I_2\hat{\sigma}_{ij}}} & \text{if} \quad i = j \\ \hat{a}_{ij} = 0 & \text{if} \quad i \neq j \end{cases} \tag{5.22}$$

Particularly under plane strain or axisymmetric conditions ($\sigma_{33} = \hat{\sigma}_{33} = \sigma_3$ or $\sigma_{13} = \sigma_{23} = 0$), a_{ij} is expressed as follows. From Figure 5.7, the angle 2α

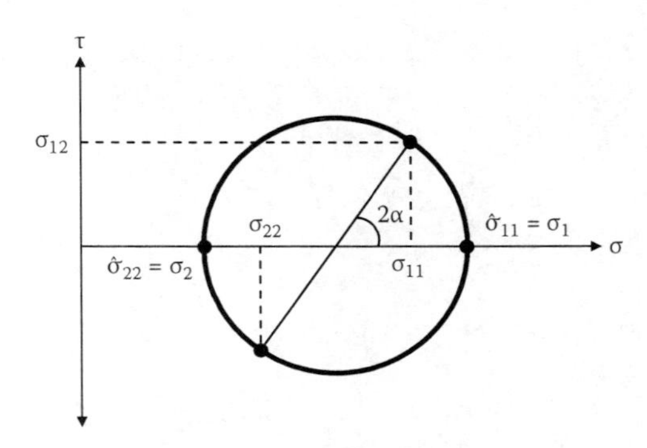

Figure 5.7 Expression of σ_{ij} and $\hat{\sigma}_{ij}$ on Mohr's stress circle.

is given using σ_{ij} and $\hat{\sigma}_{ij}$ by

$$\left.\begin{aligned} \cos 2\alpha &= \frac{\sigma_{11}-\sigma_{22}}{\sigma_1-\sigma_2} = \frac{\sigma_{11}-\sigma_{22}}{\sqrt{(\sigma_{11}-\sigma_{22})^2+4\sigma_{12}^2}} \\ \sin 2\alpha &= \frac{2\sigma_{12}}{\sigma_1-\sigma_2} = \frac{2\sigma_{12}}{\sqrt{(\sigma_{11}-\sigma_{22})^2+4\sigma_{12}^2}} \end{aligned}\right\} \tag{5.23}$$

Therefore, considering the coaxiality between σ_{ij} and a_{ij}, the following expression is obtained:

$$\left.\begin{aligned} a_{11} &= \frac{a_1+a_2}{2} + \frac{a_1-a_2}{2}\cos 2\alpha \\ a_{22} &= \frac{a_1+a_2}{2} - \frac{a_1-a_2}{2}\cos 2\alpha \\ a_{33} &= a_3 \\ a_{12} &= \frac{a_1-a_2}{2}\sin 2\alpha \\ a_{23} &= a_{13} = 0 \end{aligned}\right\} \tag{5.24}$$

Although these expressions of a_{ij} are obtained using the principal values (a_1, a_2, and a_3) defined by eq. (5.2) and the orthogonal transformation tensor Q_{ij} of σ_{ij}, the tensor a_{ij} can also be given by eq. (5.3) directly. The various steps in deriving eq. (5.3) and how it is used to determine a_{ij} are shown here. Consider the tensor r_{ij}, which is the square root of σ_{ij}. Cayley Hamilton's theorem gives the following equation:

$$r_{ik}r_{kl}r_{lj} - I_{r1}\cdot r_{ik}r_{kj} + I_{r2}\cdot r_{ij} - I_{r3}\cdot\delta_{ij} = 0 \tag{5.25}$$

Here, δ_{ij} is the unit tensor and (I_{r1}, I_{r2}, and I_{r3}) imply the first, second, and third invariants of r_{ij}, as shown in eq. (5.5). The tensor r_{ij}, which is σ_{ij} to the 1/2 power, is symmetric and is related to σ_{ij} as

$$r_{ik}r_{kj} = \sigma_{ij} \tag{5.26}$$

Equation (5.25) becomes

$$r_{ij}(\sigma_{jk} + I_{r2}\cdot\delta_{jk}) = I_{r1}\cdot\sigma_{ik} + I_{r3}\cdot\delta_{ik} \tag{5.27}$$

Thus,

$$r_{ij} = (I_{r1}\cdot\sigma_{ik} + I_{r3}\cdot\delta_{ik})(\sigma_{kj} + I_{r2}\cdot\delta_{kj})^{-1} \tag{5.28}$$

Recalling the second and third stress invariants of σ_{ij} in eq. (5.4) and r_{ij} in eq. (5.28), a_{ij} can be expressed as follows:

$$a_{ij} = \sqrt{\frac{I_3}{I_2}} \cdot r_{ij}^{-1} \tag{5.29}$$

In fact, substituting eq. (5.28) into eq. (5.29) leads to eq. (5.3). Now, the inverse of the tensor found on the right-hand side of eq. (5.3) can be easily calculated using Cramer's formula, among others, because all the tensors described here are represented by 3 × 3 matrices.

5.2.3 Meaning of the t_{ij} concept

The meaning of the t_{ij} concept is discussed here, focusing mostly on the microscopic point of view. Several researchers have shown that induced anisotropy of soils developed during stress changes is characterized by the frequency distribution of the interparticle contact angles. It has then been shown from microscopic observations (Oda 1972) and DEM simulations (Maeda, Hirabayashi, and Ohmura 2006) that, as the stress ratio increases, the average of the normal directions to the interparticle contacts gradually concentrate in the direction of the major principal stress (σ_1).

For example, Figure 5.8 shows the typical stress–strain curve for sands under triaxial compression and the rose diagrams of the observed

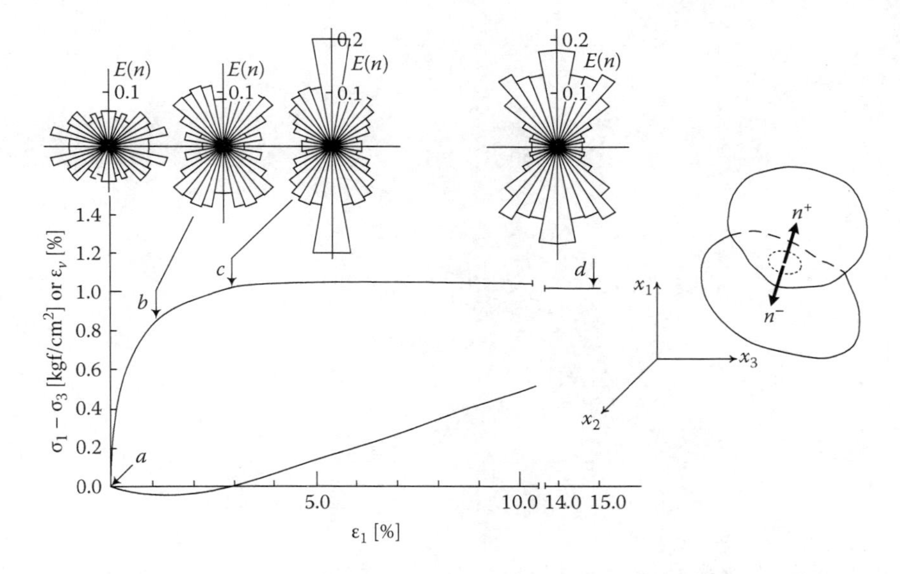

Figure 5.8 Progressive change in the distribution of contact normal directions during a triaxial compression test on a sand. (Oda, M. 1993. *Mechanics of Materials* 16 (1–2): 35–45.)

distribution of normal directions to the contacts at various loading stages from a to d on the curve (Oda 1993). Here, $E(\boldsymbol{n})$ denotes a density function of unit vectors $\boldsymbol{n}$ parallel to contact normal directions. Furthermore, based on the results of biaxial tests on a stack of photoelastic rods, Satake (1984) pointed out that the principal values (φ_1, φ_2) of the so-called fabric tensor φ_{ij}, which represents the relative distribution of the number of vectors normal to the interparticle contacts, is approximately proportional to the square root of the corresponding principal stresses:

$$\frac{\varphi_1}{\varphi_2} \simeq \left(\frac{\sigma_1}{\sigma_2}\right)^{0.5} \tag{5.30}$$

Maeda et al. (2006) also carried out two-dimensional (2D) DEM simulations of biaxial tests on 2D granular materials and obtained the same results as shown in Figure 5.9.

Employing a fabric tensor, Satake (1982) proposed the following modified stress tensor σ_{ij}^* for analyzing the behavior of granular materials:

$$\sigma_{ij}^* = \frac{1}{3}\varphi_{ik}^{-1}\sigma_{kj} \tag{5.31}$$

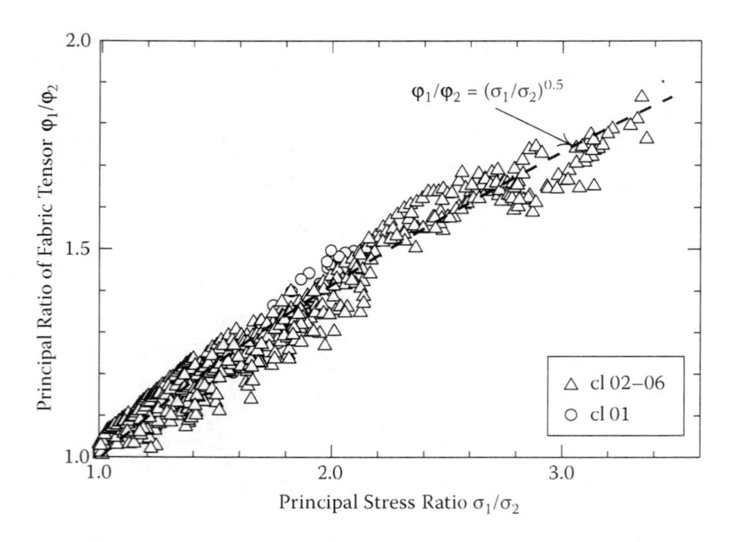

Figure 5.9 Change in the ratio of principal values φ_1/φ_2 of the fabric tensor φ_{ij} during DEM simulation of biaxial tests on two-dimensional granular materials. (Replotted from data in Maeda, K. et al. 2006. *Proceedings of Geomechanics and Geotechnics of Particulate Media,* Yamaguchi, 173–179.)

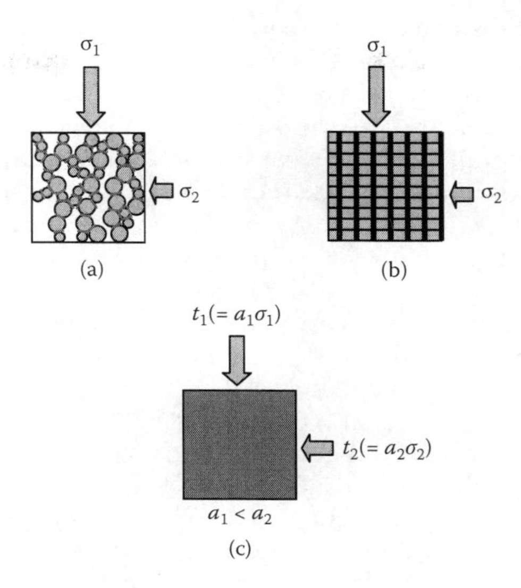

Figure 5.10 Anisotropy and its expression.

Figure 5.10(a) shows schematically the distribution of interparticle contact normal directions in 2D conditions in a granular material. Considering an equivalent continuum, such a material exhibits anisotropy with rearrangement of contact normal directions since the stiffness in the σ_1 direction becomes larger than that in the σ_2 direction with an increase of stress ratio as shown in diagram (b). Within the framework of elastoplastic theory, it is reasonable to treat the soil as an isotropic material by introducing the modified stress t_{ij} in which any induced anisotropy is already taken into consideration. Just as when analyzing seepage in anisotropic soils, the transformation of the stress space into a modified one allows us to revert to isotropic conditions where the normality rule applies, with the direction of plastic flow being normal to the yield surface (plastic potential). As can be deduced from eq. (5.2), the principal values of a_{ij} are inversely proportional to the square root of the respective principal stresses; therefore,

$$a_1 : a_2 = \frac{1}{\sqrt{\sigma_1}} : \frac{1}{\sqrt{\sigma_2}} \tag{5.32}$$

Interpreting the relationship in eq. (5.29) in terms of principal values, we find that the tensor a_{ij} corresponds to the inverse of the fabric tensor φ_{ij}. As a result, the modified stress tensor t_{ij} as defined in eq. (5.6) correlates with the one introduced by Satake (1982) in eq. (5.31). As shown in diagram (c) of Figure 5.10 referring to an equivalent isotropic material, the stress ratio t_1/t_2 in the modified stress space is smaller than the stress ratio σ_1/σ_2 in the ordinary stress space, because a_1 is smaller than a_2 as a consequence of eq. (5.32). Then, it is reasonable to assume that the flow rule (normality condition) does not hold in the σ_{ij} space, but rather in the t_{ij} space because the condition of the anisotropic material under anisotropic stress ratio in diagram (b) can be considered to be the same as that of an equivalent isotropic material under lower modified stress ratio in diagram (c) of Figure 5.10.

Although this explanation of the modified stress holds for the case of the development of anisotropy under monotonic loading without rotation of principal stress axes, a method to determine the modified stress under general cyclic loading with rotation of principal stress axes is described in the paper of Kikumoto, Nakai, and Kyokawa (2009).

Next, we demonstrate the rationale of relating the stress and strain increment invariants to normal and in-plane components of the respective measures in the SMP within the t_{ij} concept instead of the octahedral plane used in traditional models. The failure behavior of highly cohesive materials such as metals is governed mostly by the deviatoric stress (or shear stress) alone, so it is generally accepted to analyze failure in the plane where the shear stress is maximized—that is, a 45° plane (in 2D condition)—or the average plane where the shear stress arising from two principal stresses is maximized—that is, the octahedral plane (in 3D condition).

On the other hand, turning to geomaterials such as soils, failure is governed by Coulomb's friction between particles, which relates to the shear-to-normal stress ratio. Therefore, attention should be paid to the plane where the shear-to-normal stress ratio is maximized—that is, the mobilized plane (in 2D conditions)—or the combination of the three planes where the shear-to-normal stress ratio between two respective principal stresses is maximized—that is, the SMP (in 3D conditions). From this viewpoint, it is natural to formulate constitutive models using the normal and in-plane components of the stress and strain increments referred in the SMP. As a consequence, the influence of the intermediate principal stress on induced anisotropy and on the frictional resistance of geomaterials is naturally introduced when the t_{ij} concept is adopted.

5.3 THREE-DIMENSIONAL MODELING OF NORMALLY CONSOLIDATED SOILS BASED ON THE t_{ij} CONCEPT

5.3.1 Formulation of model

From the preceding developments, it becomes clear that the t_{ij} concept lends itself well to the consideration of the intermediate principal stress influence on the deformation and strength of soils. Thus, in the constitutive modeling of soils, all that we need is to formulate the yield function using the stress

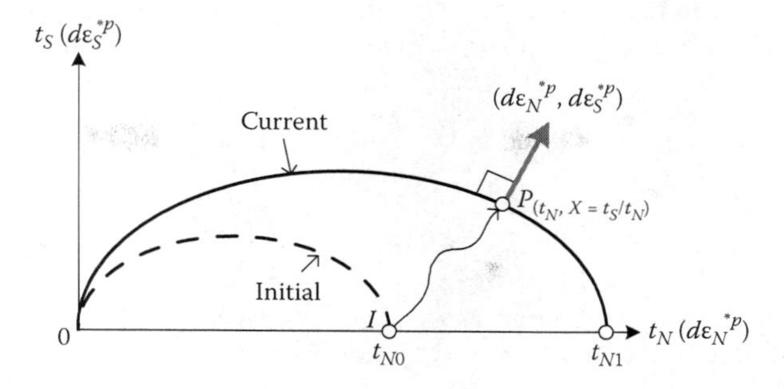

Figure 5.11 Initial and current yield surfaces in the t_N–t_S plane and direction of plastic flow for the model based on the t_{ij} concept.

invariants (t_N and t_S) instead of p and q and to assume the flow rule in the modified stress t_{ij} space instead of the Cauchy (ordinary) σ_{ij} space (Nakai and Mihara 1984), as described in Section 5.2.1.

Figure 5.11 shows the yield surfaces of an elastoplastic model, based on the t_{ij} concept and represented in the t_N–t_S space. The broken curve and solid curve indicate the initial and current yield surfaces, respectively, when the stress evolves from the initial state I ($t_N = t_{N0}$) to the current state P ($t_N = t_N$, $X = t_S/t_N$) with ensuing elastoplastic deformations. Here, t_{N0} and t_{N1} on the t_N-axis represent the size of the initial and current yield surfaces, respectively, similarly to p_0 and p_1 in the ordinary model as shown in Figure 4.7 in Chapter 4. The yield function for the model based on the t_{ij} concept is hereby defined by simply replacing σ_0 and σ in the 1D models with t_{N0} and t_{N1} (rather than p_0 and p_1), respectively:

$$F = H \quad \text{or} \quad f = F - H = 0 \tag{5.33}$$

$$F = (\lambda - \kappa)\ln\frac{t_{N1}}{t_{N0}} = (\lambda - \kappa)\left\{\ln\frac{t_N}{t_{N0}} + \varsigma(X)\right\} \quad (\text{where} \quad X = t_S/t_N) \tag{5.34}$$

$$H = (-\Delta e)^p = (1 + e_0)\cdot\varepsilon_v^p \tag{5.35}$$

where $\zeta(X)$ is an increasing function of the stress ratio $X(= t_S/t_N)$ that satisfies the condition $\zeta(0) = 0$, in the same way as for $\zeta(\eta)$ in the ordinary model described in Chapter 4. Then, the plastic strain increment is calculated using an associated flow rule in t_{ij} space (see Figure 5.11) as follows:

$$d\varepsilon_{ij}^p = \Lambda\frac{\partial F}{\partial t_{ij}} = \Lambda\left(\frac{\partial F}{\partial t_N}\frac{\partial t_N}{\partial t_{ij}} + \frac{\partial F}{\partial X}\frac{\partial X}{\partial t_{ij}}\right) \tag{5.36}$$

The proportionality constant Λ in this equation is obtained from the consistency condition ($df = 0$) in the same way as that in the ordinary models:

$$\Lambda = \frac{dF}{(1+e_0)\dfrac{\partial F}{\partial t_{kk}}} = \frac{dF}{h^p} \qquad \left(\text{where} \quad dF = \frac{\partial F}{\partial \sigma_{ij}} d\sigma_{ij}\right) \tag{5.37}$$

The full expressions of the derivative of F with respect to t_{ij} and σ_{ij} are developed in Section 5.3.2.

The elastic strain increment $d\varepsilon_{ij}^e$ is usually given by eqs. (4.29) and (4.30) in terms of elastic parameters and Cauchy stresses. Following the t_{ij} treatment used for calculating plastic strains, the elastic volumetric strain is assumed to be controlled by the mean modified stress t_N, rather than the usual mean stress p. Then, the elastic strain increment can be expressed by the following equation:

$$d\varepsilon_{ij}^e = \frac{1+\nu_e}{E_e} d\left(\frac{\sigma_{ij}}{1+X^2}\right) - \frac{\nu_e}{E_e} d\left(\frac{\sigma_{mm}}{1+X^2}\right)\delta_{ij} = C_{ijkl}^e d\sigma_{kl} = \left[D_{ijkl}^e\right]^{-1} d\sigma_{kl} \tag{5.38}$$

As described in Chapter 4, C_{ijkl}^e and D_{ijkl}^e are called the elastic compliance tensor and the elastic stiffness tensor, respectively. This equation derives from the fact that the following relationship always holds between p and t_N:

$$t_N = \frac{p}{1+X^2} \tag{5.39}$$

This can be readily deduced from eqs. (5.12) and (5.14). Furthermore, the elastic modulus E_e is expressed in terms of the swelling index κ, herein as the slope of the unloading/reloading line in the e–$\ln t_N$ space, and Poisson's ratio ν_e as

$$E_e = \frac{3(1-2\nu_e)(1+e_0)t_N}{\kappa} \tag{5.40}$$

Additionally, in the elastoplastic region, the calculated stress–strain relation does not depend so much on whether eq. (4.29) or eq. (5.38) is used to compute the elastic component. The constitutive model for normally consolidated soils presented here essentially draws directly from the t_{ij} clay model developed by Nakai and Matsuoka (1986).

Now, as in recent models (Chowdhury and Nakai 1998; Nakai and Hinokio 2004), the following functional form is adopted for defining the stress ratio $\zeta(X)$ in eq. (5.34):

$$\varsigma(X) = \frac{1}{\beta}\left(\frac{X}{M^*}\right)^{\beta} \tag{5.41}$$

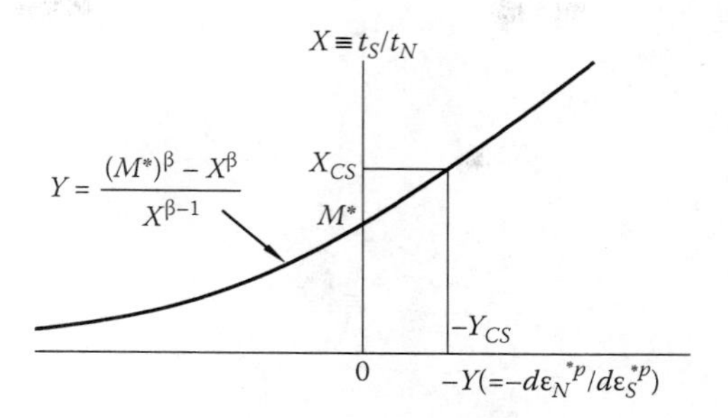

Figure 5.12 Stress–dilatancy relation of recent t_{ij} model.

where β (≥1) is a parameter that controls the shape of the yield function. When β = 1, the shape of the yield function is the same as that of the original Cam clay model. From eqs. (5.34), (5.36), and (5.41), the following stress–dilatancy relation holds:

$$\frac{d\varepsilon_N^{*p}}{d\varepsilon_S^{*p}} = \frac{\dfrac{\partial F}{\partial t_N} + \dfrac{\partial F}{\partial X}\dfrac{\partial X}{\partial t_N}}{\dfrac{\partial F}{\partial X}\dfrac{\partial X}{\partial t_S}} = \frac{1-\zeta'(X)\cdot X}{\zeta'(X)} = \frac{(M^*)^\beta - X^\beta}{X^{\beta-1}} \quad (5.42)$$

Figure 5.12 illustrates this equation in terms of the relation between $X = t_S/t_N$ and $Y = d\varepsilon_N{}^{*p}/d\varepsilon_S{}^{*p}$. Here, M^* implies the intercept with the vertical axis and is given by the following equation:

$$M^* = \left(X_{CS}^\beta + X_{CS}^{\beta-1} Y_{CS}\right)^{1/\beta} \quad (5.43)$$

when making both the stress ratio X_{CS} and plastic strain increment Y_{CS} at critical state ($d\varepsilon_v{}^p = 0$) satisfy eq. (5.42).

The values of X_{CS} and Y_{CS} are expressed as follows using the principal stress ratio at critical state under triaxial compression $R_{CS} = (\sigma_1/\sigma_3)_{CS(comp)}$ (Nakai and Mihara 1984):

$$X_{CS} = \frac{\sqrt{2}}{3}\left(\sqrt{R_{CS}} - \frac{1}{\sqrt{R_{CS}}}\right) \quad (5.44)$$

$$Y_{CS} = \frac{1-\sqrt{R_{CS}}}{\sqrt{2}\left(\sqrt{R_{CS}} + 0.5\right)} \quad (5.45)$$

The derivations of X_{CS} and Y_{CS} are given in Section 5.3.2.

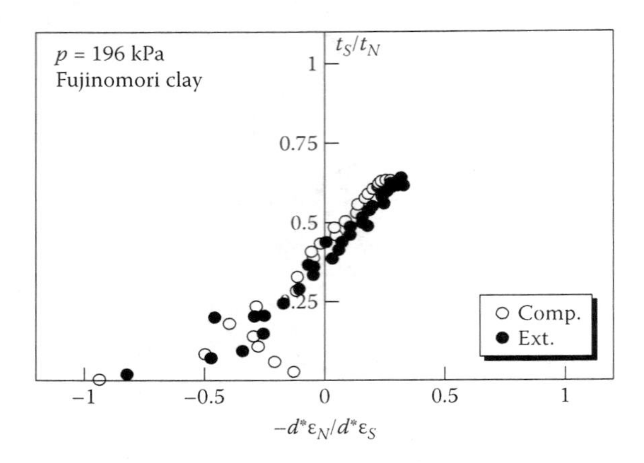

Figure 5.13 t_S/t_N versus $d\varepsilon_N^*/d\varepsilon_S^*$ relation of drained triaxial compression and extension tests on a normally consolidated clay.

Figure 5.13 shows the experimental data of the stress–dilatancy relation for the normally consolidated clay reported in Figure 4.21 (Chapter 4), replotted in terms of the stress ratio t_S/t_N and the strain increment ratio $d\varepsilon_N^*/d\varepsilon_S^*$. It can be seen that the stress–dilatancy relation based on the t_{ij} concept is independent of the intermediate principal stress, whereas a strong dependency is found when expressed in terms of q/p and $d\varepsilon_v/d\varepsilon_d$, as illustrated in Figure 4.21.

Figure 5.14 shows a view of the yield surface of the t_{ij} model in the principal spaces of σ_{ij} and t_{ij} (Pedroso, Farias, and Nakai 2005). The shape of the yield surface on the octahedral plane in the σ_{ij} space is a rounded triangle and corresponds to that of the SMP criterion shown in Figure 5.5 (Matsuoka and Nakai 1974); see diagram (c). The shape of the yield surface in the t_{ij} space is also oval, though a little more rounded, but it is not a circle; see diagram (d).

Figure 5.15 shows the shape of yield surfaces under different stress ratios ($X = 0.2$, 0.4, and 0.63) on the octahedral plane in the principal stress space, together with the direction of the plastic strain increment $d\varepsilon_d^p$ (arrows) calculated by the t_{ij} model. Here, the dotted lines indicate the normal vectors to the yield surfaces in the σ_{ij} space. As can be seen from this figure, although the directions of the plastic strain increments inevitably coincide with the direction of the corresponding deviatoric strain increments (i.e., radial directions) under triaxial compression ($\theta = 0°$) and extension ($\theta = 60°$) conditions, they deviate leftward from the direction of shear stress (radial direction) with the increase in stress ratio under three different principal stresses ($0° < \theta < 60°$), and they lie between the directions normal to the yield surfaces and the radial directions. These deviations of the calculated plastic strain increments are in good agreement with many reported results of true triaxial tests

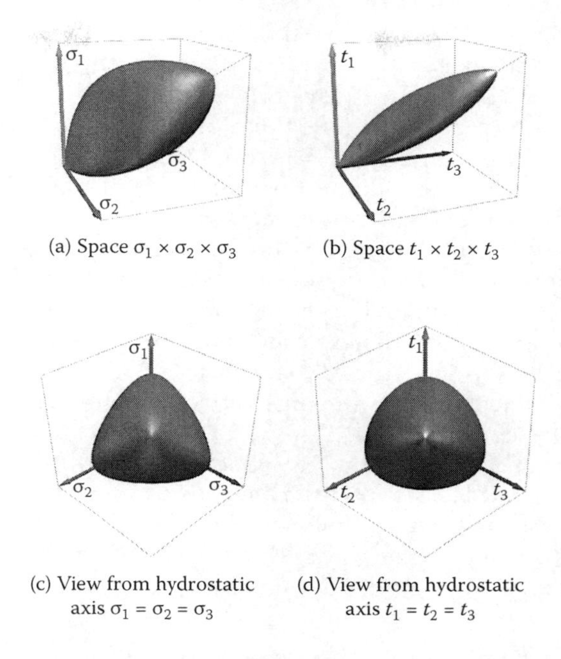

Figure 5.14 Shape of yield surfaces in the principal spaces of σ_{ij} and t_{ij}. (Pedroso, D. M. et al. 2005. *Soils and Foundations* 45 (4): 61–77.)

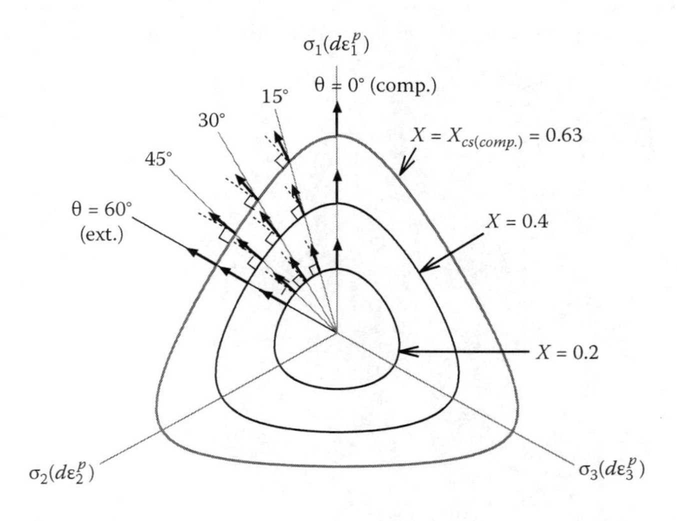

Figure 5.15 Yield surfaces and directions of plastic flow of t_{ij} model on the octahedral plane in the σ_{ij} space.

(Yong and Mckyes 1971; Lade and Musante 1978) as well as those in Figure 4.23 in Chapter 4.

On the other hand, if the flow rule were assumed in the σ_{ij} space but not in the t_{ij} space, the directions of the plastic flow would follow the dotted lines in Figure 5.15, and hence there would be even more deviation from the radial direction. Therefore, the t_{ij} concept can describe not only the observed uniqueness of the stress–dilatancy relation (Figure 5.13) but also the observed deviations of the directions of the plastic flow from the direction of shear stress on the octahedral plane (Figure 4.23). Employing the t_{ij} concept (i.e., the stress term F in the yield function is formulated in such a form as that in eq. (5.34)) and assuming the flow rule in the t_{ij} space as in eq. (5.36), the influence of intermediate principal stress in constitutive modeling can be automatically taken into consideration.

As mentioned before, the Cam clay model for normally consolidated soils can be easily extended to a model valid under three different principal stresses using the t_{ij} concept. Then, the following loading condition is employed for the present model in the same way as that in eq. (4.32) in Chapter 4 because there is no strain-softening behavior in normally consolidated soils:

$$\begin{cases} d\varepsilon_{ij}^{p} \neq 0 \quad \text{if } f = F(t_N, X) - H(-\Delta e^{p}) = 0 \quad \& \quad dF = \dfrac{\partial F}{\partial \sigma_{ij}} d\sigma_{ij} \geq 0 \\ d\varepsilon_{ij}^{p} = 0 \quad \text{otherwise} \end{cases} \tag{5.46}$$

Figure 5.16 shows the yield surface of the original Cam clay model in the p–q plane under triaxial compression (upper half) and triaxial

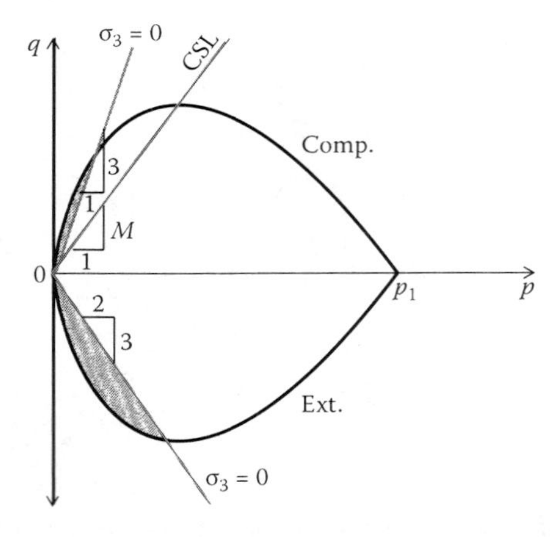

Figure 5.16 Yield surface of the Cam clay model and tension zone on the p–q plane.

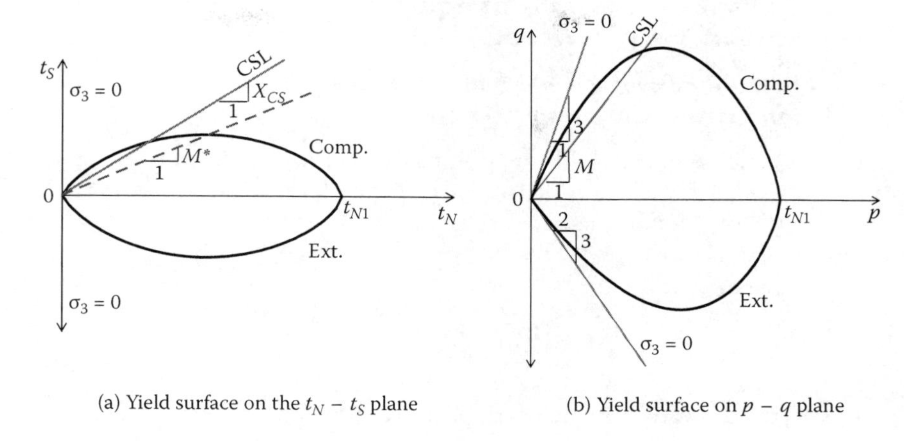

(a) Yield surface on the $t_N - t_S$ plane

(b) Yield surface on $p - q$ plane

Figure 5.17 Yield surface of the t_{ij} model.

extension (lower half) conditions. Figure 5.17 shows the yield surface of the present model in (a) the t_N–t_S plane and (b) the p–q plane. It can be observed that the yield surface of the model based on the t_{ij} concept is symmetric with respect to the t_N-axis but not symmetric with respect to the p-axis. Also, it is noted that the yield surface of the proposed model is smooth over the whole domain, whereas the one in the original Cam clay model is not smooth at the tip on the p-axis. Such smoothness of the present yield surface ensures numerical stability in computations in the same way as the modified Cam clay model (Roscoe and Burland 1968) does.

Now, the lines for which the minor principal stress σ_3 is zero are indicated in every figure. It can be seen that models formulated using p and q such as the Cam clay model have tension zones (shaded area in Figure 5.16) on and inside the yield surface. This means that the stress condition may enter the tension zone ($\sigma_3 < 0$) during elastic deformation in the numerical simulations of the ground behavior because the yield surface is fixed during elastic deformation in conventional elastoplasticity. On the other hand, there is no tension zone in the yield surface formulated using t_N and t_S because the $\sigma_3 = 0$ condition is satisfied on the vertical axis (t_S-axis) in Figure 5.17(a). Hence, models based on the t_{ij} concept not only are capable of describing properly the influence of the intermediate principal stress but also have the benefit of numerical computations associated with the tension zone.

5.3.2 Explicit expressions of equations used in the t_{ij} concept

5.3.2.1 *Partial derivatives of yield function and stress variables*

The stress term F in the yield function based on the t_{ij} concept is expressed by eq. (5.34) as a function of the mean stress t_N and the stress ratio X, with the various definitions of tensors and scalars related to the t_{ij} concept shown in Table 5.1. Thus,

$$F = (\lambda - \kappa)\ln\frac{t_{N1}}{t_{N0}} = (\lambda - \kappa)\left\{\ln\frac{t_N}{t_{N0}} + \varsigma(X)\right\} \quad (\text{where } X = t_S/t_N) \quad (5.34\text{bis})$$

Firstly, the derivatives of F with respect to the modified stress t_{ij} are shown:

$$\frac{\partial F}{\partial t_{ij}} = \frac{\partial F}{\partial t_N}\frac{\partial t_N}{\partial t_{ij}} + \frac{\partial F}{\partial X}\frac{\partial X}{\partial t_{ij}} \quad (5.47)$$

$$\frac{\partial F}{\partial t_N} = (\lambda - \kappa)\frac{1}{t_N} \quad (5.48)$$

$$\frac{\partial t_N}{\partial t_{ij}} = \frac{\partial (t_{kl}a_{kl})}{\partial t_{ij}} = a_{ij} \quad (5.49)$$

$$\frac{\partial F}{\partial X} = (\lambda - \kappa)\zeta'(X) \quad (5.50)$$

$$\frac{\partial X}{\partial t_{ij}} = \frac{\partial\left(\sqrt{x_{mn}x_{mn}}\right)}{\partial x_{kl}}\frac{\partial x_{kl}}{\partial t_{ij}} = \frac{1}{X\cdot t_N}\left(x_{ij} - X^2 a_{ij}\right) \quad (5.51)$$

where

$$x_{ij} = \frac{t'_{ij}}{t_N} = \frac{t_{ij} - t_N a_{ij}}{t_N} = \frac{t_{ij}}{t_N} - a_{ij} \quad (5.52)$$

$$\frac{\partial X}{\partial x_{kl}} = \frac{\partial\left(\sqrt{x_{mn}x_{mn}}\right)}{\partial x_{kl}} = \frac{x_{kl}}{X} \quad (5.53)$$

$$\frac{\partial x_{kl}}{\partial t_{ij}} = \frac{\partial}{\partial t_{ij}}\left(\frac{t_{kl}}{t_N} - a_{kl}\right) = \frac{1}{t_N}\{\delta_{ik}\delta_{jl} - (x_{kl} + a_{kl})a_{ij}\} \quad (5.54)$$

The mean stress t_N and the stress ratio X are also given using the ordinary stress σ_{ij} by eqs. (5.12) and (5.14). Then, the derivative of F with respect to the ordinary stress σ_{ij} is expressed as follows:

$$\frac{\partial F}{\partial \sigma_{ij}} = \frac{\partial F}{\partial t_N}\frac{\partial t_N}{\partial \sigma_{ij}} + \frac{\partial F}{\partial X}\frac{\partial X}{\partial \sigma_{ij}} \tag{5.55}$$

$$\frac{\partial t_N}{\partial \sigma_{ij}} = \frac{\partial}{\partial \sigma_{ij}}\left(3\frac{I_3}{I_2}\right) = -3\frac{I_3}{I_2^2}\frac{\partial I_2}{\partial \sigma_{ij}} + 3\frac{1}{I_2}\frac{\partial I_3}{\partial \sigma_{ij}} \tag{5.56}$$

$$\frac{\partial X}{\partial \sigma_{ij}} = \frac{\partial}{\partial \sigma_{ij}}\left(\sqrt{\frac{I_1 I_2}{9I_3}-1}\right) = \frac{1}{2X}\left(\frac{I_2}{9I_3}\frac{\partial I_1}{\partial \sigma_{ij}} + \frac{I_1}{9I_3}\frac{\partial I_2}{\partial \sigma_{ij}} - \frac{I_1 I_2}{9I_3^2}\frac{\partial I_3}{\partial \sigma_{ij}}\right) \tag{5.57}$$

where

$$\frac{\partial I_1}{\partial \sigma_{ij}} = \frac{\partial \sigma_{kk}}{\partial \sigma_{ij}} = \delta_{ij} \tag{5.58}$$

$$\frac{\partial I_2}{\partial \sigma_{ij}} = \frac{\partial}{\partial \sigma_{ij}}\left(\frac{(\sigma_{kk})^2 - \sigma_{lm}\sigma_{ml}}{2}\right) = \sigma_{kk}\delta_{ij} - \sigma_{ij} \tag{5.59}$$

$$\frac{\partial I_3}{\partial \sigma_{ij}} = \frac{\partial}{\partial \sigma_{ij}}\left(\frac{e_{klm}e_{opq}\sigma_{ko}\sigma_{lp}\sigma_{mq}}{6}\right) = \frac{1}{2}e_{ilm}e_{jpq}\sigma_{lp}\sigma_{mq} \tag{5.60}$$

Referring to this formulation, the plastic strain increment $d\varepsilon_{ij}{}^p$ can be calculated using eqs. (5.36) and (5.37). These equations are sufficient for the calculation of isotropic hardening models based on the t_{ij} concept, in which the function F is given by the invariants of t_{ij} such as $F(t_N, X)$.

For reference, the partial derivatives of the tensors (a_{ij} and t_{ij}) with respect to the ordinary stress σ_{kl} follow for the elaborate calculation of other hardening models (e.g., kinematic hardening, rotational hardening, and others) based on the t_{ij} concept.

Since a_{ij} is given by eq. (5.3), the following equation holds:

$$\begin{aligned}\frac{\partial a_{ij}}{\partial \sigma_{kl}} &= \frac{\partial a_{ij}}{\partial I_2}\frac{\partial I_2}{\partial \sigma_{kl}} + \frac{\partial a_{ij}}{\partial I_3}\frac{\partial I_3}{\partial \sigma_{kl}} + \frac{\partial a_{ij}}{\partial r_{mn}}\frac{\partial r_{mn}}{\partial \sigma_{kl}} \\ &= -\frac{a_{ij}}{2I_2}\frac{\partial I_2}{\partial \sigma_{kl}} + \frac{a_{ij}}{2I_3}\frac{\partial I_3}{\partial \sigma_{kl}} + \sqrt{\frac{I_3}{I_2}}\frac{\partial r_{ij}^{-1}}{\partial r_{mn}}\frac{\partial r_{mn}}{\partial \sigma_{kl}}\end{aligned} \tag{5.61}$$

The derivatives $(\partial I_2/\partial \sigma_{kl})$ and $(\partial I_3/\partial \sigma_{kl})$ are given by eqs. (5.59) and (5.60), whereas the other derivatives in eq. (5.61) are calculated as follows using the condition $r_{ki}r_{ij}^{-1} = \delta_{kj}$:

$$\frac{\partial\left(r_{ki}r_{ij}^{-1}\right)}{\partial r_{mn}} = \delta_{km}\delta_{in}r_{ij}^{-1} + r_{ki}\frac{\partial r_{ij}^{-1}}{\partial r_{mn}} = \delta_{km}r_{nj}^{-1} + r_{ki}\frac{\partial r_{ij}^{-1}}{\partial r_{mn}} = 0 \tag{5.62}$$

Then,

$$\frac{\partial r_{ij}^{-1}}{\partial r_{mn}} = -\delta_{km} r_{nj}^{-1} r_{ki}^{-1} = -r_{nj}^{-1} r_{mi}^{-1} \tag{5.63}$$

Equation (5.26) leads to

$$\frac{\partial \sigma_{kl}}{\partial r_{mn}} = \frac{\partial (r_{ko} r_{ol})}{\partial r_{mn}} = \delta_{km}\delta_{on} r_{ol} + r_{ko}\delta_{om}\delta_{ln} = \delta_{km} r_{nl} + r_{km}\delta_{ln} \tag{5.64}$$

Then,

$$\frac{\partial r_{mn}}{\partial \sigma_{kl}} = (\delta_{km} r_{nl} + r_{km}\delta_{nl})^{-1} \tag{5.65}$$

Therefore, substituting eqs. (5.59), (5.60), (5.63), and (5.65) into eq. (5.61), the partial derivatives of a_{ij} with respect to the ordinary stress σ_{kl} can be expressed as

$$\begin{aligned}\frac{\partial a_{ij}}{\partial \sigma_{kl}} &= -\frac{a_{ij}}{2I_2}(\sigma_{mm}\delta_{kl} - \sigma_{kl}) + \frac{a_{ij}}{2I_3}\left(\frac{1}{2} e_{kpq} e_{lst} \sigma_{ps} \sigma_{qt}\right) \\ &\quad - \sqrt{\frac{I_3}{I_2}} r_{nj}^{-1} r_{mi}^{-1} (\delta_{km} r_{nl} + r_{km}\delta_{nl})^{-1} \\ &= -\frac{a_{ij}}{2I_2}(\sigma_{mm}\delta_{kl} - \sigma_{kl}) + \frac{a_{ij}}{2I_3}\left(\frac{1}{2} e_{kpq} e_{lst} \sigma_{ps} \sigma_{qt}\right) \\ &\quad - \sqrt{\frac{I_3}{I_2}} (r_{ik}\sigma_{jl} + \sigma_{ik} r_{jl})^{-1}\end{aligned} \tag{5.66}$$

Also, the partial derivatives of t_{ij} with respect to the ordinary stress σ_{kl} can be calculated as

$$\begin{aligned}\frac{\partial t_{ij}}{\partial \sigma_{kl}} &= \frac{\partial (a_{im}\sigma_{mj})}{\partial \sigma_{kl}} = \frac{\partial}{\partial \sigma_{kl}}\left(\sqrt{\frac{I_3}{I_2}} r_{ij}\right) = \frac{\partial t_{ij}}{\partial I_2}\frac{\partial I_2}{\partial \sigma_{kl}} + \frac{\partial t_{ij}}{\partial I_3}\frac{\partial I_3}{\partial \sigma_{kl}} + \frac{\partial t_{ij}}{\partial r_{mn}}\frac{\partial r_{mn}}{\partial \sigma_{kl}} \\ &= -\frac{t_{ij}}{2I_2}(\sigma_{mm}\delta_{kl} - \sigma_{kl}) + \frac{t_{ij}}{2I_3}\left(\frac{1}{2} e_{kpq} e_{lst} \sigma_{ps} \sigma_{qt}\right) \\ &\quad - \sqrt{\frac{I_3}{I_2}} \delta_{im}\delta_{jn} (\delta_{km} r_{nl} + r_{km}\delta_{nl})^{-1} \\ &= -\frac{t_{ij}}{2I_2}(\sigma_{mm}\delta_{kl} - \sigma_{kl}) + \frac{t_{ij}}{2I_3}\left(\frac{1}{2} e_{kpq} e_{lst} \sigma_{ps} \sigma_{qt}\right) \\ &\quad - \sqrt{\frac{I_3}{I_2}} (\delta_{ik} r_{jl} + r_{ik}\delta_{jl})^{-1}\end{aligned} \tag{5.67}$$

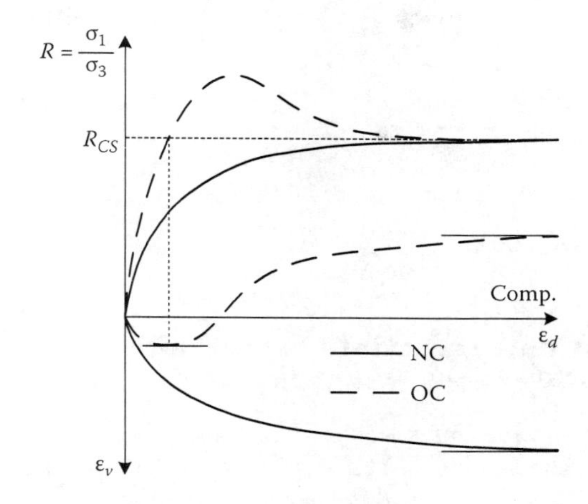

Figure 5.18 Definition of R_{CS} in triaxial compression condition for normally and overconsolidated soils. (a) Shape on the t_N–t_S plane; (b) shape on the p–q plane.

5.3.2.2 *Derivation of X_{CS} and Y_{CS}*

Under the triaxial compression condition ($\sigma_1 > \sigma_2 = \sigma_3$ and $d\varepsilon_1 > d\varepsilon_2 = d\varepsilon_3$), as shown in Figure 5.18, and noting eqs. (5.2), (5.14), (5.10), and (5.11), the stress ratio $X = t_S/t_N$ and the plastic strain increment ratio $Y = d\varepsilon_N^*/d\varepsilon_S^*$ are expressed as follows:

$$X = \frac{t_S}{t_N} = \frac{\sqrt{2}}{3}\left(\sqrt{\frac{\sigma_1}{\sigma_3}} - \sqrt{\frac{\sigma_3}{\sigma_1}}\right) \tag{5.68}$$

$$Y = \frac{d\varepsilon_N^{*p}}{d\varepsilon_S^{*p}} = \frac{d\varepsilon_1^p a_1 + 2d\varepsilon_3^p a_3}{\sqrt{2}\left(d\varepsilon_1^p a_3 - d\varepsilon_3^p a_1\right)} = \frac{1 + 2\sqrt{\frac{\sigma_1}{\sigma_3}} \cdot \frac{d\varepsilon_3^p}{d\varepsilon_1^p}}{\sqrt{2}\left(\sqrt{\frac{\sigma_1}{\sigma_3}} - \frac{d\varepsilon_3^p}{d\varepsilon_1^p}\right)} \tag{5.69}$$

Since the plastic volumetric strain increment $d\varepsilon_v{}^p$ is zero at critical state, the plastic strain increment ratio $d\varepsilon_3{}^p/d\varepsilon_1{}^p$ in triaxial compression condition is expressed as

$$\frac{d\varepsilon_3^p}{d\varepsilon_1^p} = \frac{1}{2}\left(\frac{d\varepsilon_v^p}{d\varepsilon_1^p} - 1\right) = -\frac{1}{2} \tag{5.70}$$

Therefore, X_{CS} and Y_{CS} take the following form using the principal stress ratio $R_{CS} = (\sigma_1/\sigma_3)_{CS(comp)}$ at the critical state under triaxial compression condition:

$$X_{CS} = \frac{\sqrt{2}}{3}\left(\sqrt{R_{CS}} - \frac{1}{\sqrt{R_{CS}}}\right) \quad (5.44\text{bis})$$

$$Y_{CS} = \frac{1-\sqrt{R_{CS}}}{\sqrt{2}\left(\sqrt{R_{CS}}+0.5\right)} \quad (5.45\text{bis})$$

5.3.3 Validation by test data for remolded normally consolidated clay

The validity of the present model is next confirmed by examining experimental data of shear tests on saturated, remolded, normally consolidated Fijinomori clay (F-clay). Physical properties of the F-clay are as follows: liquid limit $w_L = 44.7\%$, plastic limit $w_P = 24.7\%$, and specific gravity $G_S = 2.65$. Details regarding methods of sample preparation are described in Nakai and Matsuoka (1986). Table 5.2 shows the various material parameters for saturated Fujinomori clay. As indicated in the table, the parameters are fundamentally the same as those of the Cam clay model. The parameter β, which represents the shape of the yield surface (see eq. 5.41), can be determined from the observed stress–strain–dilatancy curve of a shear test, as described later.

The sensitivity of β to the stress–strain–dilatancy relation and others is discussed here. Figures 5.19(a) and (b) show the shape of the yield surfaces with different values of β ($\beta = 1.1$, 1.5, and 2.0) on the t_N–t_S and p–q planes. The curves are drawn using the material parameters in Table 5.2 except for the value of β. The critical state line (CSL) in triaxial extension condition is marginally influenced by the value of β.

Figure 5.20 shows the stress–dilatancy relations corresponding to the yield surfaces (plastic potentials) in Figure 5.19(a). The thin broken lines represent the condition of no plastic volume change in triaxial compression and extension. It is also seen from this figure that the stress ratio X at critical state in triaxial extension is slightly larger than that in triaxial

Table 5.2 Values of material parameters for remolded normally consolidated Fujinomori clay

λ	0.090	
k	0.020	
$N(e_N$ at p = 98 kPa)	0.83	Same parameters as Cam clay model
$R_{CS} = (\sigma_1/\sigma_3)_{CS\,(comp)}$	3.5	
ν_e	0.2	
β	1.5	Shape of yield surface (same as original Cam clay if β = 1)

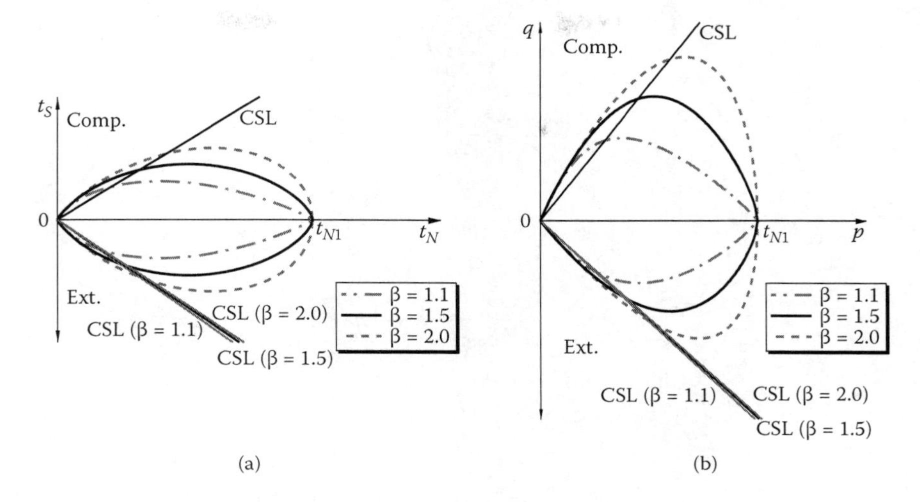

Figure 5.19 Yield surfaces with different values of parameter β and critical state line under triaxial compression and extension conditions.

compression. The critical state conditions on the octahedral plane with different values of β are illustrated in Figure 5.21, together with the SMP criterion and the Mohr–Coulomb criterion.

Figure 5.22 shows the calculated stress–strain–dilatancy curves of triaxial compression test under constant mean principal stress ($p = 196$ kPa) using different values of β, together with the observed data in Figure 4.21. It can be seen that the calculated curve of β = 1.5 shows the best agreement with the observed results, and the model describes higher stiffness and less volume contraction with increasing values of β. Figure 5.23 shows the calculated stress–strain curves and the effective

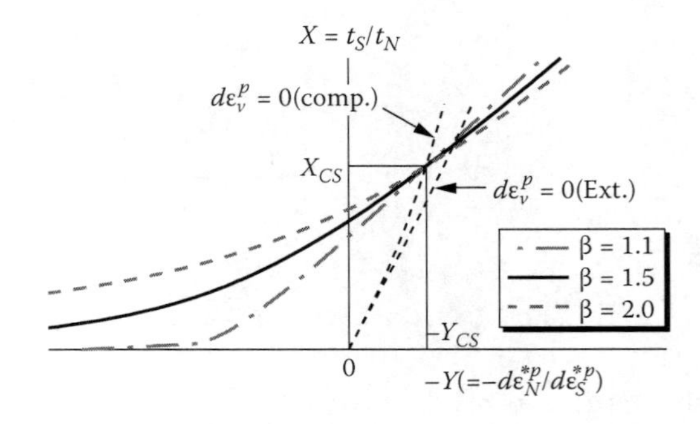

Figure 5.20 Stress–dilatancy relations with different values of β.

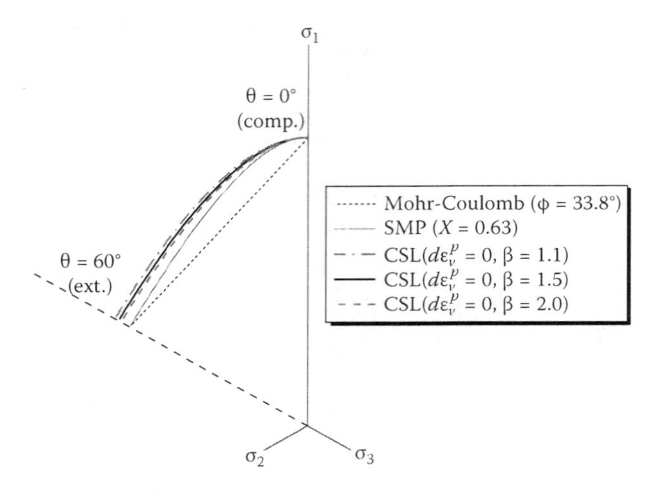

Figure 5.21 Critical state condition based on the t_{ij} concept, SMP criterion, and Mohr–Coulomb criterion on the octahedral plane.

stress paths for values of β = 1.1, 1.5, and 2.0 in an undrained triaxial compression test.

Figure 5.24 plots the calculated change in void ratio for a constant mean principal triaxial compression test as well as the change in effective mean principal stress for an undrained triaxial compression test on the e–ln$p$

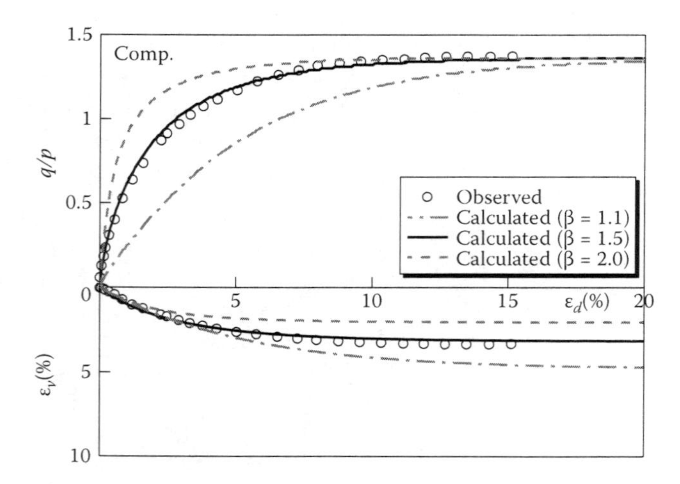

Figure 5.22 Calculated stress–strain curves with different values of β and corresponding observed results during constant mean principal (196 kPa) triaxial compression test.

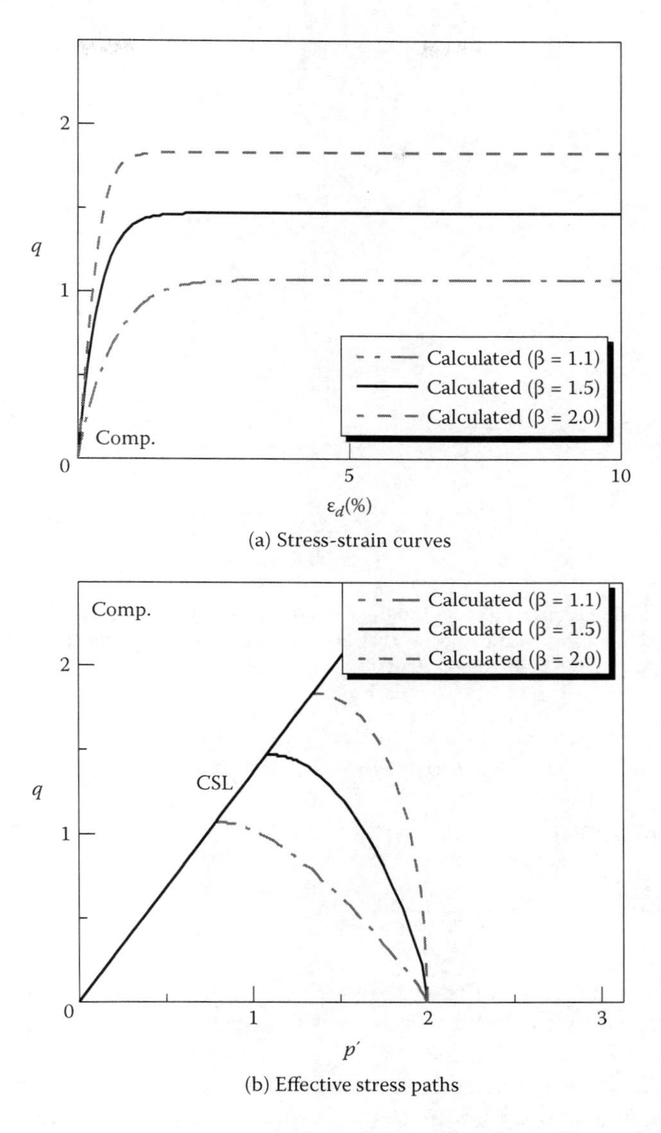

Figure 5.23 Calculated results of undrained triaxial compression test with different values of β.

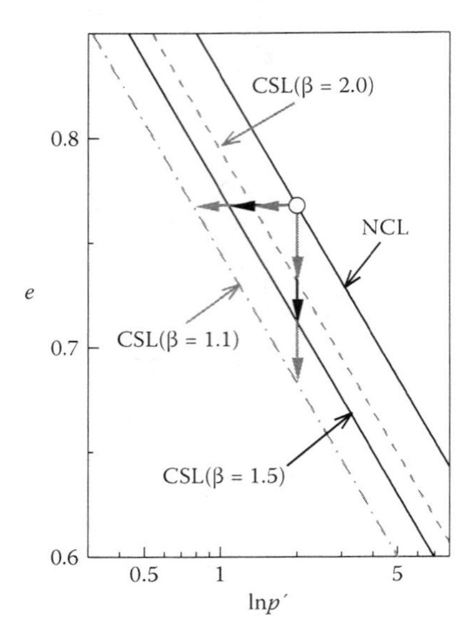

Figure 5.24 Critical state lines with different values of β and change in void ratio during constant mean principal test and change of effective mean principal stress during undrained compression test.

plane, together with the critical state lines for different values of β. Here, from eqs. (5.34), (5.41), and (5.43), the difference in void ratio between the normal consolidation line (NCL) and the critical state line (CSL) at the same mean principal stress is expressed as

$$(\lambda - \kappa)\cdot \varsigma(X_{CS}) = (\lambda - \kappa)\frac{1}{\beta}\left(\frac{X_{CS}}{M^*}\right)^{\beta} = \frac{\lambda - \kappa}{\beta}\frac{X_{CS}}{X_{CS} + Y_{CS}} \tag{5.71}$$

Therefore, the value of the parameter β can be determined by fitting the computed results to the observed ones arranged in terms of the relations described in Figures 5.21–5.24.

Figure 5.25 shows the stress–strain–dilatancy results (symbols) compared to calculated curves in triaxial compression and extension tests on the normally consolidated Fujinomori clay under constant mean principal stress (p = 196 kPa). It is seen that the present model can describe well both the deformation and strength of normally consolidated clay in triaxial compression and extension tests. Here, the test data are the same as those in Figure 4.19. Figure 5.26 shows the observed (symbols) and calculated (curves) results of the undrained triaxial compression and extension tests. The model simulates well the undrained behavior of clay, including the difference in strength between triaxial compression and extension conditions.

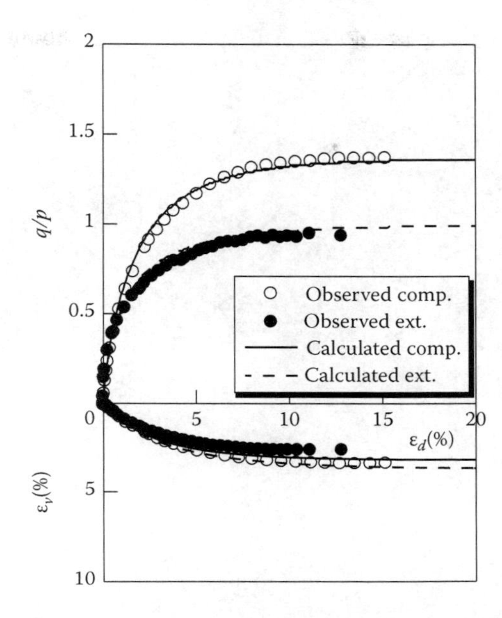

Figure 5.25 Observed and calculated results of triaxial compression and extension tests on normally consolidated Fujinomori clay under monotonic loading.

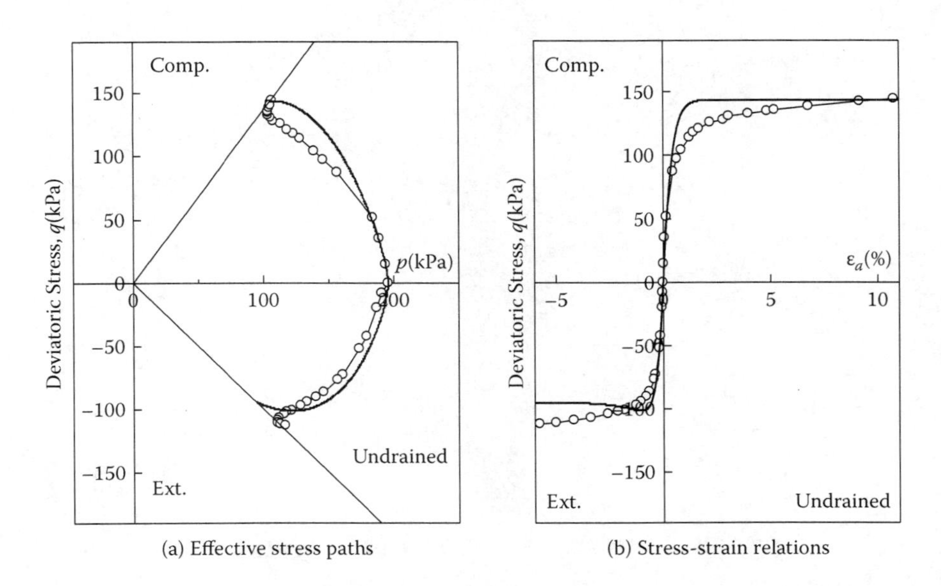

Figure 5.26 Observed and calculated effective stress paths and stress–strain relations during undrained triaxial compression and extension tests on normally consolidated clay. (a) Effective stress paths; (b) stress–strain relations.

(a) Whole view of the apparatus

(b) Specimen with horizontal loading device

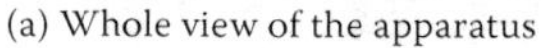

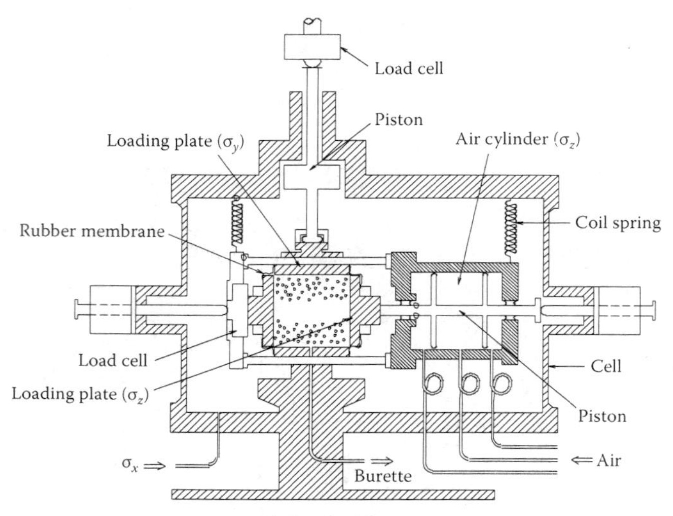

(c) Sketch of the apparatus

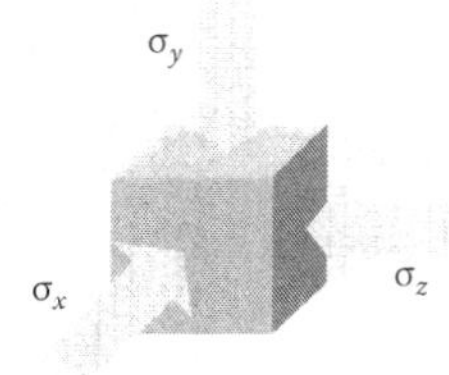

(d) Explanation of three principal stresses system

Figure 5.27 System of chamber type true triaxial test apparatus. (a) Whole view of the apparatus; (b) specimen with horizontal loading device; (c) sketch of the apparatus; (d) explanation of three principal stresses system.

Figure 5.27 shows details of the true triaxial apparatus, which is of the chamber type. Photo (a) shows a whole view of the complete apparatus together with the controlling devices, and photo (b) gives a close-up view of the specimen (10 × 10 × 7 cm) with the loading device in the horizontal direction. Diagram (c) shows a schematic picture of the interior part of the true triaxial chamber, whereas diagram (d) illustrates the specimen subjected to three principal stresses. As is seen in this figure, σ_x is applied by chamber water pressure while the other two principal stresses (σ_y and σ_z) are applied by rigid plates. To reduce the effect of friction on the loading plate (σ_z) in the horizontal direction, the loading device in the horizontal direction is suspended and floated by coil springs during the test, as shown in diagram (c).

Also, in order to apply three different principal stresses to the specimen, a rubber membrane surrounds not only the cap and pedestal in the vertical direction but also the loading plate in the horizontal direction. In the ordinary chamber type true triaxial testing apparatus (Shibata and Karube 1965; Green 1971; Lade and Duncan 1973), the loading plate or the pressure bag that applies one of the two horizontal stresses is outside the membrane, so the principal stress, which is less than the chamber water pressure, cannot be applied in the horizontal direction. Details of the present apparatus are described by Nakai et al. (1986).

Figure 5.28 shows the observed (symbols) and calculated (curves) variations of the three principal strains (ε_1, ε_2, and ε_3) and the volumetric strain (ε_v) against the stress ratio q/p in true triaxial tests (θ = 0°, 15°, 30°, 45°, and 60°) on normally consolidated clay under constant mean principal effective stress (p = 196 kPa). As shown in Figure 5.29, θ denotes the angle between the σ_1-axis and the corresponding radial stress path on the octahedral plane, where θ = 0° and 60° represent the stress paths under triaxial compression and extension conditions, respectively. There is the following relation between the angle θ and intermediate principal stress parameter b_σ:

$$b_\sigma = \frac{\sigma_2 - \sigma_3}{\sigma_1 - \sigma_3} = \frac{2\tan\theta}{\sqrt{3} + \tan\theta} \tag{5.72}$$

It can be seen from Figure 5.28 that the present model predicts well the stress–strain behavior of normally consolidated clay with negative dilatancy under three different principal stresses as well as axisymmetric stress conditions. Here, the two minor principal strains in triaxial compression condition—θ = 0°, diagram (a)—and two major principal strains in triaxial extension condition—θ = 60°, diagram (e)—are almost the same. It is understood that though the loading systems for horizontal principal stresses (σ_x and σ_z) are different (one is applied by chamber pressure and the other is applied by rigid plates), the apparatus has an adequate accuracy. The change in sign for the strain component ε_2 from negative to positive as

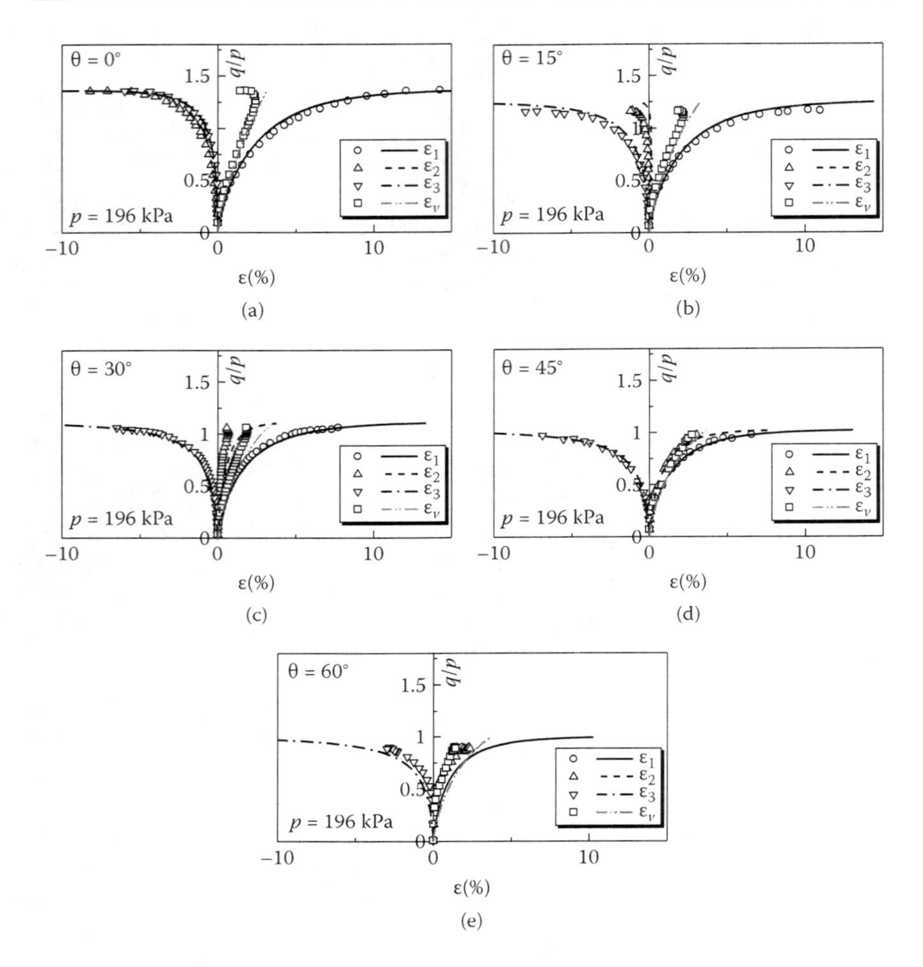

Figure 5.28 Observed and calculated results of true triaxial tests on normally consolidated Fujinomori clay.

seen in diagrams (b) and (c) suggests that the plane strain condition ($\varepsilon_2 = 0$) lies between $15° < \theta < 30°$ for clays, which is in agreement with the results reported by many researchers.

Figure 5.30 shows the stress–dilatancy relations, in which the observed data in Figure 5.28 are arranged in terms of the relation between the stress ratio t_S/t_N and the strain increment ratio $d\varepsilon_N^*/d\varepsilon_S^*$ based on the t_{ij} concept. Here, the broken line indicates the relation in the triaxial compression and extension tests using a conventional triaxial test apparatus (see Figure 5.13). It can be seen that the stress–dilatancy relation based on the t_{ij}

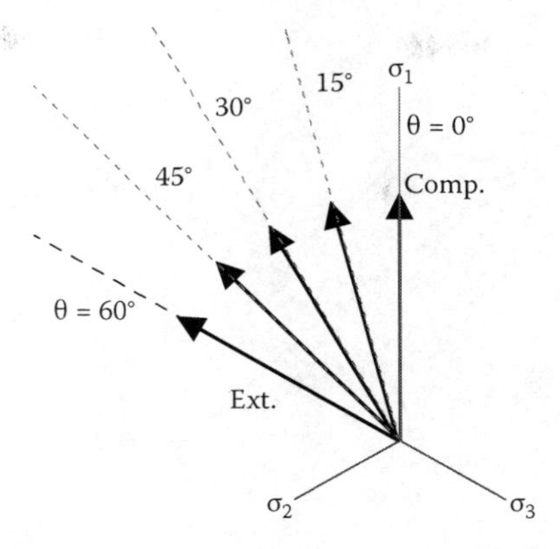

Figue 5.29 Stress paths in the octahedral plane during true triaxial tests.

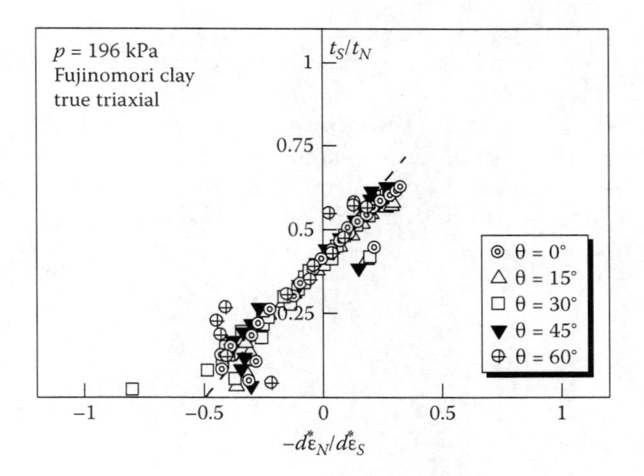

Figure 5.30 t_S/t_N versus $d\varepsilon_N^*/d\varepsilon_S^*$ relation of drained true triaxial tests on normally consolidated Fujinomori clay.

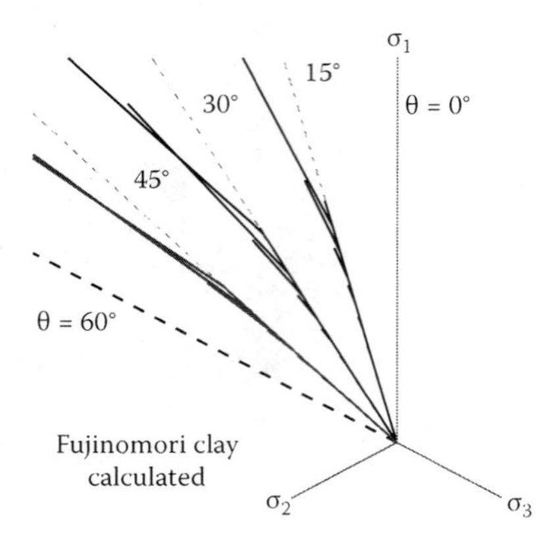

Figure 5.31 Calculated directions of $d\varepsilon_d$ on the octahedral plane under three different principal stresses ($\sigma_1 > \sigma_2 > \sigma_3$) on normally consolidated Fujinomori clay.

concept holds uniquely not only under triaxial compression and extension conditions, but also under three different principal stresses.

Figure 5.31 shows the calculated directions of strain increments on the octahedral plane for true triaxial tests on clay corresponding to the observed results in Figure 4.23 in Chapter 4. The calculated results describe well the observed directions of strain increments on the octahedral plane, including the leftward deviation from the radial direction with the increase of stress ratio.

Chapter 6

Three-dimensional modeling of various soil features based on the t_{ij} concept

6.1 INTRODUCTION

In this chapter, the advanced one-dimensional (1D) models, in which various soil features such as influence of density, bonding effect, time-dependent behavior, and others are uniquely considered. The descriptions in Chapter 3 are extended to three-dimensional (3D) models using the t_{ij} concept explained in the previous chapter (Nakai et al. 2011b). Also, a method to take into consideration the influence of stress path dependency on the direction of plastic flow, which is one of the features in multidimensional conditions, is presented and the explicit expression of the constitutive models based on the t_{ij} concept is shown, in order to apply these models to 2D and 3D geotechnical boundary value problems. The validities of the 3D models are confirmed by various kinds of element tests (triaxial tests, true triaxial tests, plane strain tests, and hollow cylinder tests) on clay and sand. Throughout these verifications, unified material parameters are used for clay and sand.

6.2 THREE-DIMENSIONAL MODELING OF OVERCONSOLIDATED SOILS—ADVANCED ELASTOPLASTIC MODELING AT STAGE I

6.2.1 Formulation of model

The elastoplastic model based on the t_{ij} concept for normally consolidated soils is extended here to overconsolidated soils, introducing the subloading surface concept proposed by Hashiguchi (1980) and revising it. It is assumed that the stress condition changes from point I to point P as in

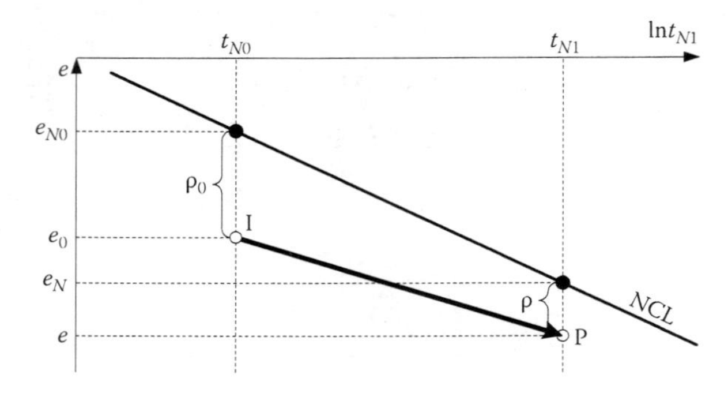

Figure 6.1 Change of void ratio in an over consolidated soil.

Figure 5.11. Since t_{N1} moves from t_{N0} to t_{N1}, the plastic change of void ratio $(-\Delta e)_{NC}{}^p$ for a normally consolidated soil is given by

$$\begin{aligned}(-\Delta e)^p_{NC} &= (e_{N0} - e_N) - (-\Delta e)^e \\ &= (-\Delta e)_{NC} - (-\Delta e)^e = (\lambda - \kappa)\ln\frac{t_{N1}}{t_{N0}}\end{aligned} \tag{6.1}$$

As shown in Figure 6.1, the initial and current void ratios for overconsolidated soils are expressed as e_0 and e, whereas the state variable ρ, which represents the influence of density, is defined as $\rho = e_N - e$ in the same way as in the 1D advanced model (stage I) described in Chapter 3. Then, the corresponding plastic change in void ratio $(-\Delta e)^p$ for overconsolidated soils is given by

$$\begin{aligned}(-\Delta e)^p &= (-\Delta e)^p_{NC} - (\rho_0 - \rho) \\ &= (\lambda - \kappa)\ln\frac{t_{N1}}{t_{N0}} - (\rho_0 - \rho)\end{aligned} \tag{6.2}$$

From this equation, the yield function for overconsolidated soils can be written using F and H defined by eqs. (5.34) and (5.35) in the same form as that in the 1D model (see eq. 3.14):

$$F + \rho = H + \rho_0 \quad \text{or} \quad f = F - \{H + (\rho_0 - \rho)\} = 0 \tag{6.3}$$

The consistency condition ($df = 0$) and the flow rule in eq. (5.36) give

$$df = dF - \{dH - d\rho\}$$
$$= dF - \{d(-e)^p - d\rho\}$$
$$= dF - \left\{(1+e_0)\Lambda \frac{\partial F}{\partial t_{ii}} - d\rho\right\} = 0 \tag{6.4}$$

It can be assumed that the positive value of ρ decreases and finally becomes zero with the development of plastic strains in the same way as in the 1D model. The proportionality constant $\Lambda(>0)$, which represents the magnitude of the plastic deformation, has the same dimension as stress because F is a dimensionless function. Then, the derivative $(\partial F/\partial t_{ii})$ has the dimension of the inverse of stress. To satisfy these conditions, the following evolution rule of ρ is given using a monotonically increasing function $G(\rho)$ that satisfies $G(0) = 0$:

$$d\rho = -(1+e_0)\frac{G(\rho)}{t_N}\Lambda \tag{6.5}$$

As mentioned regarding the 1D model, any increasing function $G(\rho)$ that satisfies $G(0) = 0$ can be employed. As described in the 1D model, the variable ρ may become negative, particularly when considering the bonding effect and other effects. Then, the function $G(\rho)$ in the present model is given by the following equation so as to be an increasing function, even if $\rho < 0$:

$$G(\rho) = sign(\rho)a\rho^2 \tag{6.6}$$

Substituting eq. (6.5) into eq. (6.4), the proportionality constant Λ is expressed as

$$\Lambda = \frac{dF}{(1+e_0)\left\{\frac{\partial F}{\partial t_{ii}} + \frac{G(\rho)}{t_N}\right\}} = \frac{dF}{h^p} \quad \left(\text{where} \quad dF = \frac{\partial F}{\partial \sigma_{kl}} d\sigma_{kl}\right) \tag{6.7}$$

In the modeling based on the subloading surface concept (Hashiguchi 1980), it is assumed that the current stress point always passes over the yield surface (subloading surface) whether plastic deformation occurs or not. Also, the loading condition is given by

$$\begin{cases} d\varepsilon_{ij}^{p} \neq 0 & \text{if} \quad \Lambda = \dfrac{dF}{h^{p}} > 0 \\ d\varepsilon_{ij}^{p} = 0 & \text{otherwise} \end{cases} \tag{6.8}$$

Then, the plastic strain increment is calculated as

$$d\varepsilon_{ij}^{p} = \langle\Lambda\rangle \frac{\partial F}{\partial t_{ij}} = \left\langle \frac{dF}{h^{p}} \right\rangle \frac{\partial F}{\partial t_{ij}} \qquad \left(\text{where} \quad dF = \frac{\partial F}{\partial \sigma_{kl}} d\sigma_{kl} \right) \tag{6.9}$$

where the symbol < > denotes the Macaulay bracket—that is, $<A> = A$ if $A > 0$; otherwise, $<A> = 0$. The present model, in which the evolution rule of ρ is given by eqs. (6.5) and (6.6), coincides with the subloading t_{ij} model by Nakai and Hinokio (2004).

Finally, the total strain increment including the loading condition can be given by the following equation:

$$d\varepsilon_{ij} = d\varepsilon_{ij}^{e} + d\varepsilon_{ij}^{p} = C_{ijkl}^{e} d\sigma_{kl} + \langle\Lambda\rangle \frac{\partial F}{\partial t_{ij}} = C_{ijkl}^{e} d\sigma_{kl} + \left\langle \frac{dF}{h^{p}} \right\rangle \frac{\partial F}{\partial t_{ij}} \tag{6.10}$$

Here, $C_{ijkl}^{e} = [D_{ijkl}^{e}]^{-1}$ is the elastic compliance tensor given by eq. (4.29) or eq. (5.38).

Next, an explicit expression of the incremental stress–strain relation for the model is shown. Within the elastoplastic region, the following equations hold:

$$d\sigma_{ij} = D_{ijkl}^{e} d\varepsilon_{kl}^{e} = D_{ijkl}^{e} \left(d\varepsilon_{kl} - d\varepsilon_{kl}^{p} \right) \tag{6.11}$$

Here, D_{ijkl}^{e} is the fourth order elastic stiffness tensor, which is described in eq. (4.29) or (5.38). From eqs. (6.7), (6.9), and (6.11), the following equation is obtained:

$$\Lambda h^{p} = \frac{\partial F}{\partial \sigma_{mn}} d\sigma_{mn} = \frac{\partial F}{\partial \sigma_{mn}} D_{mnop}^{e} \left(d\varepsilon_{op} - d\varepsilon_{op}^{p} \right) = \frac{\partial F}{\partial \sigma_{mn}} D_{mnop}^{e} \left(d\varepsilon_{op} - \Lambda \frac{\partial F}{\partial \sigma_{op}} \right) \tag{6.12}$$

Rearranging this equation, the proportionality constant Λ is finally obtained in terms of the total strain increment $d\varepsilon_{ij}$, but not stress increment $d\sigma_{ij}$:

$$\Lambda = \frac{\frac{\partial F}{\partial \sigma_{st}} D^e_{stkl} d\varepsilon_{kl}}{h^p + \frac{\partial F}{\partial \sigma_{mn}} D^e_{mnop} \frac{\partial F}{\partial t_{op}}} \quad (6.13)$$

Therefore, the elastoplastic stiffness tensor D^{ep}_{ijkl} is expressed as

$$\begin{aligned} d\sigma_{ij} &= D^e_{ijkl}\left(d\varepsilon_{kl} - d\varepsilon^p_{kl}\right) \\ &= \left(D^e_{ijkl} - \frac{D^e_{ijqr} \frac{\partial F}{\partial t_{qr}} \frac{\partial F}{\partial \sigma_{st}} D^e_{stkl}}{h^p + \frac{\partial F}{\partial \sigma_{mn}} D^e_{mnop} \frac{\partial F}{\partial t_{op}}} \right) d\varepsilon_{kl} = D^{ep}_{ijkl} d\varepsilon_{kl} \end{aligned} \quad (6.14)$$

Here, the denominator of Λ in eq. (6.13) is positive, even in the perfectly plastic region ($h^p = 0$) and in the strain-softening region ($h^p < 0$), because the second term $(\partial F/\partial \sigma_{mm}) D^e_{mnop} (\partial F/\partial t_{op})$ in the denominator usually has a large positive value, except for extreme cases. Therefore, the loading condition in eq. (6.8) can be rewritten as follows in terms of strain increment $d\varepsilon_{ij}$:

$$\begin{cases} d\varepsilon^p_{ij} \neq 0 & \text{if } \dfrac{\partial F}{\partial \sigma_{mm}} D^e_{mnkl} d\varepsilon_{kl} > 0 \\ d\varepsilon^p_{ij} = 0 & \text{otherwise} \end{cases} \quad (6.15)$$

6.2.2 Description of dependency of plastic flow on stress path in constitutive modeling

According to the classical theory of plasticity, the direction of plastic flow (direction of plastic strain increments) is independent of the direction of stress increments. This means that the stress–dilatancy relation is not influenced by the stress path. However, it was experimentally pointed out that the stress–dilatancy relation depends on the stress paths except for the stress condition near or at failure (El-Shohby 1969; Tatsuoka 1978). Figures 6.2(a) and (b) show the observed stress–dilatancy relations plotted as $\eta = q/p$ versus $d\varepsilon_v/d\varepsilon_d$ for normally consolidated Fujinomori clay and for medium dense Toyoura sand, respectively. More precisely, they pertain to triaxial compression and extension tests under p = constant, σ_3 = constant, σ_1 = constant, and $R(= \sigma_1/\sigma_3)$ = constant.

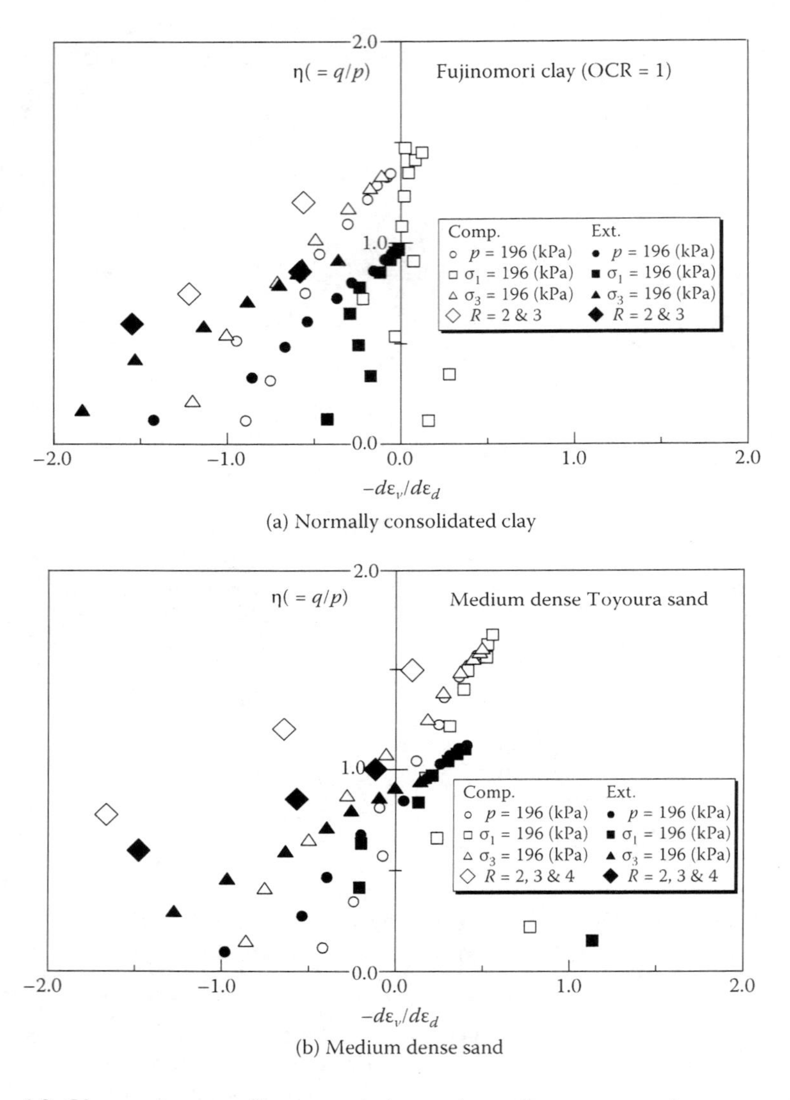

Figure 6.2 Observed stress–dilatancy relation under various stress paths, arranged in terms of the relation between q/p and $d\varepsilon_v/d\varepsilon_d$.

The stress–dilatancy relations for the clay and sand based on ordinary stress and strain increment invariants are influenced not only by the intermediate principal stress but also by the stress paths. Figure 6.3 shows the same observed results of the clay and sand plotted using the stress ratio $X = t_S/t_N$ against the strain increment ratio $d\varepsilon_N^*/d\varepsilon_S^*$ based on the t_{ij} concept.

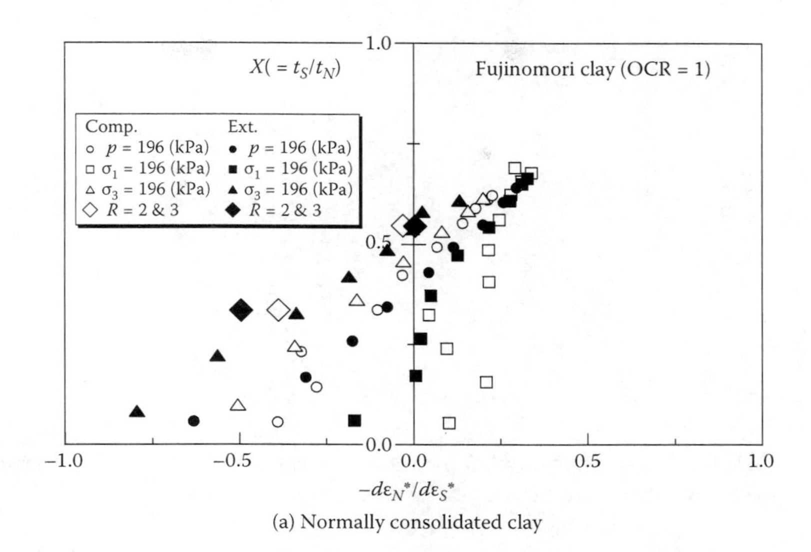

(a) Normally consolidated clay

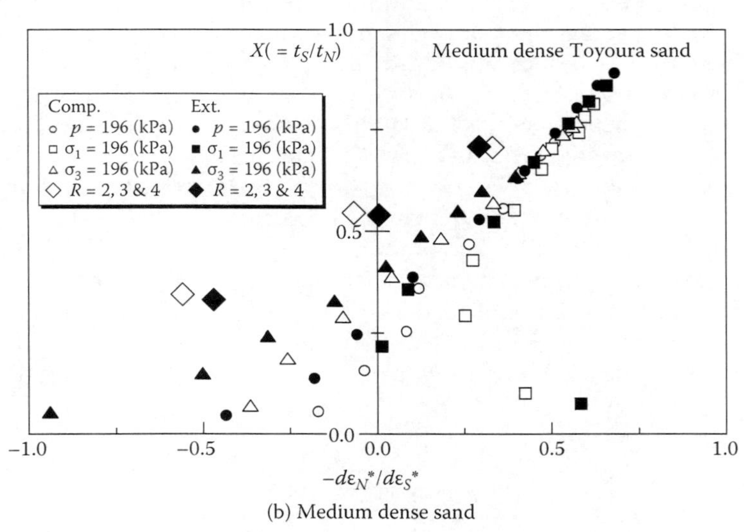

(b) Medium dense sand

Figure 6.3 Observed stress–dilatancy relation under various stress paths, arranged in terms of the relation between t_S/t_N and $d\varepsilon_N^*/d\varepsilon_S^*$.

The newly plotted data show that, even using the t_{ij} concept, the strain increment ratio depends on the stress path, except near and at peak strength, even though there is not much difference between the results of the triaxial compression and extension tests. Therefore, it can be deduced that the direction of plastic flow is generally influenced by the direction of stress increments except at peak strength.

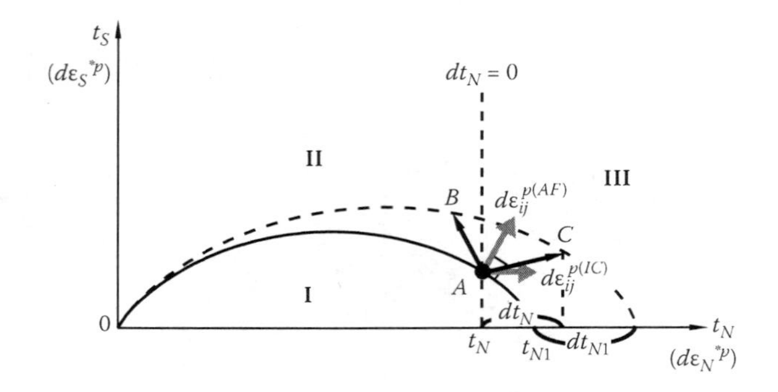

Figure 6.4 Explanation of strain increment split into $d\varepsilon_{ij}^{p(AF)}$ and $d\varepsilon_{ij}^{p(IC)}$.

To describe this stress path dependency of the stress–dilatancy relation (the direction of plastic flow), the plastic strain increment is divided into two components: the plastic strain increment $d\varepsilon_{ij}^{p(AF)}$ satisfying the associated flow rule in the t_{ij} space and the isotropic plastic strain increment $d\varepsilon_{ij}^{p(IC)}$ under increasing mean stress t_N—in spite of using one yield function in eq. (6.3) and one strain-hardening parameter $(-\Delta e)^p$ (or ε_v^p) (Nakai and Hinokio 2004). Such plastic strain decomposition has been previously employed in other models (Nakai and Matsuoka 1986; Nakai 1989).

Figure 6.4 shows the yield surface $f = 0$ (solid curve) and its subsequent yield surface (broken curve), where point A is the current stress state and points B and C are the stress states on the same subsequent yield surface during strain hardening. When the stress state moves to point B in region II ($dF > 0$ and $dt_N \leq 0$), the plastic strain increment is only $d\varepsilon_{ij}^{p(AF)}$ given by eq. (6.9). However, when it moves to point C in region III ($dF > 0$ and $dt_N > 0$), it is assumed that the plastic strain increment is expressed by the summation of $d\varepsilon_{ij}^{p(AF)}$ and $d\varepsilon_{ij}^{p(IC)}$.

Now, from eqs. (6.7) and (6.9), plastic volumetric strain under isotropic compression is given by

$$d\varepsilon_v^p = d\varepsilon_{ii}^p = \Lambda \frac{\partial F}{\partial t_{ii}} = \frac{dF}{(1+e_0)\left(\frac{\partial F}{\partial t_{kk}} + \frac{G(\rho)}{t_N}\right)} \cdot \frac{\partial F}{\partial t_{ii}} \tag{6.16}$$

Under the isotropic compression condition ($X = 0$), the following equations hold:

$$t_{N1} = t_N = p \tag{6.17}$$

$$\frac{\partial F}{\partial t_{ii}} = \frac{\lambda - \kappa}{t_N} a_{ii} \tag{6.18}$$

Then, eq. (6.16) is expressed as

$$d\varepsilon_v^p = \frac{\lambda-\kappa}{(1+e_0)\left(1+\frac{G(\rho)}{(\lambda-\kappa)a_{ii}}\right)}\cdot\frac{dt_N}{t_N} \tag{6.19}$$

It is assumed that the plastic volumetric isotropic strain increment $d\varepsilon_v{}^{p(IC)}$ in general stress conditions, which occurs when $dt_N > 0$, is a fraction t_N/t_{N1} of the plastic volumetric strain $d\varepsilon_v{}^p$ given by eq. (6.19) in the same way as in the t_{ij} clay model (Nakai and Matsuoka 1986). Thus,

$$\begin{aligned} d\varepsilon_{ij}^{p(IC)} &= \Lambda^{(IC)}\frac{\delta_{ij}}{3} = \frac{\lambda-\kappa}{(1+e_0)\left(1+\frac{G(\rho)}{(\lambda-\kappa)a_{kk}}\right)}\cdot\frac{\langle dt_N\rangle}{t_N}\cdot\frac{t_N}{t_{N1}}\cdot\frac{\delta_{ij}}{3} \\ &= \frac{\frac{\lambda-\kappa}{t_{N1}}\langle dt_N\rangle}{(1+e_0)\left(1+\frac{G(\rho)}{(\lambda-\kappa)a_{kk}}\right)}\cdot\frac{\delta_{ij}}{3} = \frac{\frac{\lambda-\kappa}{t_{N1}}\langle dt_N\rangle}{h^{p(IC)}}\cdot\frac{\delta_{ij}}{3} \end{aligned} \tag{6.20}$$

Here, as mentioned before, the symbol < > denotes the Macaulay bracket—that is, $\langle A\rangle = A$ if $A > 0$; otherwise, $\langle A\rangle = 0$. From eq. (5.34), t_{N1} is expressed as

$$t_{N1} = t_N \cdot \exp(\varsigma(X)) \tag{6.21}$$

As mentioned in Chapter 4, Henkel's (1960) experimental results show that the current void ratio (or plastic volumetric strain) of normally consolidated soils can be determined by the current stress condition alone and is independent of the past stress paths. Then, it can be considered that in the same way as for the t_{ij} clay model (Nakai and Matsuoka, 1986), the present model satisfies the unique relation between the plastic strain increment $\varepsilon_v{}^p$ and stresses at normally consolidated state ($\rho = 0$), even though the total strain increment consists of the two components. Under normally consolidated state ($\rho = 0$), consistency condition ($df = 0$) gives

$$\begin{aligned} dF = dH = d(-e)^p &= (1+e_0)d\varepsilon_v^p = (1+e_0)\left(d\varepsilon_v^{p(AF)} + d\varepsilon_v^{p(IC)}\right) \\ &= (1+e_0)\left(\Lambda_{(\rho=0)}^{(AF)}\frac{\partial F}{\partial t_{ii}} + \frac{\lambda-\kappa}{1+e_0}\frac{1}{t_{N1}}\langle dt_N\rangle\right) \end{aligned} \tag{6.22}$$

From this equation, the proportionality constant $\Lambda_{(\rho=0)}^{(AF)}$ of $d\varepsilon_{ij}{}^{p(AF)}$ at normally consolidated states ($\rho = 0$) is given as follows:

$$\Lambda_{(\rho=0)}^{(AF)} = \frac{dF - \frac{\lambda-\kappa}{t_{N1}}\langle dt_N\rangle}{(1+e_0)\frac{\partial F}{\partial t_{kk}}} \tag{6.23}$$

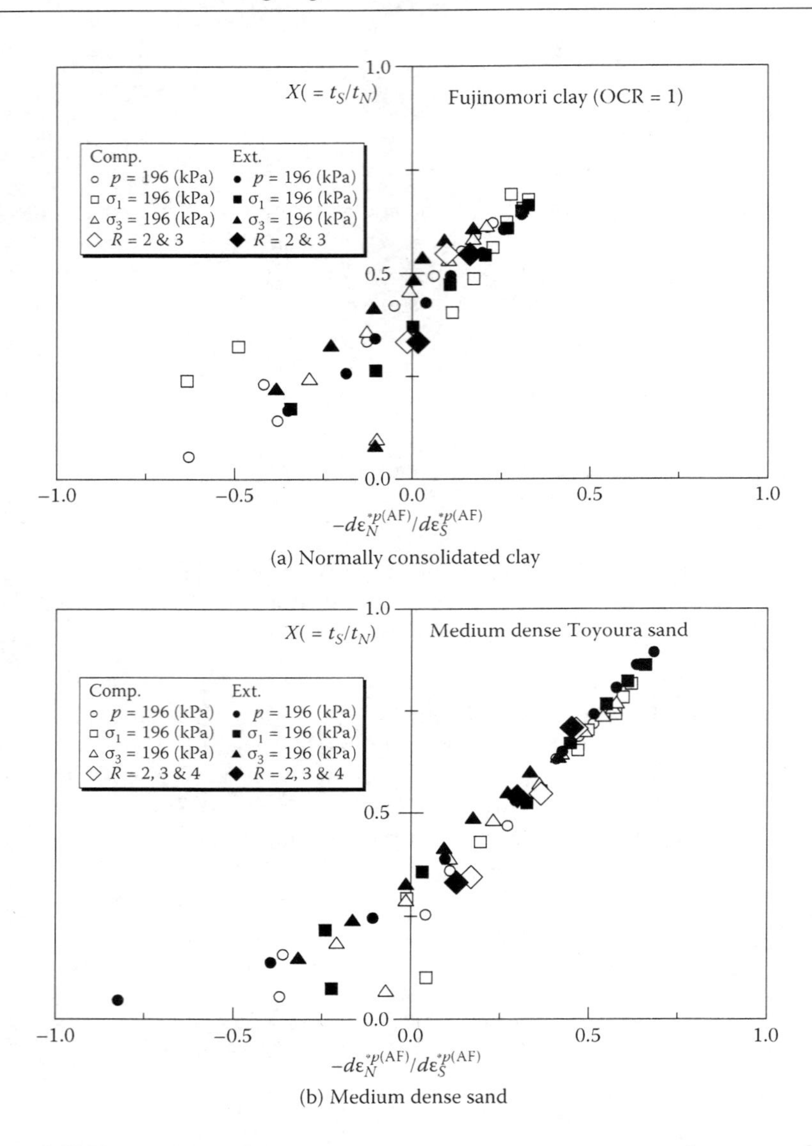

(a) Normally consolidated clay

(b) Medium dense sand

Figure 6.5 Observed stress–dilatancy relation under various stress paths, arranged in terms of the relation between t_S/t_N and $d\varepsilon_N^{*p(AF)}/d\varepsilon_S^{*p(AF)}$.

Then, the strain increment $d\varepsilon_{ij}^{p(AF)}$ at overconsolidated states as well as normally consolidated states is defined in the following form, considering that $\Lambda^{(AF)}$ becomes $\Lambda_{(NC)}^{(AF)}$ at normally consolidated states ($\rho = 0$) and referring to eqs. (6.7), (6.9), and (6.23):

$$d\varepsilon_{ij}^{p(AF)} = \Lambda^{(AF)} \frac{\partial F}{\partial t_{ij}} = \frac{dF - \frac{\lambda - \kappa}{t_{N1}} \langle dt_N \rangle}{(1+e_0)\left(\frac{\partial F}{\partial t_{kk}} + \frac{G(\rho)}{t_N}\right)} \cdot \frac{\partial F}{\partial t_{ij}}$$

$$= \frac{dF - \frac{\lambda-\kappa}{t_{N1}} \langle dt_N \rangle}{h_p} \cdot \frac{\partial F}{\partial t_{ij}} \qquad \left(\text{where} \quad dF = \frac{\partial F}{\partial \sigma_{mn}} d\sigma_{mn}\right) \tag{6.24}$$

Hence, the strain increments, in which dependency of plastic flow on the stress path is considered, are summarized as follows:

1. Elastic region $\left(\frac{\partial F}{\partial \sigma_{mn}} D_{mnkl}^e d\varepsilon_{kl} < 0\right)$:

$$d\varepsilon_{ij} = d\varepsilon_{ij}^e$$

$$= \frac{1+\upsilon_e}{E_e} d\sigma_{ij} - \frac{\upsilon_e}{E_e} d\sigma_{mm} \delta_{ij} \tag{6.25}$$

$$= [D_{ijkl}^e]^{-1} d\sigma_{kl}$$

2. Elastoplastic region with strain hardening $\left(\frac{\partial F}{\partial \sigma_{mn}} D_{mnkl}^e d\varepsilon_{kl} > 0 \quad \& \quad h^p > 0\right)$:

$$d\varepsilon_{ij} = d\varepsilon_{ij}^e + d\varepsilon_{ij}^{p(AF)} + d\varepsilon_{ij}^{p(IC)}$$

$$= \frac{1+\nu_e}{E_e} d\sigma_{ij} - \frac{\nu_e}{E_e} d\sigma_{kk} \delta_{ij} + \frac{dF - \frac{\lambda-\kappa}{t_{N1}} \langle dt_N \rangle}{(1+e_0)\left(\frac{\partial F}{\partial t_{mm}} + \frac{G(\rho)}{t_N}\right)} \cdot \frac{\partial F}{\partial t_{ij}} + \frac{\frac{\lambda-\kappa}{t_{N1}} \langle dt_N \rangle}{(1+e_0)\left(1 + \frac{G(\rho)}{(\lambda-\kappa) a_{kk}}\right)} \cdot \frac{\delta_{ij}}{3}$$

$$= \left[D_{ijkl}^e\right]^{-1} d\sigma_{kl} + \frac{dF - \frac{\lambda-\kappa}{t_{N1}} \langle dt_N \rangle}{h^p} \cdot \frac{\partial F}{\partial t_{ij}} + \frac{\frac{\lambda-\kappa}{t_{N1}} \langle dt_N \rangle}{h^{p(IC)}} \cdot \frac{\delta_{ij}}{3} \tag{6.26}$$

In the case that the stress condition moves from point A to region II ($dt_N < 0$), this equation results in eq. (6.10) (described in Section 6.2.1) because $\langle dt_N \rangle = 0$. Furthermore, when the stress increment is applied to various directions, as shown in Figure 6.4, the development of strain increment continuously changes even at the boundary ($dt_N = 0$) between region II and region III.

3. Elastoplastic region with strain softening

$$\left(\frac{\partial F}{\partial \sigma_{mn}} D^e_{mnkl} d\varepsilon_{kl} > 0 \quad \& \quad h^p < 0 \right):$$

$$\begin{aligned} d\varepsilon_{ij} &= d\varepsilon^e_{ij} + d\varepsilon^p_{ij} \\ &= \frac{1+\nu_e}{E_e} d\sigma_{ij} - \frac{\nu_e}{E_e} d\sigma_{kk}\delta_{ij} + \frac{dF}{(1+e_0)\left(\frac{\partial F}{\partial t_{mm}} + \frac{G(\rho)}{t_N}\right)} \cdot \frac{\partial F}{\partial t_{ij}} \\ &= \left[D^e_{ijkl}\right]^{-1} d\sigma_{kl} + \frac{dF}{h^p} \cdot \frac{\partial F}{\partial t_{ij}} \end{aligned} \tag{6.27}$$

This is because, as shown in Figure 6.3, the direction of the strain increment at and after peak strength is independent of the direction of stress increment.

Figure 6.5 shows the arrangement of the plastic strain increments satisfying the associated flow rule $d\varepsilon_{ij}^{p(AF)}$ of the normally consolidated clay and medium dense sand in the same relation as Figure 6.3. Here, the component $d\varepsilon_{ij}^{p(AF)}$ is obtained by removing the plastic component $d\varepsilon_{ij}^{p(IC)}$ evaluated in eq. (6.20) and the elastic component $d\varepsilon_{ij}^{e}$ obtained as a nonlinear elastic body (eqs. 4.29 and 4.30) from the observed total strain increment $d\varepsilon_{ij}$. It can be seen that the present formulation shown in eq. (6.26) is effective in considering stress path influence on plastic flow not only for normally consolidated soils ($\rho = 0$) but also for overconsolidated soils ($\rho > 0$) because there are unique relations between the stress ratio $X = t_S/t_N$ and the strain increment ratio $d\varepsilon_N^{*p(AF)}/d\varepsilon_S^{*p(AF)}$ derived from the component $d\varepsilon_{ij}^{p(AF)}$ under various stress conditions and stress paths.

Figure 6.6 shows the calculated relation between the stress ratio $X = t_S/t_N$ and the plastic strain increment ratio $Y = d\varepsilon_N^{*p}/d\varepsilon_S^{*p}$ for constant mean principal tests (solid curve) and constant stress ratio tests (symbols) on normally consolidated and overconsolidated clays. Here, the solid curve is the same as that in Figure 5.12. In this relation, negative values in the abscissa axis denote volumetric contraction while positive values denote dilatancy under the same deviatoric strain increment. It is seen from this figure that

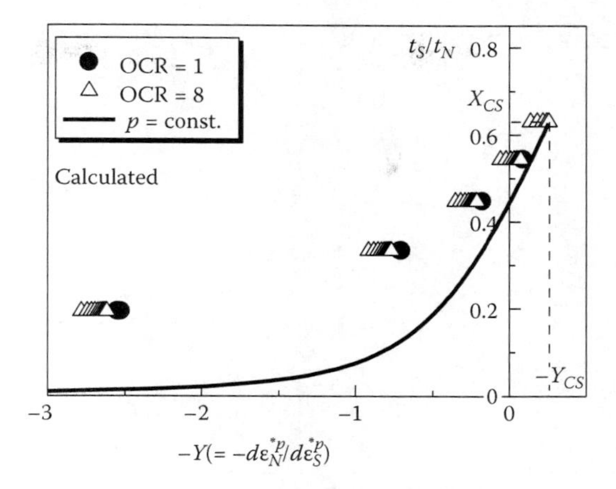

Figure 6.6 Calculated relation between stress ratio and plastic strain increment ratio under constant mean principal stress and under constant stress ratio.

the development of the incremental plastic strains under constant stress ratio is more contractive under low stress ratio and converges to that under constant mean stress with the increase of the stress ratio. This tendency corresponds to the observed results in Figures 6.2 and 6.3. Here, although the observed results in Figures 6.2 and 6.3 are arranged with respect to the total strain increments including the elastic component, the magnitude of the elastic strain increments is relatively small.

For reference, Figure 6.7 shows the comparison between the stress–dilatancy relations based on plastic and total strain components for shear tests. It can be seen that the elastic components (eqs. 4.29 and 4.30) hardly influence the stress–dilatancy equation. Most elastoplastic models such as the Cam clay model do not take into consideration the previously mentioned influence of the stress path on the plastic flow. When using such models in deformation analyses of real ground, various results, such as the coefficient of earth pressure at rest ($K_0 = \sigma_3/\sigma_1$) and the lateral displacement of the embankment foundation, among others, are overestimated because deviatoric strains under stress paths with increasing mean stress are also overestimated (Nakai and Matsuoka 1987).

The explicit expression of the elastoplastic incremental stress–strain relation with the split of plastic strain increment turns out to be somewhat complex. Since the explicit expression for the case of the elastoplastic stiffness tensor under $dt_N \le 0$ (region II in Figure 6.4) and that under strain softening are already given by eq. (6.14), the expression for the case of the elastoplastic stiffness tensor under $dt_N > 0$ (region III in Figure 5.35) will be described.

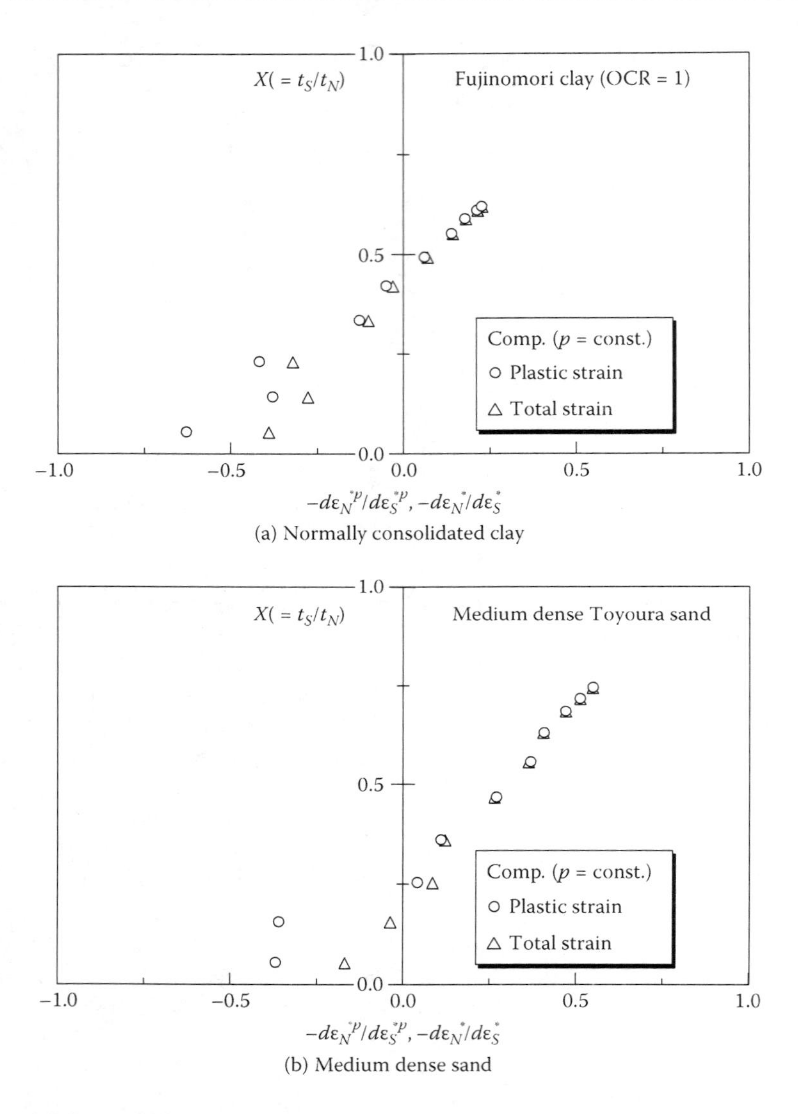

Figure 6.7 Stress–dilatancy relations during shear tests on clay and sand, arranged with respect to the plastic components and the total components.

Using elastic stiffness tensor $D_{ijkl}{}^{e}$, the stress increment $d\sigma_{ij}$ is expressed as

$$d\sigma_{ij} = D^{e}_{ijkl} d\varepsilon^{e}_{kl} = D^{e}_{ijkl}\left(d\varepsilon_{kl} - d\varepsilon^{p}_{kl}\right) = D^{e}_{ijkl}\left(d\varepsilon_{kl} - d\varepsilon^{p(AF)}_{kl} - d\varepsilon^{p(IC)}_{kl}\right) \tag{6.28}$$

Here, $d\varepsilon_{ij}{}^{(AF)}$ and $d\varepsilon_{ij}{}^{p(IC)}$ are given by eqs. (6.24) and (6.20) using dF and dt_N, respectively. Therefore, the following equations hold between dF and dt_N:

$$\begin{aligned} dF &= \frac{\partial F}{\partial \sigma_{ij}} d\sigma_{ij} \\ &= \frac{\partial F}{\partial \sigma_{ij}} D^{e}_{ijkl} d\varepsilon_{kl} - \frac{\partial F}{\partial \sigma_{ij}} D^{e}_{ijkl} \frac{\partial F}{\partial t_{kl}} \frac{dF - \frac{\lambda-\kappa}{t_{N1}} dt_N}{h^p} - \frac{\partial F}{\partial \sigma_{ij}} D^{e}_{ijkl} \frac{\delta_{kl}}{3} \frac{\frac{\lambda-\kappa}{t_{N1}} dt_N}{h^{p(IC)}} \end{aligned} \tag{6.29}$$

$$\begin{aligned} dt_N &= \frac{\partial t_N}{\partial \sigma_{ij}} d\sigma_{ij} \\ &= \frac{\partial t_N}{\partial \sigma_{ij}} D^{e}_{ijkl} d\varepsilon_{kl} - \frac{\partial t_N}{\partial \sigma_{ij}} D^{e}_{ijkl} \frac{\partial F}{\partial t_{kl}} \frac{dF - \frac{\lambda-\kappa}{t_{N1}} dt_N}{h^p} - \frac{\partial t_N}{\partial \sigma_{ij}} D^{e}_{ijkl} \frac{\delta_{kl}}{3} \frac{\frac{\lambda-\kappa}{t_{N1}} dt_N}{h^{p(IC)}} \end{aligned} \tag{6.30}$$

Rearrangement of eqs. (6.29) and (6.30) leads to

$$\begin{cases} P_1 \cdot dF + P_2 \cdot dt_N = \dfrac{\partial F}{\partial \sigma_{ij}} D^{e}_{ijkl} d\varepsilon_{kl} \\[2ex] Q_1 \cdot dF + Q_2 \cdot dt_N = \dfrac{\partial t_N}{\partial \sigma_{ij}} D^{e}_{ijkl} d\varepsilon_{kl} \end{cases} \tag{6.31}$$

Here,

$$\begin{cases} P_1 = 1 + \dfrac{\partial F}{\partial \sigma_{ij}} D^{e}_{ijkl} \left(\dfrac{\partial F}{\partial t_{kl}} \cdot \dfrac{1}{h^p} \right) = 1 + \dfrac{\partial F}{\partial \sigma_{ij}} D^{e}_{ijkl} L^{(1)}_{kl} \\[2ex] P_2 = \dfrac{\partial F}{\partial \sigma_{ij}} D^{e}_{ijkl} \left(\dfrac{\partial F}{\partial t_{kl}} \cdot \dfrac{-1}{h^p} + \dfrac{\delta_{kl}}{3} \cdot \dfrac{1}{h^{p(IC)}} \right) \dfrac{\lambda - \kappa}{t_{N1}} = \dfrac{\partial F}{\partial \sigma_{ij}} D^{e}_{ijkl} L^{(2)}_{kl} \\[2ex] Q_1 = \dfrac{\partial t_N}{\partial \sigma_{ij}} D^{e}_{ijkl} \left(\dfrac{\partial F}{\partial t_{kl}} \cdot \dfrac{1}{h^p} \right) = \dfrac{\partial t_N}{\partial \sigma_{ij}} D^{e}_{ijkl} L^{(1)}_{kl} \\[2ex] Q_2 = 1 + \dfrac{\partial t_N}{\partial \sigma_{ij}} D^{e}_{ijkl} \left(\dfrac{\partial F}{\partial t_{kl}} \cdot \dfrac{-1}{h^p} + \dfrac{\delta_{kl}}{3} \cdot \dfrac{1}{h^{p(IC)}} \right) \dfrac{\lambda - \kappa}{t_{N1}} = 1 + \dfrac{\partial t_N}{\partial \sigma_{ij}} D^{e}_{ijkl} L^{(2)}_{kl} \\[2ex] \text{where,} \quad L^{(1)}_{kl} = \dfrac{\partial F}{\partial t_{kl}} \cdot \dfrac{1}{h^p}, \quad L^{(2)}_{kl} = \left(\dfrac{\partial F}{\partial t_{kl}} \cdot \dfrac{-1}{h^p} + \dfrac{\delta_{kl}}{3} \cdot \dfrac{1}{h^{p(IC)}} \right) \dfrac{\lambda - \kappa}{t_{N1}} \end{cases} \tag{6.32}$$

From eq. (6.31), both dF and dt_N are expressed using total strain increment $d\varepsilon_{ij}$ as

$$\begin{cases} dF = \dfrac{Q_2 \frac{\partial F}{\partial \sigma_{ij}} - P_2 \frac{\partial t_N}{\partial \sigma_{ij}}}{P_1 Q_2 - P_2 Q_1} D^e_{ijkl} d\varepsilon_{kl} \\ dt_N = \dfrac{P_1 \frac{\partial t_N}{\partial \sigma_{ij}} - Q_1 \frac{\partial F}{\partial \sigma_{ij}}}{P_1 Q_2 - P_2 Q_1} D^e_{ijkl} d\varepsilon_{kl} \end{cases} \tag{6.33}$$

Finally, from eqs. (6.20), (6.24), (6.28), and (6.33), the explicit expression of the elastoplastic stiffness tensor emerges as

$$d\sigma_{ij} = \left[D^e_{ijkl} - D^e_{ijmn} L^{(1)}_{mn} \frac{Q_2 \frac{\partial F}{\partial \sigma_{op}} - P_2 \frac{\partial t_N}{\partial \sigma_{op}}}{P_1 Q_2 - P_2 Q_1} D^e_{opkl} - D^e_{ijmn} L^{(2)}_{mn} \frac{P_1 \frac{\partial t_N}{\partial \sigma_{op}} - Q_1 \frac{\partial F}{\partial \sigma_{op}}}{P_1 Q_2 - P_2 Q_1} D^e_{opkl} \right] d\varepsilon_{kl} \tag{6.34}$$

Efficient substepping integration schemes of the present model with splitting of the plastic strain increment are explained by Pedrose, Farias, and Nakai (2005) and Farias, Pedroso, and Nakai (2009).

6.2.3 Model validation using test data for remolded normally consolidated and overconsolidated clays

The 3D advanced model (stage I), which has been named "subloading t_{ij} model," is verified against various types of shear tests on Fujinomori clay. Only one material parameter a in Table 6.1, which relates to the degradation of ρ, is added to the material parameters for normally consolidated soils in Table 5.2. The value of a can be determined by fitting the calculated stress–strain curve and/or the strength of an overconsolidated clay with the experimentally observed ones after the values of the material parameters for normally consolidated clay (see Table 5.2) have been fixed. Here, the relation between the stress ratio X_f at peak strength and the density variable

Table 6.1 Additional material parameter for Fujinomori clay in advanced model at stage I considering influence of density

$a/(\lambda - \kappa)$	500	Influence of density

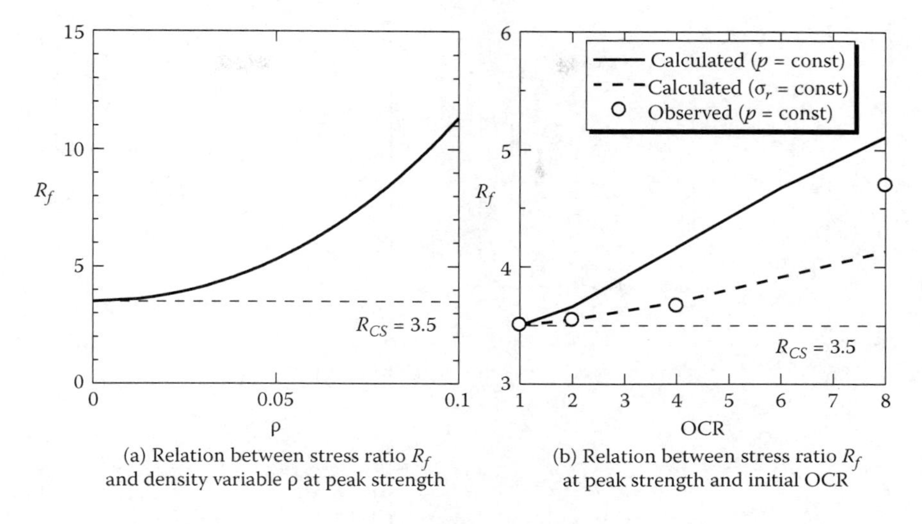

(a) Relation between stress ratio R_f and density variable ρ at peak strength

(b) Relation between stress ratio R_f at peak strength and initial OCR

Figure 6.8 Calculated strength of Fujinomori clay under triaxial compression conditions.

ρ is obtained from the condition of $h^p = 0$. This condition leads to

$$\frac{\partial F}{\partial t_{ii}} + \frac{G(\rho)}{t_N} = \frac{\partial F}{\partial t_{ii}} + \frac{a\rho^2}{t_N} = 0 \tag{6.35}$$

where

$$\frac{\partial F}{\partial t_{ii}} = \frac{\lambda - \kappa}{t_N}\left(a_{ii} + \frac{X^{\beta-2}}{M^{*\beta}}(x_{ii} - X^2 a_{ii})\right) \tag{6.36}$$

From this condition, the relationship between the peak strength $R_f = (\sigma_1/\sigma_3)_{f(comp)}$ and the density variable ρ is obtained as shown in Figure 6.8(a). Figure 6.8(b) gives the variation of peak strength with initial overconsolidation ratio for Fujinomori clay, which was also calculated using the present model. This kind of relation provides a method to determine the value of parameter *a* by fitting the computed values to the observed ones.

6.2.3.1 Conventional triaxial tests under monotonic and cyclic loadings

Figure 6.9 shows the results of triaxial compression and extension tests on Fujinomori clay with different overconsolidation ratios (OCR = 1, 2, 4, and 8). Here, tests with OCR = 8 were carried out under p = 98 kPa,

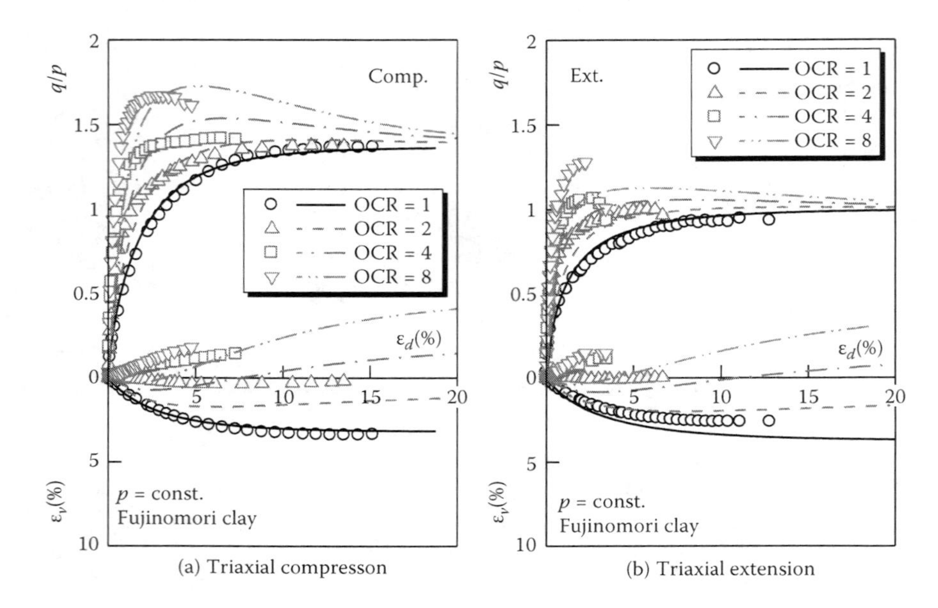

Figure 6.9 Observed and calculated results of triaxial compression and extension tests with different overconsolidation ratio under monotonic loading.

and the other tests were conducted under p = 196 kPa. The observed (symbols) and calculated (curves) results in these figures are plotted in terms of stress ratio q/p, deviatoric strain ε_d, and volumetric strain ε_v. Diagrams (a) and (b) show results under triaxial compression and extension conditions, respectively. The model predicts well not only the influence of overconsolidation ratio (density) on the deformation, dilatancy, and strength of clay but also the influence of intermediate principal stress on the latter.

Figures 6.10 and 6.11 show the observed (symbols) and calculated (curves) variations of (a) deviatoric strain ε_d and (b) volumetric strain ε_v against the stress ratio q/p in drained cyclic triaxial tests on normally consolidated clay. The stress path of each test is drawn in diagram (c). As shown in the stress path of diagram (c), Figure 6.10 refers to the cyclic constant mean principal stress test under constant amplitude of stress ratio, and Figure 6.11 gives the results of the cyclic constant mean principal stress test under increasing stress ratio for each cycle.

It is seen that in spite of using an isotropic hardening law, the model can describe the cyclic behavior of clays. This is due to the adoption of the subloading surface concept and the loading condition in eq. (6.8) or (6.15). The present model can describe the behavior of clays as they become stiffer with

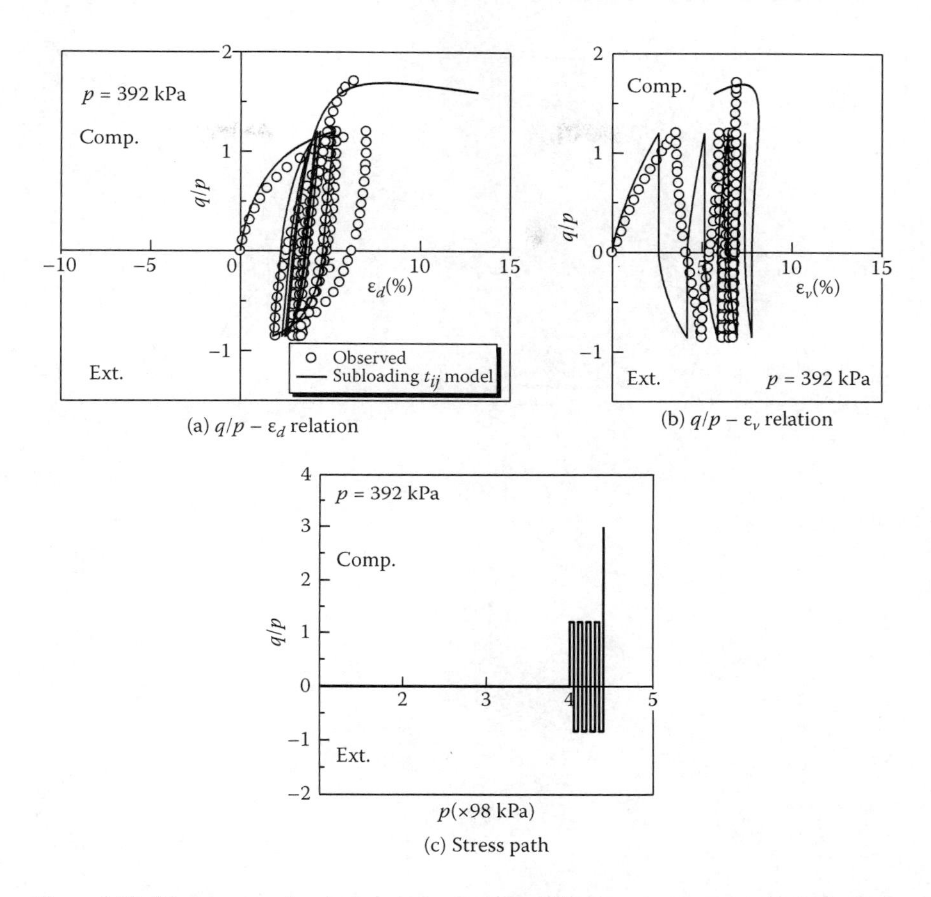

Figure 6.10 Observed and calculated results of cyclic constant mean principal stress tests with constant amplitude of stress ratio.

an increasing number of cycles because the state variable ρ increases under cyclic loadings, even if the clay is initially normally consolidated ($\rho_0 = 0$). Also, Figure 6.12 shows the results of cyclic triaxial tests under constant radial stress. In the calibration under cyclic loading with increasing and decreasing of mean stress (radial stress σ_r = const.), the influence of stress path on the plastic flow is considered using the formulation described in Section 6.2.2.

6.2.3.2 True triaxial tests under cyclic loadings

Real ground conditions normally involve three different principal stresses that may vary cyclically, such as under earthquake loading. To investigate the behavior of clay subjected to general 3D stress conditions, both

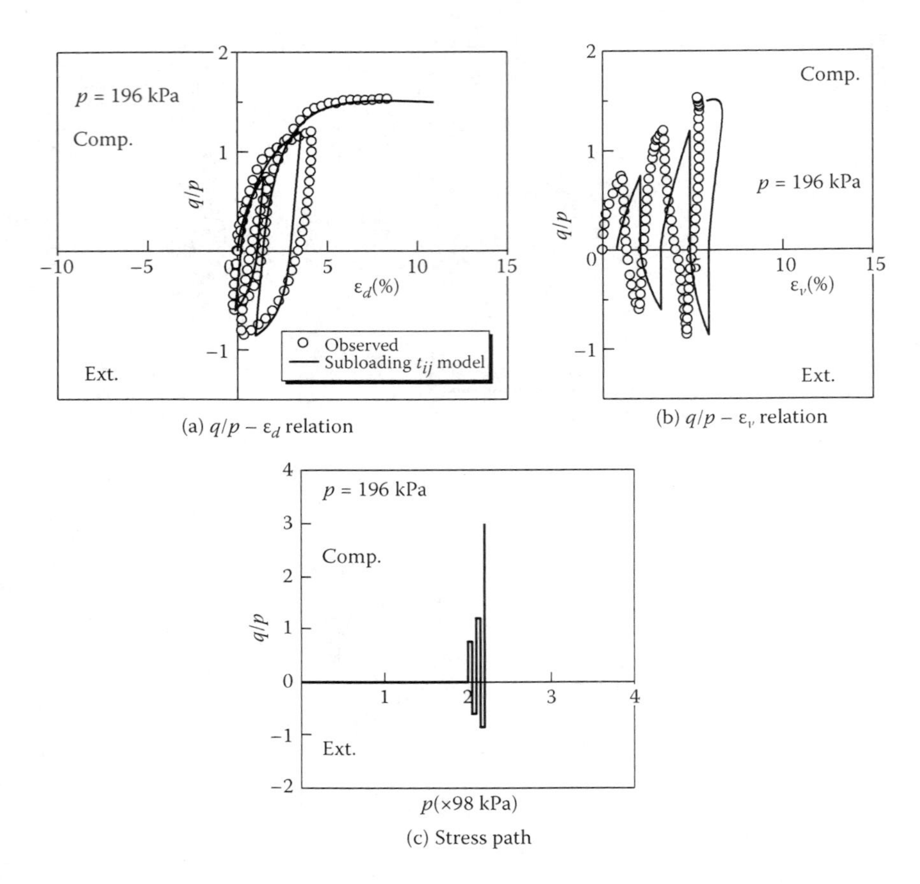

Figure 6.11 Observed and calculated results of cyclic constant mean principal stress tests with increasing amplitude of stress ratio.

monotonic and cyclic true triaxial tests were carried out on normally consolidated clay. A particular interest is to examine the changing of directions of shear stress and/or mean principal stress during cyclic loading. The comparisons between experimental observations (symbols) and numerical simulations (curves) under cyclic loading are herein discussed. The apparatus and the testing methods pertaining to the cyclic true triaxial tests were described in Section 5.3.3.

Figures 6.13(a) and (b) show the stress path of the cyclic true triaxial test and the corresponding material response in terms of variations of principal strains and volumetric strain with the major/minor principal stress ratio. The stress path describes the shape of a bow tie on the octahedral

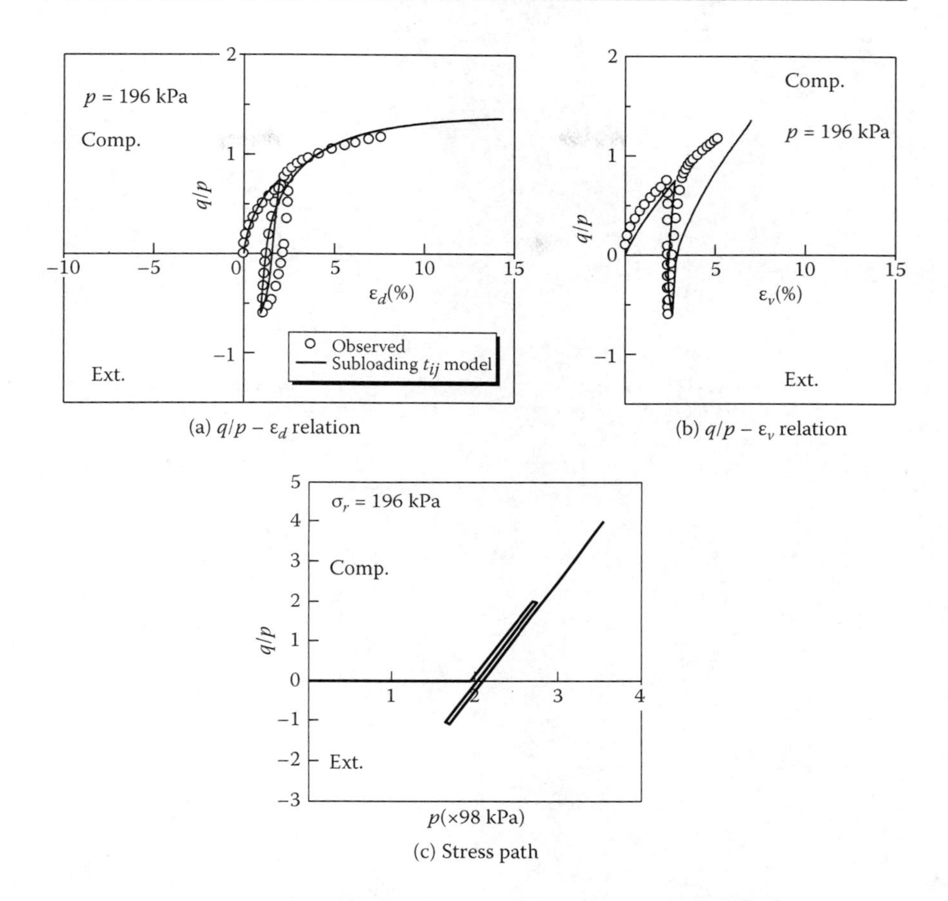

Figure 6.12 Observed and calculated results of cyclic constant radial stress tests.

plane under constant mean principal stress. The same shape of the stress path, but with varying mean principal stress, is also shown in Figure 6.14, together with numerical simulations. It can be seen from these figures that, as a whole, the present model describes quite well the cyclic behavior of clay under three different principal stresses.

6.2.3.3 Plane strain tests on Ko consolidated clay

Plane strain tests were carried out using the chamber type true triaxial apparatus shown in Figure 5.27 by fixing the horizontal loading plates (σ_z direction). The sample is consolidated one-dimensionally before shearing

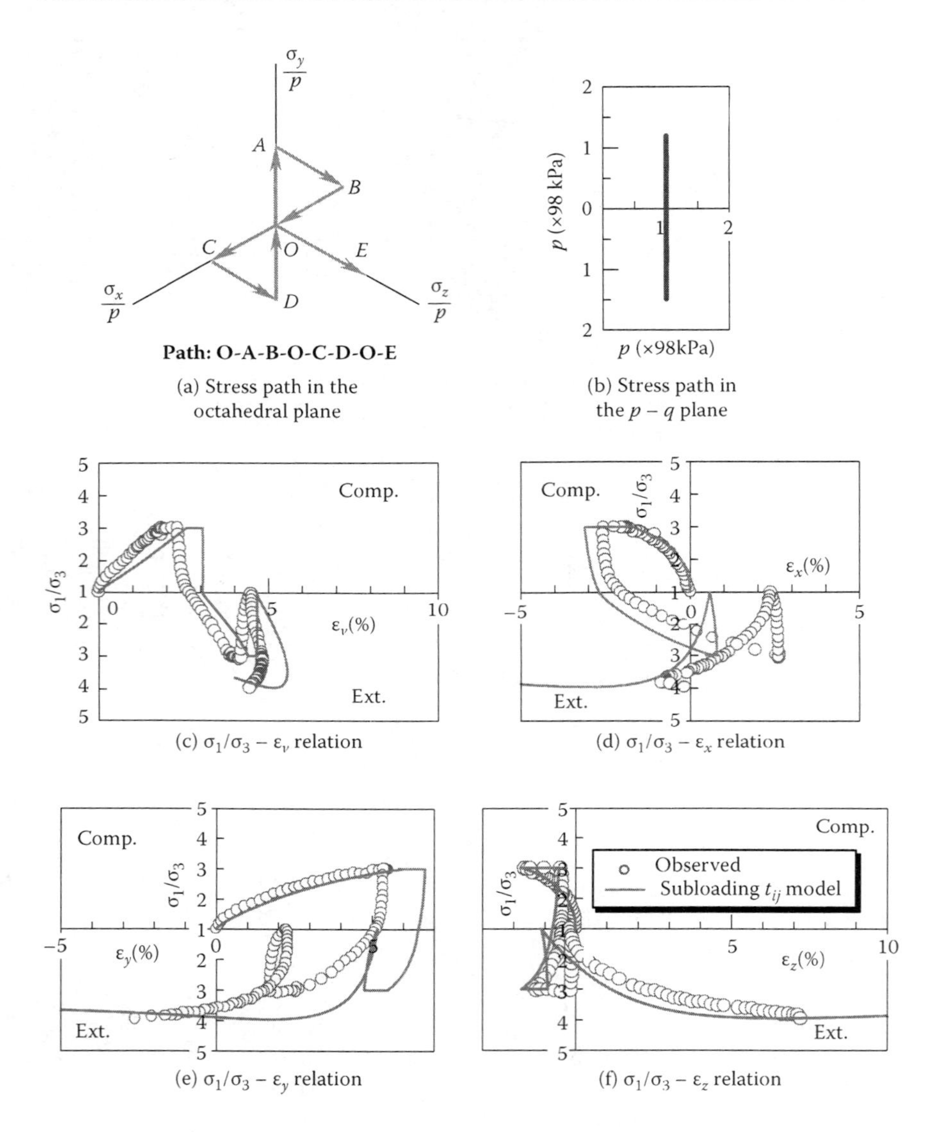

Figure 6.13 Observed and calculated results of cyclic true triaxial test under constant mean principal stress.

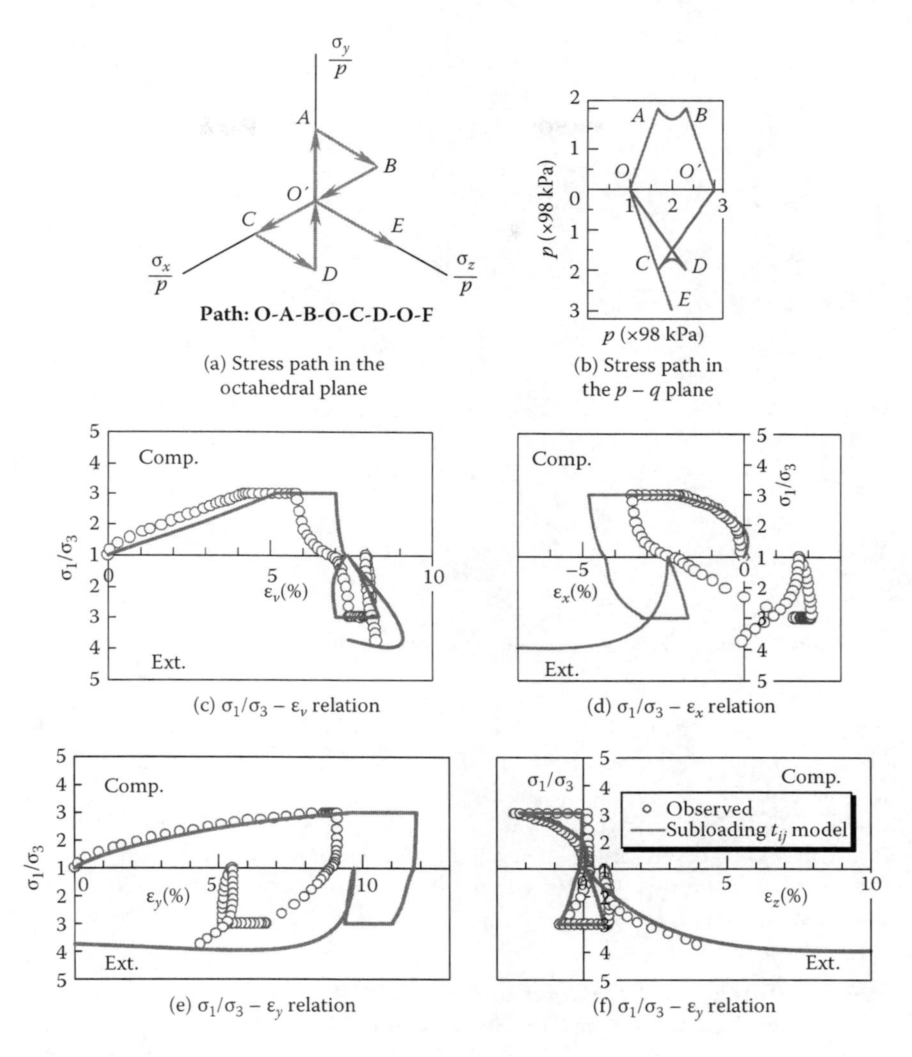

Figure 6.14 Observed and calculated results of cyclic true triaxial tests under varying mean principal stress.

by controlling the vertical stress, σ_y, and one of the horizontal stresses, σ_x. Figure 6.15 shows schematically the specimen under plane strain conditions and four stress paths (tests 1–4) carried out starting from a normally *Ko* consolidated condition ($\sigma_y = 196$ kPa, $\sigma_z \cong \sigma_x = 98$ kPa).

Comparisons of numerical simulations (curves) with experimental results (symbols) for all these stress paths are shown for constant minor

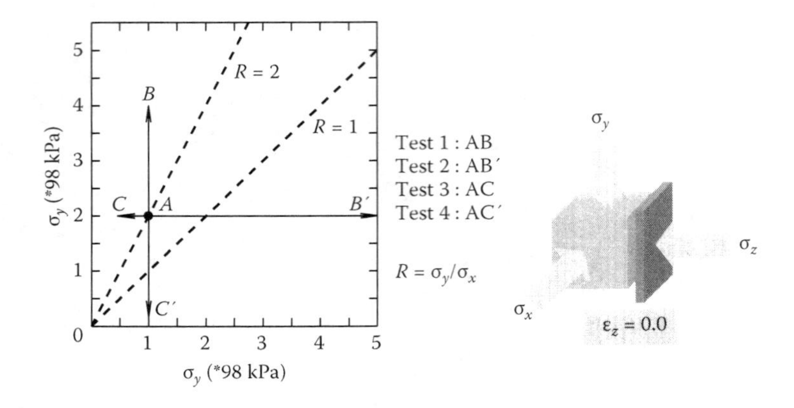

Figure 6.15 Stress paths and specimen in plane strain tests.

(σ_3) and major (σ_1) principal stresses in Figures 6.16 and 6.17, respectively. The corresponding stress paths in the octahedral plane are also given. As a whole, it can be seen that the proposed model describes well the plane strain behavior of clay under both active and passive states. However, there are some discrepancies between observed and calculated responses in test 4, where the direction of the major principal stress changes while the mean stress decreases.

6.2.3.4 Torsional shear tests on isotropically and anisotropically consolidated clays

To investigate the influence of the continuous rotation of principal stress axes on the behavior of clay, including soil dilatancy during shear, cyclic torsional shear tests were carried out. The chamber type hollow cylinder apparatus shown in Figure 6.18 was used. Starting from an isotropic consolidated state, the principal stress axes are made to rotate continuously during cyclic loading; the stress state is chosen to be the same as that in the true triaxial test, where $b_\sigma = (\sigma_2 - \sigma_3)/(\sigma_1 - \sigma_3) = 0.5$. Among all results presented in Figure 6.19, it is found that there is generally a good agreement between numerical simulations and experimental observations with

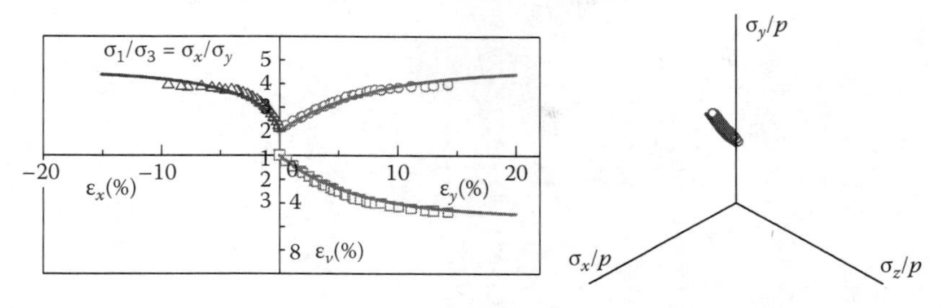

(a) Stress-strain relation and stress path in the octahedral plane during Test 1 (active side: path - AB)

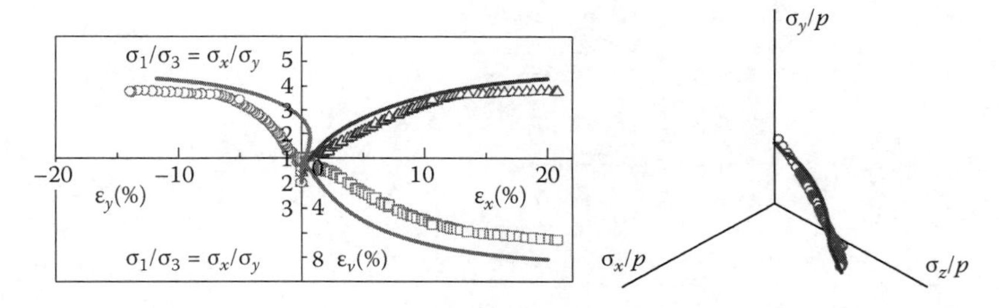

(b) Stress-strain relation and stress path in the octahedral plane during Test 2 (passive side: path - AB′)

Figure 6.16 Observed and calculated results of plane strain tests under constant minor principal stress.

the variation of volumetric strain ε_v; principal stress ratio σ_1/σ_3—diagram (a)—is the most difficult to match.

The next loading program starts from an isotropic stress state from which the axial stress is increased, keeping the radial stresses constant until an anisotropic state of stress where $\sigma_1/\sigma_3 = 2.5$ is reached. This is referred to as process A; it is then followed by process B, where cyclic torsional stresses are applied to the specimen so as to induce continuous rotation of principal axes (see Figure 6.20). It can be seen that there is a good agreement for the $\tau_{a\theta}$–$\gamma_{a\theta}$ relation in diagram (c); however, the model overpredicts the deviatoric strain ε_d in the process B in which the principal axes rotate continuously from the anisotropic stress condition. This is because the isotropic hardening model cannot fully take into account the anisotropy. However, as a whole, the proposed model describes the behavior of clay quite well under such complex stress conditions even though isotropic hardening is assumed. It is also worth mentioning that a unified set of material parameters was used in the modeling.

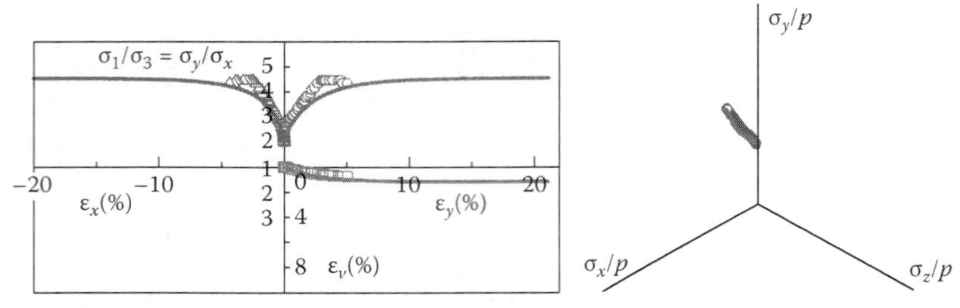

(a) Stress-strain relation and stress path in the octahedral plane during Test 3 (active side: path - AC)

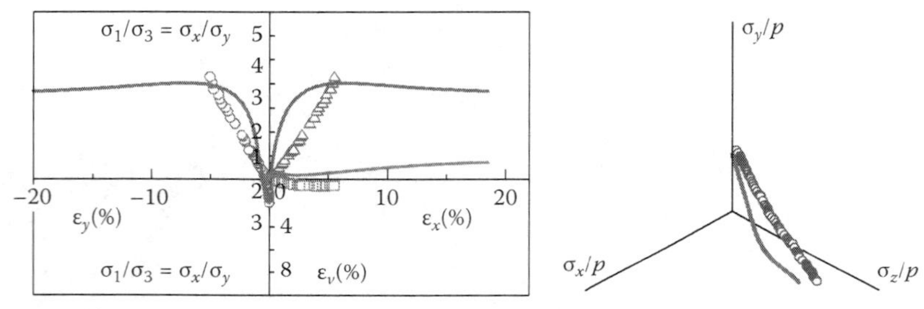

(b) Stress-strain relation and stress path in the octahedral plane during Test 4 (passive side: path - AC′)

Figure 6.17 Observed and calculated results of plane strain tests under constant major principal stress.

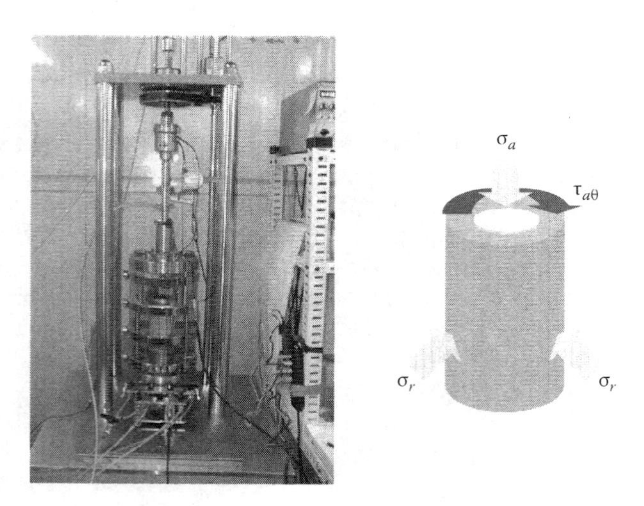

Figure 6.18 Chamber type hollow cylinder apparatus and stress condition of the specimen under torsional shear loading.

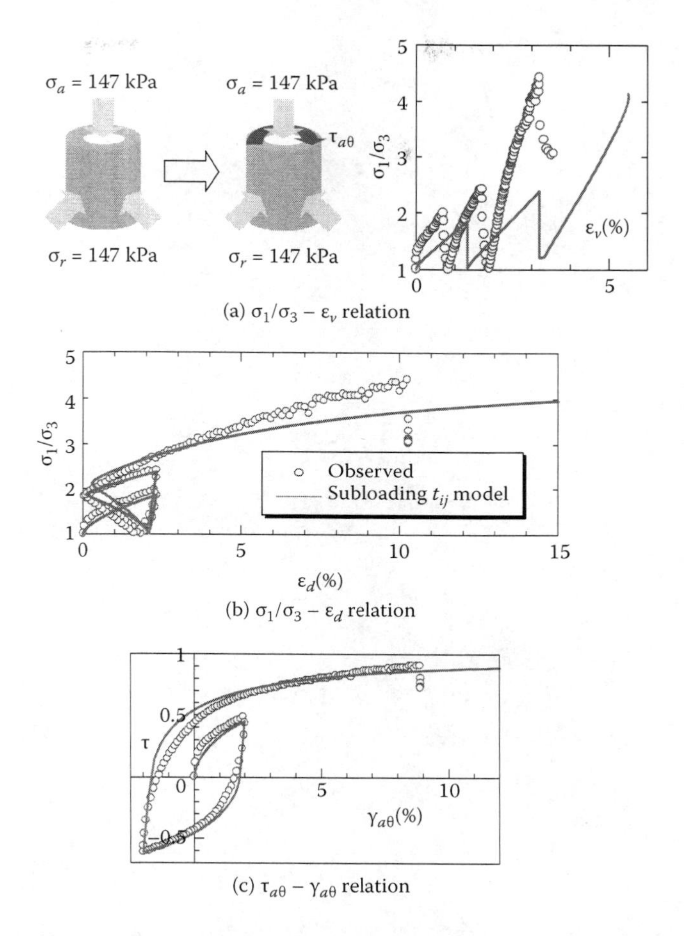

Figure 6.19 Observed and calculated results of cyclic torsional shear tests starting from an isotropic stress condition.

6.2.4 Model validation using test data for sand

6.2.4.1 Conventional triaxial tests under monotonic and cyclic loadings

In this exercise, we use Toyoura sand, which has the following characteristics: mean diameter D_{50} = 0.2 mm, uniformity coefficient U_c = 1.3, specific gravity G_s = 2.65, maximum void ratio $e_{\max}$ = 0.95, and minimum void ratio $e_{\min}$ = 0.58. Two kinds of procedures were used for preparing specimens of sand with different initial void ratios. A dense specimen was prepared by pouring the saturated sand into a mold in several layers and

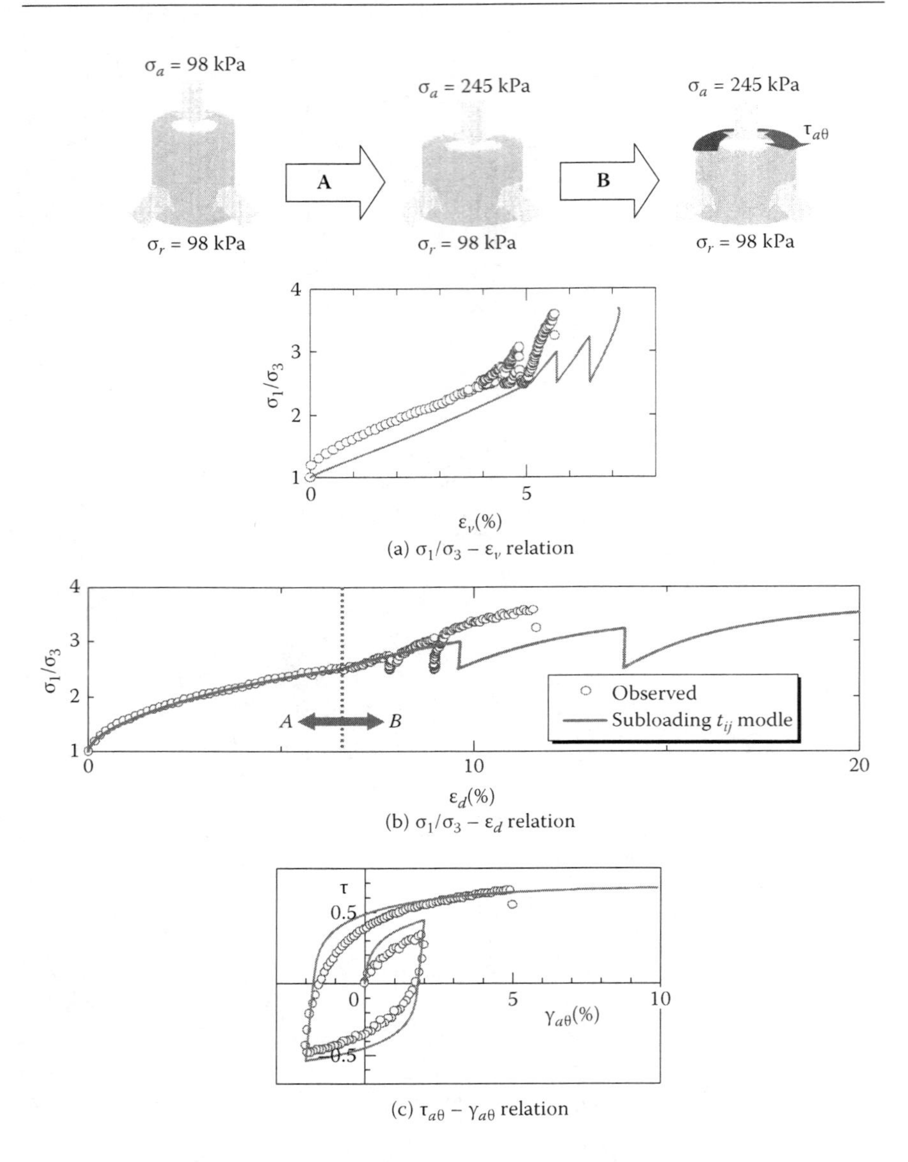

Figure 6.20 Observed and calculated results of cyclic torsional shear tests starting from an anisotropic stress condition.

compacting each layer with a 6 mm diameter rod so that the specimen reached a desired void ratio ($e_{\text{initial}} \approx 0.68$). On the other hand, a loose specimen ($e_{\text{initial}} \approx 0.92$) was achieved by depositing the saturated sand slowly in de-aired water using a funnel with an opening of 3 mm. Both specimens prepared in these ways have quasi-isotropic structures. Every specimen was

Table 6.2 Values of material parameters for Toyoura sand

λ		0.070	
κ		0.0045	
N(e_N at p = 98 kPa)		1.10	Same parameters as Cam clay model
$R_{CS} = (\sigma_1/\sigma_3)_{CS(comp.)}$		3.2	
ν_e		0.2	
β		2.0	Shape of yield surface (same as original Cam clay if β = 1)
$a/(\lambda - \kappa)$	$a^{(AF)}/(\lambda - \kappa)$	30	Influence of density
	$a^{(IC)}/(\lambda - \kappa)$	500	

then consolidated isotropically to the prescribed stress state and thereafter sheared and/or consolidated along the given stress paths.

Table 6.2 shows the values of material parameters for Toyoura sand. Unlike clay, the values of λ and N cannot be determined directly from isotropic consolidation tests on sand, because it is difficult to consolidate the sand to similar states as in remolded normally consolidated clays. The dots in Figure 6.21 are the observed results of isotropic compression tests on dense and loose sands plotted in void ratio e versus confining pressure

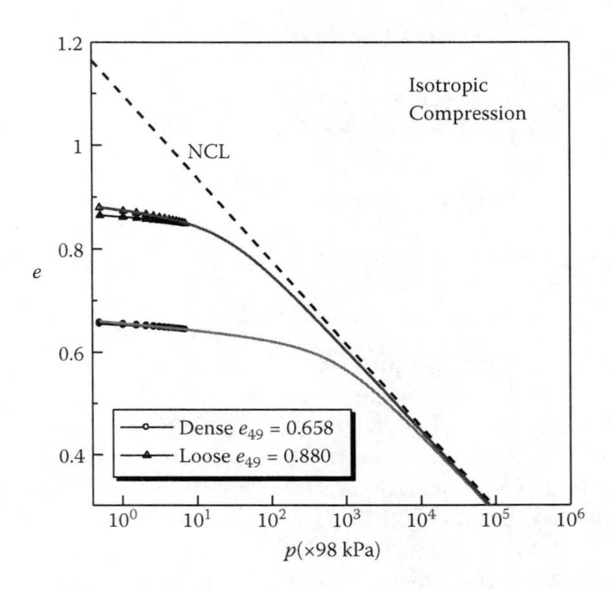

Figure 6.21 Observed and calculated results of isotropic compression tests.

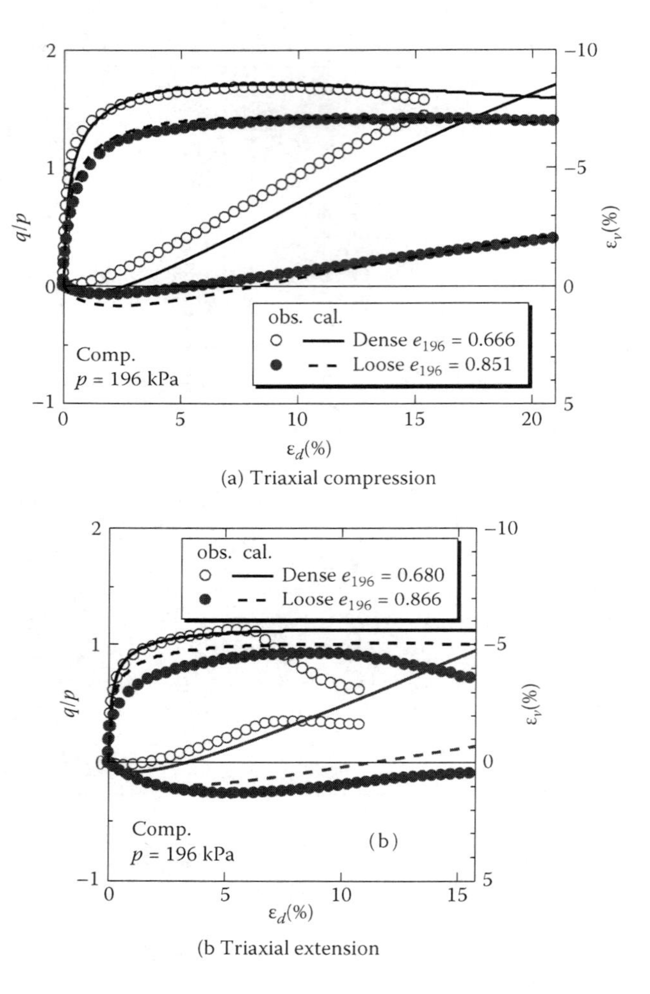

Figure 6.22 Observed and calculated results of constant mean principal tests on dense and loose Toyoura sand.

p in logarithmic scale. Assuming the normal consolidation line (NCL) to be the broken line shown in Figure 6.21 and using the relation between the density variable ρ, the peak strength X_f, and the parameter a (see eqs. 6.35 and 6.36) for conventional triaxial compression tests in the same way as overconsolidated clay, we can determine the stress–strain curves that fit the observed experimental data.

After an iterative process, a set of material parameters for the sand is determined. Herein, to get better fittings of the calculated curves with shear and consolidation test data for Toyoura sand, different values of a ($a^{(AF)}$ and

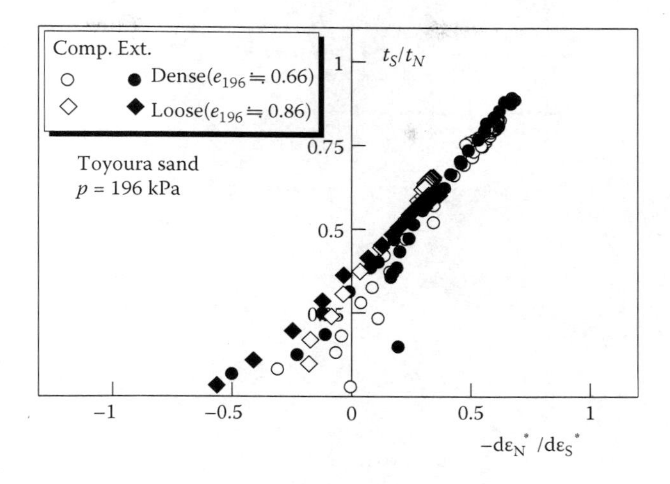

Figure 6.23 t_S/t_N versus $d\varepsilon_N^*/d\varepsilon_S^*$ relation of triaxial compression and extension tests on dense and loose sands under constant mean principal stress.

$a^{(IC)}$), which are the parameters of $G(\rho)$, are used for the two components ($d\varepsilon_{ij}^{p(AF)}$ and $d\varepsilon_{ij}^{p(IC)}$) in eq. (6.26), respectively. The reason to use different values of a for two components and the method to determine the values are described later.

Figure 6.22 shows the observed and calculated results of constant mean principal tests on dense and loose sands with reference to the stress ratio q/p, deviatoric strain ε_d, and volumetric strain ε_v. It is observed that the present model can describe the stress–strain–strength behavior for sands under triaxial compression and extension conditions from dense state to loose states with a unified set of material parameters, in the same way as for clays.

Figure 6.23 gives the observed stress–dilatancy relation for the previously mentioned tests by plotting the stress ratio t_S/t_N as a function of the strain increment ratio $d\varepsilon_N^*/d\varepsilon_S^*$. Although the relation between q/p and $d\varepsilon_v/d\varepsilon_d$ in Figure 4.22 is strongly influenced by the intermediate principal stress, the stress–dilatancy relation based on the t_{ij} concept is independent of the density of sand as well as the intermediate principal stress. As mentioned before, it is difficult to carry out the tests on sand under perfectly normally consolidated states. Then, the parameter a (or $a^{(AF)}$) can be determined by fitting the calculated stress–strain curves to the observed ones in shear tests under different densities and/or different confining stresses. The value of $a^{(AF)}$ in Table 6.2 is determined from the observed results in Figure 6.22 because the isotropic component $d\varepsilon_{ij}^{p(IC)}$ is not included in the total strain increments there.

Figures 6.24 and 6.25 show the observed results (symbols) and the calculated curves of constant mean principal tests, constant major principal

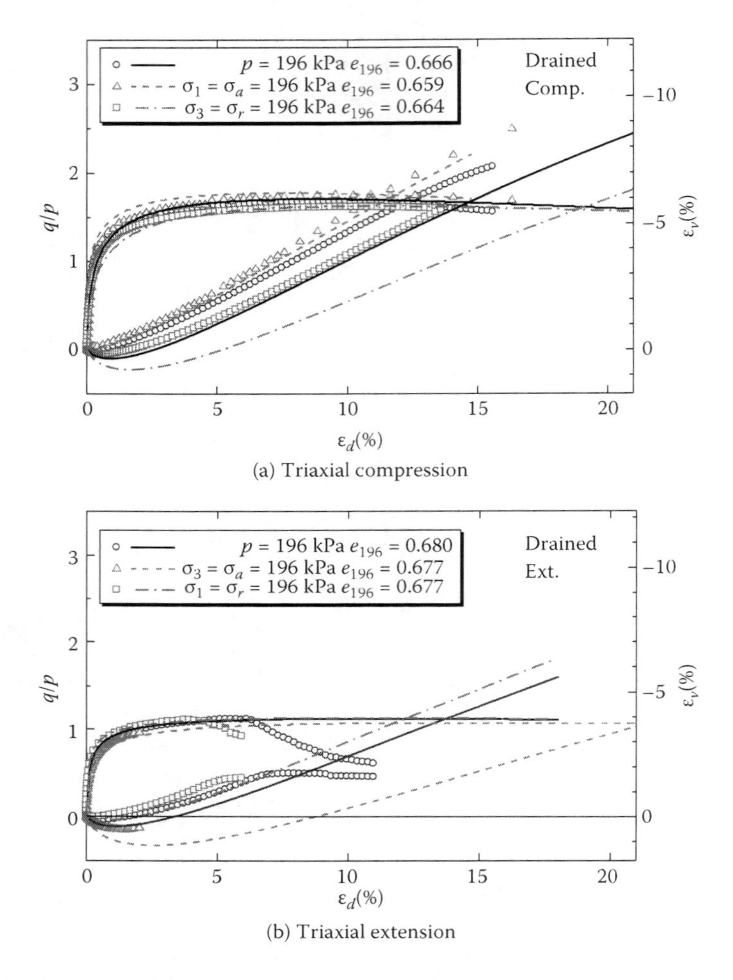

Figure 6.24 Observed and calculated results of triaxial compression and extension tests on dense sand under various stress paths.

stress tests, and constant minor principal stress tests on dense and loose sands, respectively. In each of the figures, diagram (a) refers to results under triaxial compression conditions, whereas diagram (b) relates to those under triaxial extension conditions. It can be seen from these figures that the present model can describe the stress–strain–strength behavior, including the influence of stress paths, for sands under triaxial compression and extension conditions from dense state to loose states.

The observed and calculated results of isotropic and anisotropic consolidation tests on dense and loose Toyoura sands are shown in Figures 6.26–6.29.

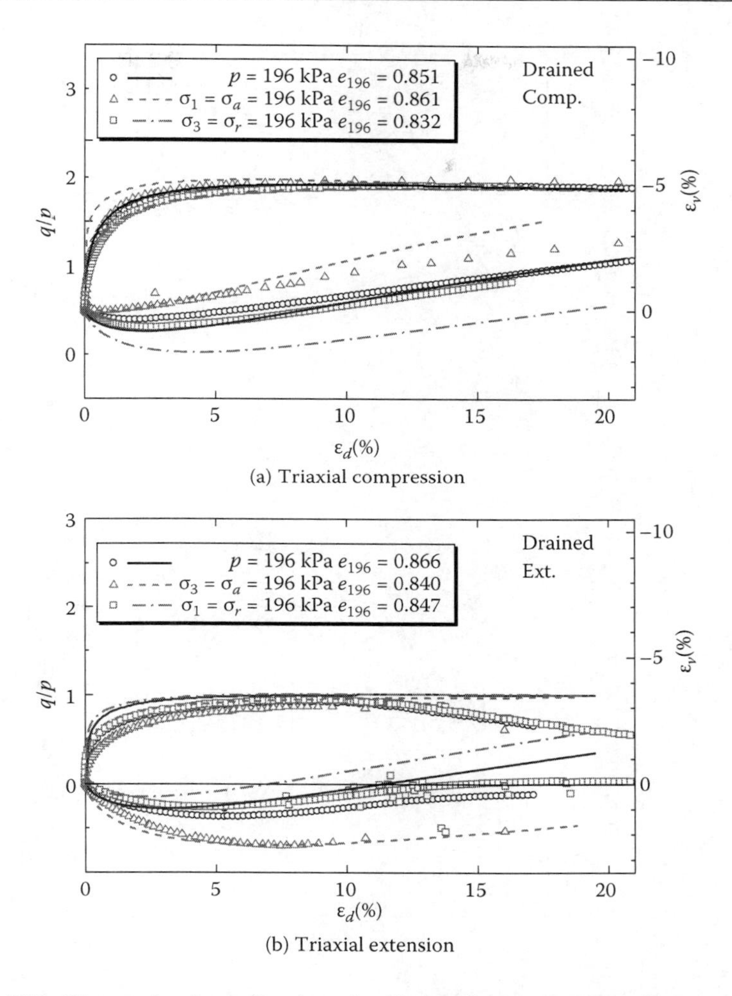

Figure 6.25 **Observed and calculated results of triaxial compression and extension tests on loose sand under various stress paths.**

The solid curves in Figures 6.26 and 6.28 denote the calculated e–log$p$ relations in isotropic compression on dense and loose sands, respectively. The value of $a^{(IC)}$ in Table 6.2 is determined for the curves to agree with the observed results represented by symbols (open dots) because the calculated results using the value of $a^{(AF)}$, which is determined from shear tests, do not describe the observed results properly in the isotropic compression tests on dense and loose Toyoura sands.

Figures 6.26 and 6.27 show (a) the ε_v–ε_1 relation and (b) the ε_v–log$p$ relation in the isotropic and anisotropic consolidation tests on dense

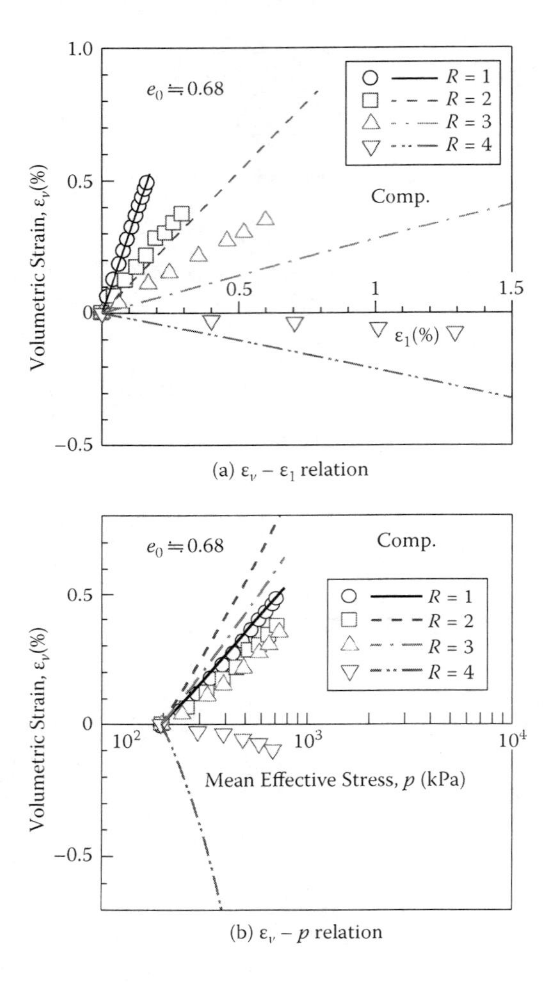

Figure 6.26 Observed and calculated results of isotropic and anisotropic consolidation tests on dense sand under triaxial compression conditions.

sand under triaxial compression and extension conditions, respectively. In these figures, R denotes the major/minor principal stress ratio σ_1/σ_3. The lines show the calculated results, and the symbols show the observed ones. Figures 6.28 and 6.29 show the same relations for loose sand. It can be seen from these figures that the model can describe the deformation behavior under anisotropic consolidation, although positive dilatancy at large stress ratio ($R = 4$) is overestimated, as shown in Figures 6.26 and 6.27. If the stress path influence on the direction of plastic flow is not

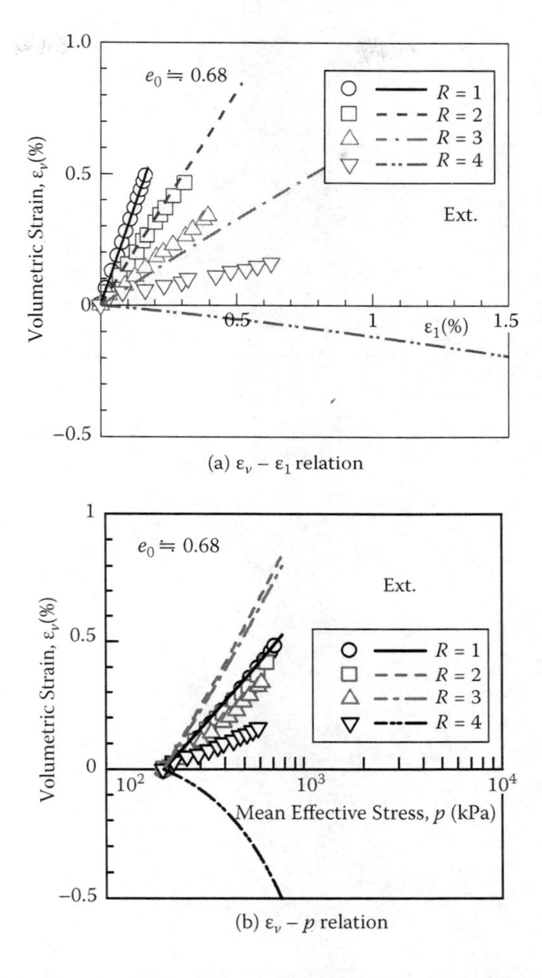

Figure 6.27 Observed and calculated results of isotropic and anisotropic consolidation tests on dense sand under triaxial extension conditions.

taken into consideration, the dilatancy under anisotropic consolidation is inevitably much more overestimated—for example, overestimation of *K*o value in 1D consolidation, as mentioned in Section 6.2.2.

Figure 6.30 shows the observed (symbols) and calculated curves of the cyclic constant mean principal stress test on dense sand under increasing stress ratio for each cycle (the amplitude of the principal stress ratio $R = \sigma_1/\sigma_3$ under cyclic loading increases in order of 2, 3, and 4 and, finally, the sand fails under triaxial compression). Figures 6.31 and 6.32 give the

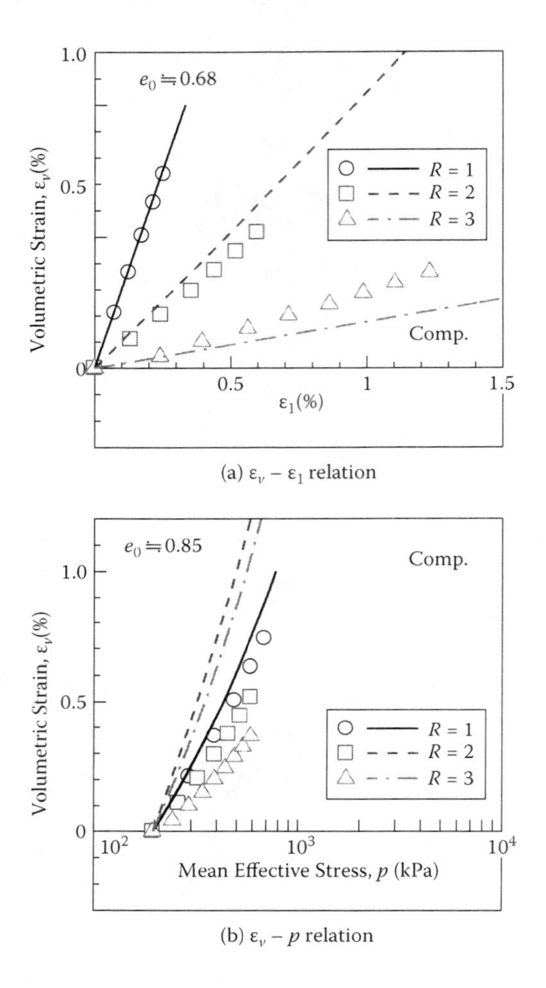

Figure 6.28 Observed and calculated results of isotropic and anisotropic consolidation tests on loose sand under triaxial compression conditions.

results of the constant radial stress test and constant axial stress tests, respectively. The same amplitude of the stress ratio as that in Figure 6.30 is applied to the samples. In each figure, diagram (a) represents the variations of the volumetric strain ε_v against σ_1/σ_3, and diagram (b) represents the principal stress ratio σ_1/σ_3 versus the axial strain ε_a relation. It can be seen that although the calculated volumetric strain is more contractive than the observed one, the present isotropic hardening model describes the typical cyclic behavior of sand without the need of any additional material parameter.

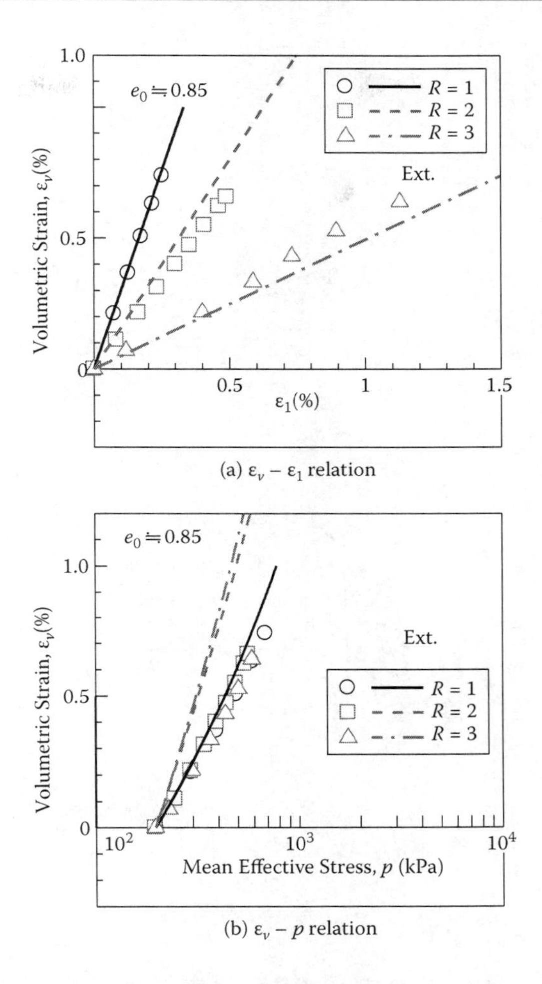

Figure 6.29 Observed and calculated results of isotropic and anisotropic consolidation tests on loose sand under triaxial extension conditions.

6.2.4.2 True triaxial tests under monotonic loading

The observed results of true triaxial tests on dense Toyoura sand were reported and calibrated by the previous model for sand, named the t_{ij} sand model (Nakai 1989). Using these experimental data, the applicability of the present model under three different principal stresses will be confirmed here. Figure 6.33 shows the observed (symbols) and calculated (curves) variations of the three principal strains (ε_1, ε_2, and ε_3) and the volumetric strain ε_v against the stress ratio q/p in true triaxial tests ($\theta = 15°$, $30°$, and $45°$) on dense sand under constant mean principal stress ($p = 196$ kPa).

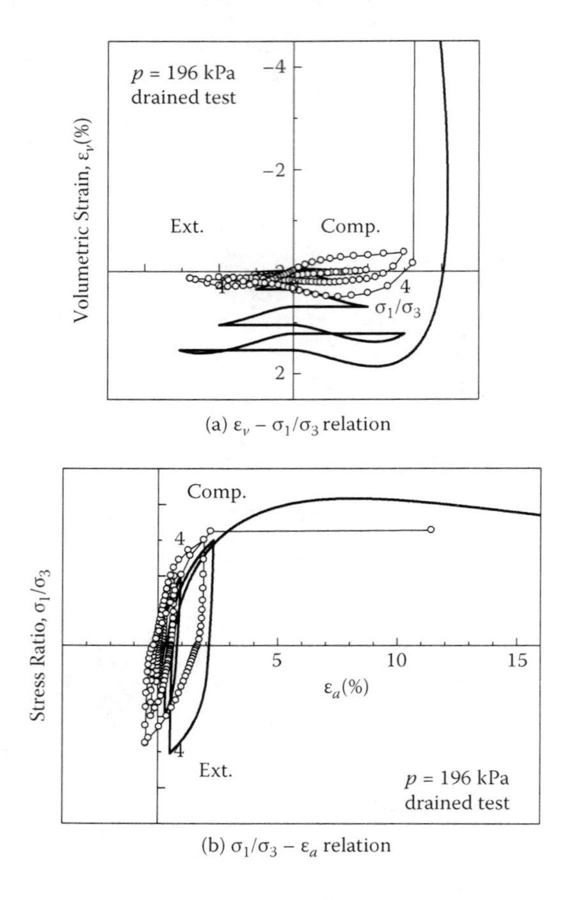

Figure 6.30 Cyclic constant mean principal stress test (p = const.) with increasing amplitude of stress ratio.

In each figure, θ denotes the angle between the σ_1-axis and the corresponding radial stress path on the octahedral plane. It can be seen that the present model predicts well the general 3D stress–strain behavior of sand, in the same way as the simulation for normally consolidated clay. Also, looking at the intermediate principal strain ε_2 in diagrams (a) and (b)—ε_2 is negative in diagram (a) but is positive in diagram (b)—it can be presumed that the stress condition θ in plane strain condition ($\varepsilon_2 = 0$) for sand also lies within $15° < \theta < 30°$.

The dots in Figure 6.34 show the observed stress–dilatancy relation for these tests, plotted in terms of the stress ratio t_S/t_N against the strain

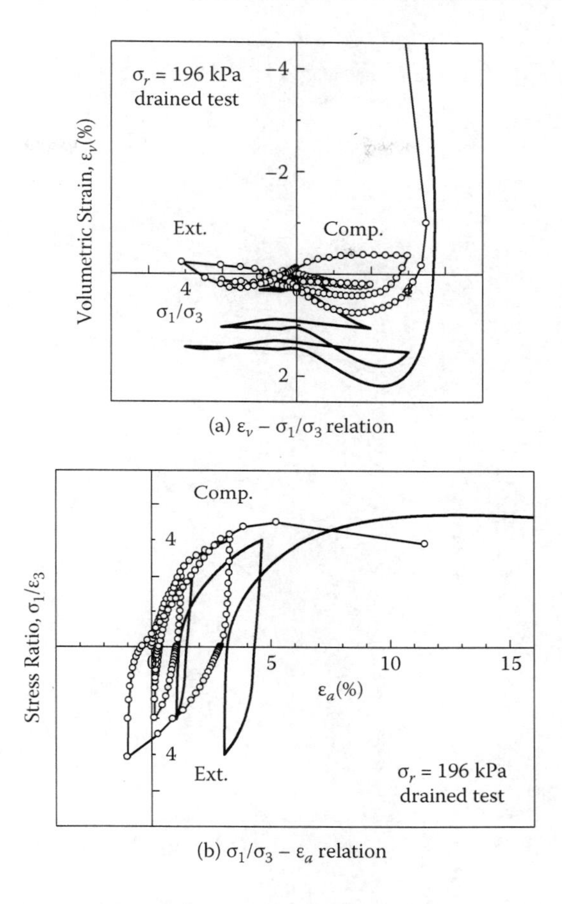

Figure 6.31 Cyclic constant radial principal stress test (σ_r = const.) with increasing amplitude of stress ratio.

increment ratio $d\varepsilon_N^*/d\varepsilon_S^*$. Here, the broken line implies the dilatancy relationship obtained from the conventional triaxial compression and extension tests (see Figure 6.23). Figure 6.35 shows the calculated directions of the strain increments on the octahedral plane for these tests. The corresponding observed results are indicated in Figure 4.24. The length of each bar is proportional to the value of the shear strain increment divided by the shear–normal stress ratio increment on the octahedral plane, similarly to the plot in Figure 4.24. It can be deduced from these figures that the present model predicts the directions of strain increments of sand in general 3D conditions uniquely in the same way as for the case of clay.

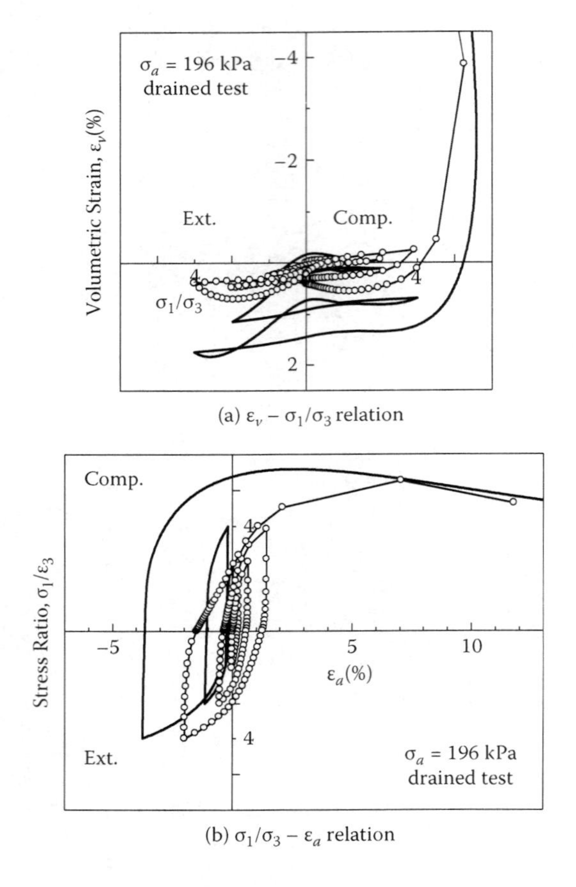

(a) $\varepsilon_v - \sigma_1/\sigma_3$ relation

(b) $\sigma_1/\sigma_3 - \varepsilon_a$ relation

Figure 6.32 Cyclic constant axial principal stress test (σ_a = const.) with increasing amplitude of stress ratio.

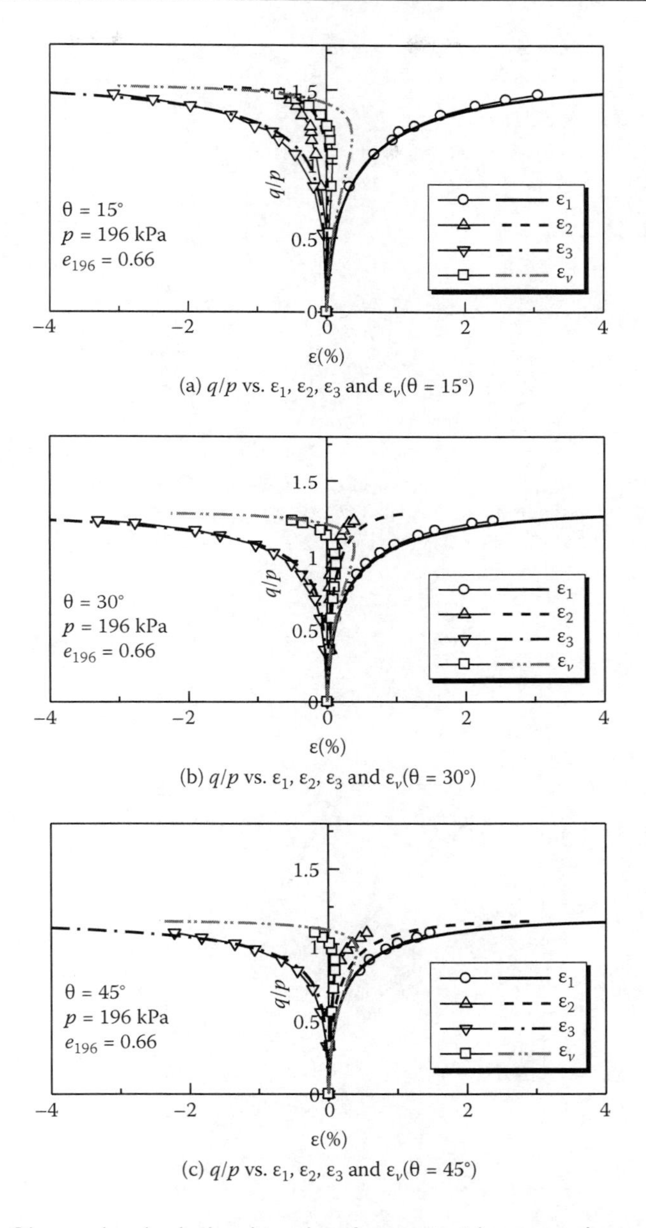

Figure 6.33 Observed and calculated results of true triaxial tests on dense sand under three different principal stresses.

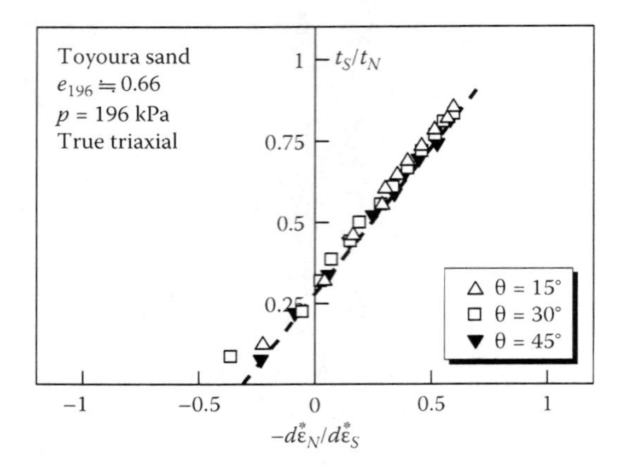

Figure 6.34 t_S/t_N versus $d\varepsilon_N^*/d\varepsilon_S^*$ relation of drained true triaxial tests on Toyoura sand under three different principal stresses.

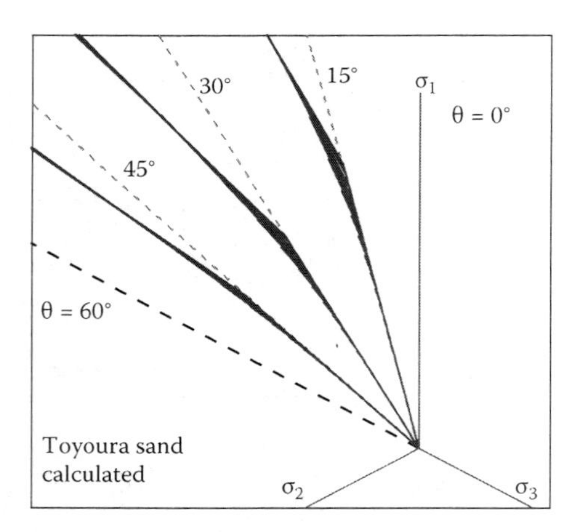

Figure 6.35 Calculated directions of $d\varepsilon_d$ on the octahedral plane under three different principal stresses ($\sigma_1 > \sigma_2 > \sigma_3$) on Toyoura sand.

6.3 THREE-DIMENSIONAL MODELING OF STRUCTURED SOILS—ADVANCED ELASTOPLASTIC MODELING AT STAGE II

6.3.1 Formulation of model

It is assumed that the stress–strain behavior of structured soils (e.g., naturally deposited soil) can be described by considering not only the effect of density as described in Section 6.2 but also the effect of bonding, in the same way as in the 1D modeling in Chapter 3. When the stress condition moves from point I to point P in Figure 5.11, the resulting change in void ratio for a structured soil is illustrated in Figure 6.1 and given by eq. (6.2). Then, the yield function is described by eq. (6.3), so that eq. (6.4) holds for structured soils as well.

As described in 1D modeling, the evolution rule of ρ with the development of plastic deformation for structured soils can be determined not only by the state variable ρ related to density but also by the state variable ω representing the bonding effect with an imaginary increase of density. Moreover, the value of the state variable ω has an additional effect on the degradation of ρ. The evolution rule of ρ is then given as follows, using not only $G(\rho)$ but also an increasing function $Q(\omega)$, which satisfies $Q(0) = 0$:

$$d\rho = -(1+e_0)\left\{\frac{G(\rho)}{t_N} + \frac{Q(\omega)}{t_N}\right\}\Lambda \tag{6.37}$$

The evolution rule of ω is given as follows, using the same function $Q(\omega)$, although it is possible to use another function:

$$d\omega = -(1+e_0)\frac{Q(\omega)}{t_N}\Lambda \tag{6.38}$$

In the present model, the following linear increasing function $Q(\omega)$ is adopted:

$$Q(\omega) = b\omega \tag{6.39}$$

The plastic strain increment is then obtained by the flow rule based on the t_{ij} concept in eq. (6.9). Substituting eq. (6.37) into eq. (6.4), the proportionality constant Λ is obtained as

$$\Lambda = \frac{dF}{(1+e_0)\left\{\frac{\partial F}{\partial t_{ii}} + \frac{G(\rho)}{t_N} + \frac{Q(\omega)}{t_N}\right\}} = \frac{dF}{h^p} \tag{6.40}$$

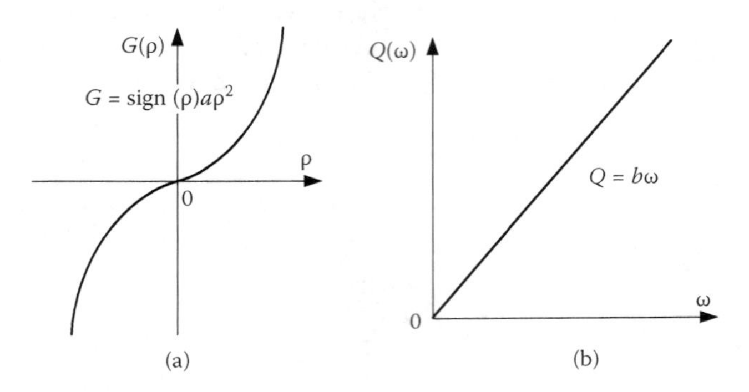

Figure 6.36 Functions $G(\rho)$ and $Q(\omega)$ for the evolution rule of variable ρ.

As described in the 1D model, since there is the possibility that the value of ρ becomes negative for the soils with bonding, the domain of the increasing function $G(\rho)$ is extended to the negative side. Therefore, a positive value of ρ has the effect of increasing the plastic modulus. On the other hand, a negative value of ρ has the effect of decreasing the plastic modulus. Figure 6.36 illustrates the functions $G(\rho)$ and $Q(\omega)$ for the evolution rule of ρ and ω. The elastic components and the loading condition are the same as those for the model at the advanced model at stage I.

6.3.2 Model validation through simulation of structured clays

Model simulations of a structured clay are carried out to further confirm the performance of the present model. All the material parameters used in the simulations are shown in Table 6.3 and have been determined for

Table 6.3 Material parameter used in simulations of structured soils using the advanced model at stage II considering bonding effects

λ	0.104	
κ	0.010	
N(e_N at p = 98 kPa)	0.83	Same parameters as Cam clay model
$R_{CS} = (\sigma_1/\sigma_3)_{CS}(comp.)$	3.5	
ν_e	0.2	
β	1.5	Shape of yield surface (same as original Cam clay if β = 1)
$a/(\lambda - \kappa)$	500	Influence of density
$b/(\lambda - \kappa)$	40	Influence of bonding

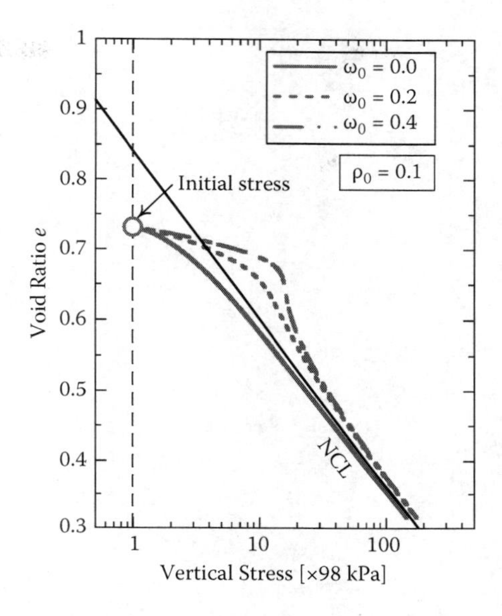

Figure 6.37 Simulation of oedometer tests on a clay with the same initial void ratio but different initial bonding parameters.

New Fujinomori clay. The values of the compression index λ and swelling index κ are the same as those employed in the simulations based on the 1D model described in Chapter 3. The other material parameters are the same as those in Tables 5.2 and 6.1 for Fujinomori clay. One single parameter b for considering the bonding effect, which assumed the value $b/(\lambda - \kappa) = 40$ as shown in Table 6.3, was added. The present model coincides with the model described in Nakai (2007), although the derivation processes of the present model are more logical.

Figure 6.37 shows the results, arranged in terms of the relations between void ratio and vertical stress in log scale, of simulations using the 3D model for oedometer tests on structured clays that have the same initial void ratio, but different initial bonding effects. Here, the solid line with $\omega_0 = 0.0$ represents the result for the nonstructured soil. It is seen that the 3D model can also describe the typical 1D consolidation behavior of structured soils simulated in Chapter 3. Figure 6.38 shows the results of simulations of constant mean principal stress tests ($p = 98$ kPa) on the same clay under triaxial compression and extension conditions. The initial stiffness, peak strength, and dilatancy become larger with the increase of the bonding effect, even if the initial void ratio is the same.

Figure 6.39 shows the results of simulations of undrained triaxial compression and extension tests ($p'_0 = 98$ kPa) on the same clay. Diagrams (a)

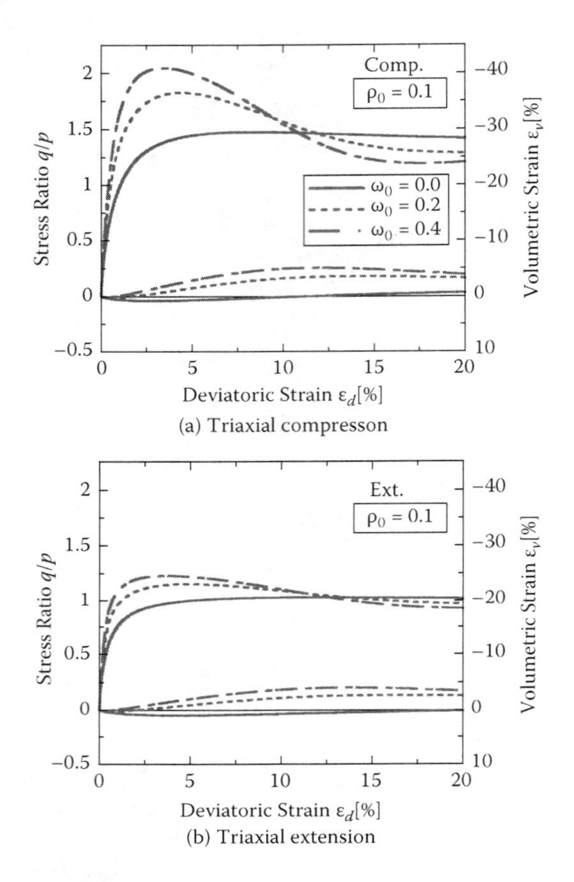

Figure 6.38 Simulation of drained triaxial compression and extension tests (p = 98 kPa) on a clay with the same initial void ratio but different initial bonding parameters.

and (b) show the results of effective stress paths and stress–strain curves, respectively. In these figures, the upper part depicts the results under triaxial compression conditions, and the lower part shows the results under triaxial extension conditions. The straight lines from the origin in diagram (a) represent the critical state lines (CSLs) in the p–q plane. Under undrained shear loadings, clays with bonding are stiffer and have higher strength than clays without bonding. It is also seen that overconsolidated clays without bonding (ω_0 = 0.0) show strain hardening with the decrease and the subsequent increase of mean stress, whereas clays with bonding (ω_0 = 0.2, 0.4) show not only strain hardening with the decrease and the subsequent increase of mean stress but also strain softening with the decrease of mean

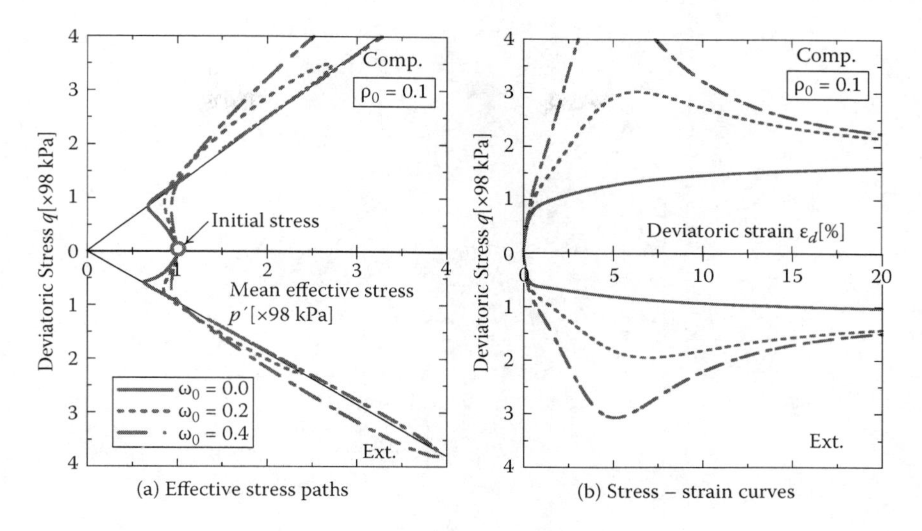

Figure 6.39 Simulation of undrained triaxial compression and extension tests on a clay with the same initial void ratio but different initial bonding parameters.

stress and deviatoric stress under undrained conditions. These represent typical undrained behaviors of structured soils.

Figure 6.40 shows numerical simulations of isotropic compression and the subsequent undrained shear tests on a structured clay. Diagram (a) refers to the consolidation curve of a structured clay (initial state: e_0 = 0.73, ($\rho_0 = 0.1$); $\omega_0 = 0.4$ at p_0 = 98 kPa) from stress condition (A). On the other hand, diagrams (b) and (c) give the results of effective stress paths and stress–strain curves for undrained triaxial compression tests on the clays, which are sheared from stress conditions (A), (B), and (C) in diagram (a).

Finally, Figure 6.41 shows the observed results of undrained shear tests on undisturbed Osaka Pleistocene clay (Ma12) published by Asaoka, Nakano, and Noda (2000b). Here, OC(e) represents the result under an initial confining pressure p_0' = 98 kPa (overconsolidation state), and NC(e) represents the result under p_0' = 490 kPa (almost the same stress as the overburden pressure in situ). The authors (Asaoka et al. 2000b; Asaoka 2003) also carried out the simulations of structured soils using the SYS Cam clay model (Asaoka et al. 2000a), in which the subloading surface concept (to increase the plastic modulus) and the superloading surface concept (to decrease the plastic modulus) are introduced in the Cam clay model.

Similarly, as can be seen from eq. (6.40) in the present model, $G(\rho)$ and $Q(\omega)$ have the effects of increasing the plastic modulus in the cases where

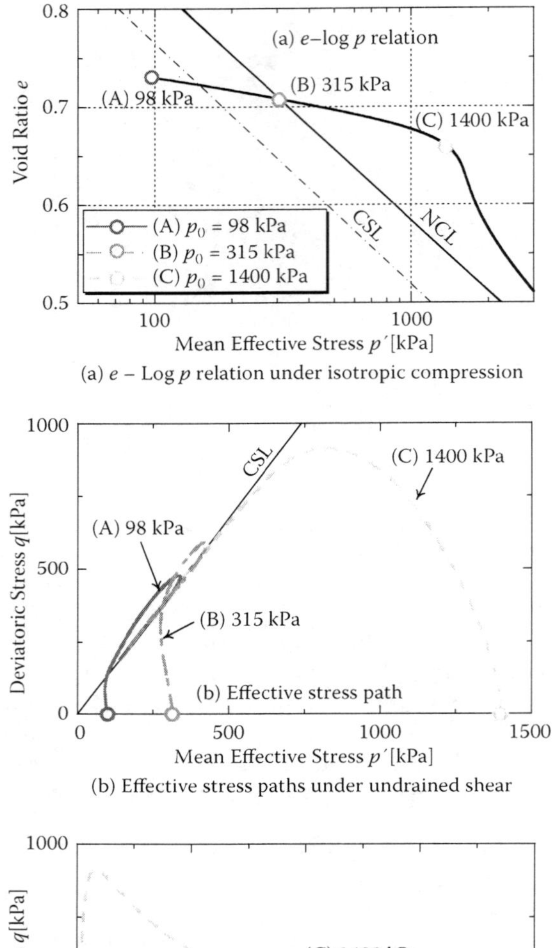

(a) e – Log p relation under isotropic compression

(b) Effective stress paths under undrained shear

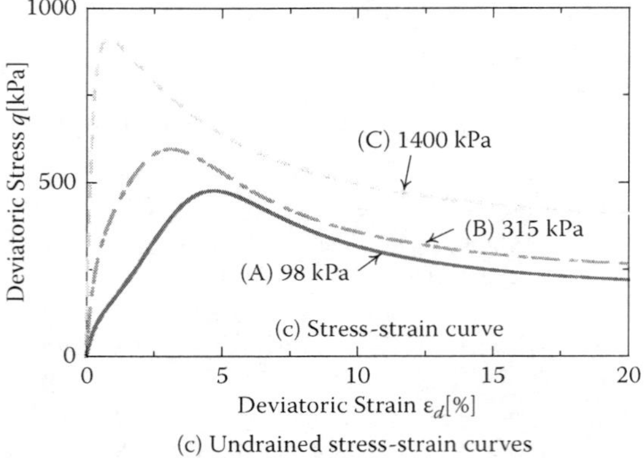

(c) Undrained stress-strain curves

Figure 6.40 Calculated results of isotropic compression and subsequent undrained shear tests on a structured clay.

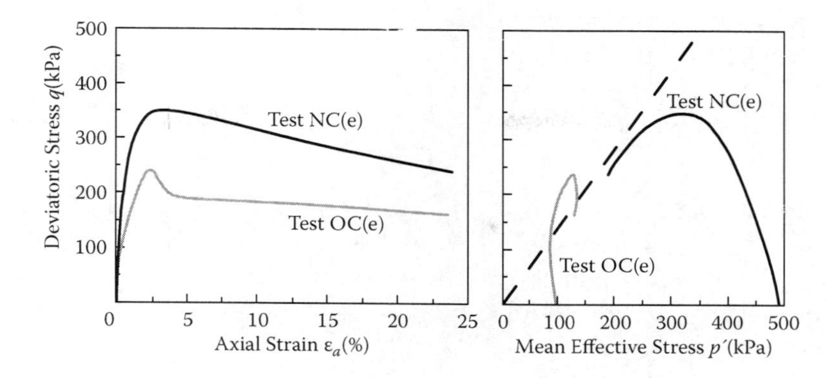

Figure 6.41 Observed results of undrained shear tests on undisturbed Osaka Pleistocene clay (Ma12). (Replotted from data in Asaoka, A. et al. 2000b. *Journal of Applied Mechanics, JSCE* 3:335–342.)

ρ and ω are positive. Only in the case when ρ becomes negative (the current void ratio is larger than that on NCL), does $G(\rho)$ have an effect of decreasing the plastic modulus. The model proposed here simulates well the typical undrained shear behavior of structured soils under different confining pressures, such as the differences of stress paths and stress–strain curves, depending on the magnitude of the confining pressure. It also reproduces the rewinding of the stress path after increasing of mean stress and deviatoric stress and other effects.

6.3.3 An alternative formulation of the model

The 3D formulation mentioned in Sections 6.2, 6.3.1, and 6.3.2 was developed in such a way that it coincides with the subloading t_{ij} model developed previously by Nakai and Hinokio (2004) and Nakai (2007). Although the validation of the model under isotropic compression (or 1D compression) is similar to that of the 1D model described in Chapter 3, there are a few quantitative differences between the results of simulations using that model and the present model. Another formulation of the 3D model in which the results of simulations coincide perfectly with those of the 1D model under isotropic compression on overconsolidated soils and structured soils is presented here.

The yield function $f = 0$ is given by eqs. (6.3), (5.34), and (5.35), and the consistency condition $df = 0$ is ensured in eq. (6.4). In the 1D model described in Chapter 3, $d\rho$ is expressed as

$$d\rho = -\{G(\rho) + Q(\omega)\} \cdot d(-e)^p \tag{6.41}$$

Under isotropic compression ($d\varepsilon_1{}^p = d\varepsilon_2{}^p = d\varepsilon_3{}^p$), there is the following relation between $d(-e)^p$ and Λ:

$$d(-e)^p = (1+e_0)d\varepsilon_v^p = \sqrt{3}(1+e_0)\,\|d\varepsilon_{ij}^p\| = \sqrt{3}(1+e_0)\Lambda\left\|\frac{\partial F}{\partial t_{ij}}\right\| \tag{6.42}$$

Here, $\|A_{ij}\|$ denotes the norm of tensor A_{ij}. Assuming that evolution of ρ under isotropic compression conditions is the same as that in the 1D model, the evolution rule of ρ in general stress conditions is obtained by replacing $d(-e)^p$ in eq. (6.41) with $\sqrt{3}(1+e_0)\Lambda\,\|\frac{\partial F}{\partial t_{ij}}\|$. Therefore, eq. (6.4) can be rewritten as

$$df = dF - (1+e_0)\frac{\partial F}{\partial t_{kk}}\Lambda - \left\{\sqrt{3}(1+e_0)\left\|\frac{\partial F}{\partial t_{ij}}\right\|\right\}G(\rho)\Lambda - \left\{\sqrt{3}(1+e_0)\left\|\frac{\partial F}{\partial t_{ij}}\right\|\right\}Q(\omega)\Lambda = 0 \tag{6.43}$$

Then, the proportional constant Λ is given by

$$\Lambda = \frac{dF}{(1+e_0)\left\{\frac{\partial F}{\partial t_{kk}} + \sqrt{3}\left\|\frac{\partial F}{\partial t_{ij}}\right\|G(\rho) + \sqrt{3}\left\|\frac{\partial F}{\partial t_{ij}}\right\|Q(\omega)\right\}} = \frac{dF}{h^p} \tag{6.44}$$

Figure 6.42 shows the comparison of the calculated results (black curves) using Λ in eq. (6.40) with the corresponding results (gray curves) using Λ in

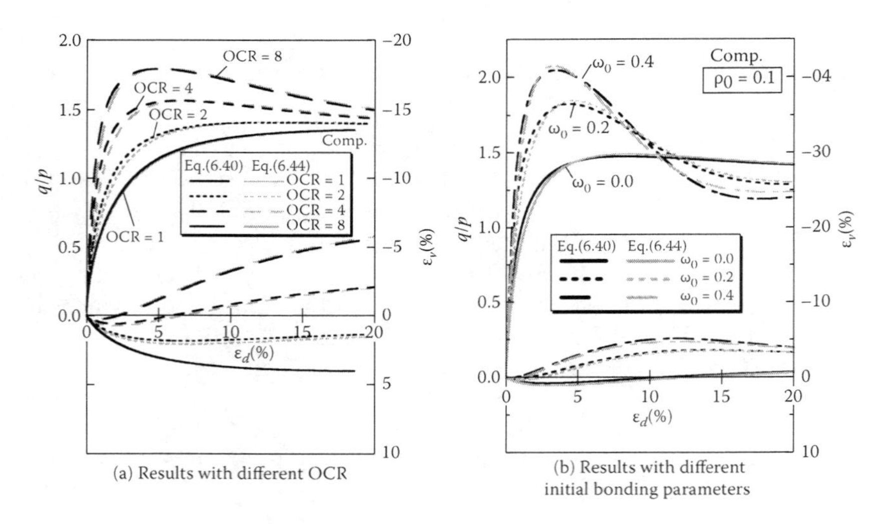

Figure 6.42 Comparison of calculated stress–strain curves using Λ from eq. (6.40) and Λ from eq. (6.44).

eq. (6.44). Here, diagram (a) shows the results for different initial density (OCR) without bonding ($\omega_0 = 0$), and diagram (b) shows the results for different initial bonding (ω_0) under the same initial density ($\rho_0 = 0.1$). The form of the equations for $G(\rho)$ and $Q(\omega)$ in eq. (6.44) are the same as those of eq. (6.40) (see Figure 6.36). Although the values of material parameters for the calculation using eq. (6.40) are shown in Table 6.3, the values of a and b alone are different for the calculation using eq. (6.44): $a = 90$ and $b = 7$. It can be seen that there is not much difference in the calculated results using either eq. (6.40) or (6.44). In this book, the simulations using 3D models are carried out based on the form of eq. (6.40).

6.4 THREE-DIMENSIONAL MODELING OF OTHER FEATURES OF SOILS—ADVANCED ELASTOPLASTIC MODELING AT STAGE III

6.4.1 General formulation of model

Experimental tests have shown that the NCL and the CSL shift on the e–ln$p$ (or e–lnt_N) plane depending on strain rate, temperature, suction (saturation), and other effects, as described in Section 3.5 in Chapter 3. To take into consideration these features in the 3D model, the e–lnt_{N1} relation in Figure 6.43 is obtained for the change of the stress condition from point I to point P in Figure 5.11. Here, ψ, the state variable to determine the position of the NCL, is related with strain rate, temperature, suction (saturation), and/or other effects in the same way as in the 1D

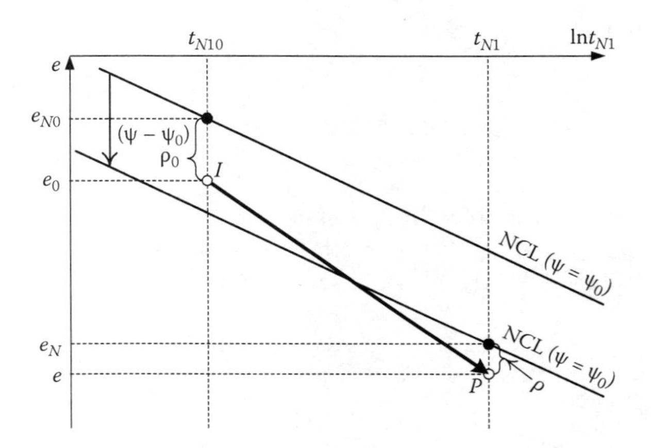

Figure 6.43 Change of void ratio in a soil with changes in variables ρ and ψ.

model; ψ_0 is the initial value of ψ. The plastic change of void ratio is then expressed as

$$(-\Delta e)^p = \left\{(e_{0N} - e_N) - (-\Delta e)^e\right\} - (\rho_0 - \rho)$$
$$= \left\{(\lambda - \kappa)\ln\frac{t_{N1}}{t_{N0}} + (\psi - \psi_0)\right\} - (\rho_0 - \rho) \tag{6.45}$$

Then the yield function can be written as follows using F and H:

$$F + \rho + \psi = H + \rho_0 + \psi_0$$

or

$$f = F - \left\{H + (\rho_0 - \rho) + (\psi_0 - \psi)\right\} = 0 \tag{6.46}$$

From the consistency condition ($df = 0$) and the flow rule in eq. (5.36), the following expression can be obtained:

$$df = dF - \{dH - d\rho - d\psi\}$$
$$= dF - \left\{d(-e)^p - d\rho - d\psi\right\} \tag{6.47}$$
$$= dF - \left\{(1 + e_0)\Lambda\frac{\partial F}{\partial t_{ii}} - d\rho - d\psi\right\} = 0$$

Since $d\rho$ is related with the proportionality constant Λ by eq. (6.37), Λ is expressed as

$$\Lambda = \frac{dF + d\psi}{(1 + e_0)\left\{\frac{\partial F}{\partial t_{kk}} + \frac{G(\rho)}{t_N} + \frac{Q(\omega)}{t_N}\right\}} = \frac{dF + d\psi}{h^p} \tag{6.48}$$

As described in the modeling at stage I and stage II, the yield surface always passes over the current stress and the plastic deformation occurs when $\Lambda > 0$. Therefore, the plastic strain increment in the model at stage III is given by

$$d\varepsilon_{ij}^p = \langle\Lambda\rangle\frac{\partial F}{\partial t_{ij}} = \left\langle\frac{dF + d\psi}{h^p}\right\rangle\frac{\partial F}{\partial t_{ij}} \qquad \left(\text{where} \quad dF = \frac{\partial F}{\partial \sigma_{kl}}d\sigma_{kl}\right) \tag{6.49}$$

As mentioned in Section 3.5, since ψ is a function of the rate of plastic void ratio change $(-\dot{e})^p$ (or plastic volumetric strain $\dot{\varepsilon}_v^p$ or other strain rates), temperature T, suction s (or degree of saturation S_r), or others, $d\psi$ is given in such forms as $d\psi = (\partial\psi/\partial t)dt$, $(\partial\psi/\partial T)dT$, $(\partial\psi/\partial s)ds$, and $(\partial\psi/\partial S_r)dS_r$.

6.4.2 Application of model to time-dependent behavior

In the 1D time-dependent model, the state variable ψ is related with the rate of plastic void ratio change $(-\dot{e})^p$, as described in Section 3.5. However, in multidimensional stress conditions, the rate of plastic void ratio change $(-\dot{e})^p$ (or the plastic volumetric strain rate $\dot{\varepsilon}_v^p$) is not necessarily positive during plastic deformation because of soil dilatancy, and then the rate of plastic void ratio change (or plastic volumetric strain rate) is not suitable for the measure of time-dependent behavior. Since the norm of plastic strain rate $\|\dot{\varepsilon}_{ij}^p\|$ is always positive during plastic deformation (even in multidimensional conditions) and gives the magnitude of the plastic strain rate, it seems logical to relate ψ with some quantity using the norm of the plastic strain rate.

From experimental results of oedometer tests and undrained and drained shear tests on a clay, Leroueil and Marques (1996) show that time-dependent behavior of clays under consolidation and shear can be arranged uniquely using the norm of the strain rate as the strain rate measure. Additionally, it is known that the void ratio change for normally consolidated soils subjected to pure creep conditions under isotropic compression satisfies a linear e–ln$t$ relation with the slope of λ_α (coefficient of secondary consolidation) in the same way as that in the 1D model. Now, under isotropic compression, the norm of plastic strain rate $\|\dot{\varepsilon}_{ij}^p\|$ is expressed as follows using the rate of plastic void ratio $(-\dot{e})^p$:

$$\left\|\dot{\varepsilon}_{ij}^p\right\| = \sqrt{\dot{\varepsilon}_{ij}^p\dot{\varepsilon}_{ij}^p} = \sqrt{\dot{\varepsilon}_1^{p2} + \dot{\varepsilon}_2^{p2} + \dot{\varepsilon}_3^{p2}} = \sqrt{3}\,\dot{\varepsilon}_1^p = \frac{\dot{\varepsilon}_v^p}{\sqrt{3}} = \frac{(-\dot{e})^p}{\sqrt{3}(1+e_0)} \tag{6.50}$$

Although the state variable ψ and its initial value ψ_0 in the 1D model are related with t and t_0 or $(-\dot{e})^p$ and $(-\dot{e})_0^p$, as described in Section 3.5, ψ and ψ_0 in the 3D model are given by the same formula as that in the 1D model, using $(-\dot{e})_{(equ)}^p$ instead of $(-\dot{e})^p$:

$$\begin{cases} \psi = \lambda_\alpha \ln t \\ \psi_0 = \lambda_\alpha \ln t_0 \end{cases} \quad \text{or} \quad \begin{cases} \psi = -\lambda_\alpha \ln(-\dot{e})_{(equ)}^p \\ \psi_0 = -\lambda_\alpha \ln(-\dot{e})_{(equ)0}^p \end{cases} \tag{6.51}$$

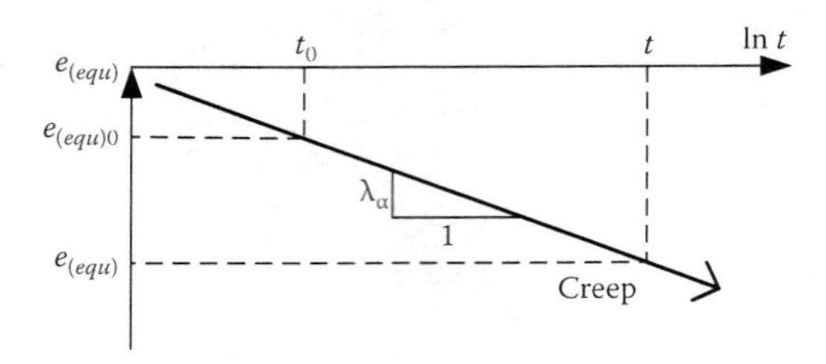

Figure 6.44 Creep characteristics of normally consolidated soils under general stress conditions.

Here, it is assumed that $(-\dot{e})^p_{(equ)}$ is an equivalent rate of plastic void ratio change, which is defined by eq. (6.50) not only under isotropic compression but also under any other stress condition:

$$(-\dot{e})^p_{(equ)} = \sqrt{3}(1+e_0)\left\|\dot{\varepsilon}^p_{ij}\right\| \tag{6.52}$$

This formulation means that for a normally consolidated soil under pure creep conditions, there is a linear relation between $e_{(equ)}$ and $\ln t$ with a slope of λ_α, regardless of stress condition, as shown in Figure 6.44. Therefore, the increment of ψ is expressed as

$$d\psi = \frac{\partial\psi}{\partial t}dt = \lambda_\alpha \frac{1}{t}dt = (-\dot{e})^p_{(equ)}dt \tag{6.53}$$

The position of the NCL shifts depending on the equivalent rate of plastic void ratio change defined by eq. (6.52) as shown in Figure 6.45.

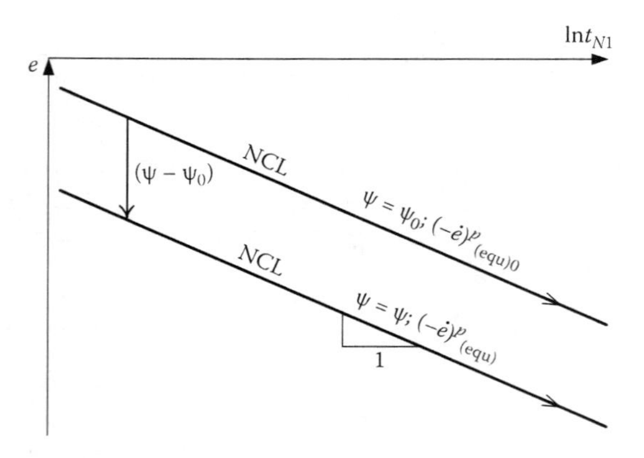

Figure 6.45 Position of the NCL depending on the rate of equivalent void ratio change.

Substituting eq. (6.53) into eq. (6.48), the proportionality constant Λ for the model considering time-dependent behavior can be obtained:

$$\Lambda = \frac{dF + (-\dot{e})^{p}_{(equ)}dt}{(1+e_0)\left\{\frac{\partial F}{\partial t_{kk}} + \frac{G(\rho)}{t_N} + \frac{Q(\omega)}{t_N}\right\}} = \frac{dF + (-\dot{e})^{p}_{(equ)}dt}{h^p} \cong \frac{dF + (-\dot{e})^{p^*}_{(equ)}dt}{h^p} \tag{6.54}$$

Here, $(-\dot{e})^{p^*}_{(equ)}$ denotes the equivalent rate of plastic void ratio change at the calculation step immediately before the current one, in the same way as in the 1D model. Also, the position of the NCL (ψ) for the next calculation step can be determined by eq. (6.51) using the updated value of $(-\dot{e})^{p}_{(equ)}$.

6.4.3 Model validation using simulations of time-dependent behavior of clays

The validity of the proposed time-dependent model is checked by performing some simulations of constant strain rate tests, creep tests, and others on normally consolidated clay, overconsolidated clay, and structured clay. The material parameters used in the simulation are the same as those of the clay in Table 6.3. The values of the added material parameters for describing time-dependent behavior are shown in Table 6.4—that is, the coefficient of secondary consolidation λ_α and the equivalent rate of plastic void ratio change at the reference state $(-\dot{e})^{p}_{(equ)ref}$. In the following simulations, the initial equivalent rate of plastic void ratio change is assumed to be the same as that at the reference state $(-\dot{e})^{p}_{(equ)0} = (-\dot{e})^{p}_{(equ)ref}$.

Figure 6.46 shows the calculated results—(a) effective stress paths, (b) stress–strain curves—of undrained triaxial compression and extension tests with different axial strain rates $\dot{\varepsilon}_a$ on a normally consolidated clay ($\rho_0 = 0.0$, $\omega_0 = 0.0$). Here, the curves labeled as "No creep" represent the results without time effect, and the thick curves indicate the results of the tests in which the strain rate changes from 2.0%/min to 0.002%/min and then to 2.0%/min during shear. The model can describe well known rate effects on the strength and development of pore water pressure: higher undrained shear strength and reduced pore pressure development with

Table 6.4 Additional material parameter in simulations using model at stage III considering time-dependent behavior

λ_α	0.003	Coefficient of secondary consolidation
$(-\dot{e})^{p}_{(equ)ref}$	1×10^{-7}/min	Equivalent rate of plastic void ratio change at reference state

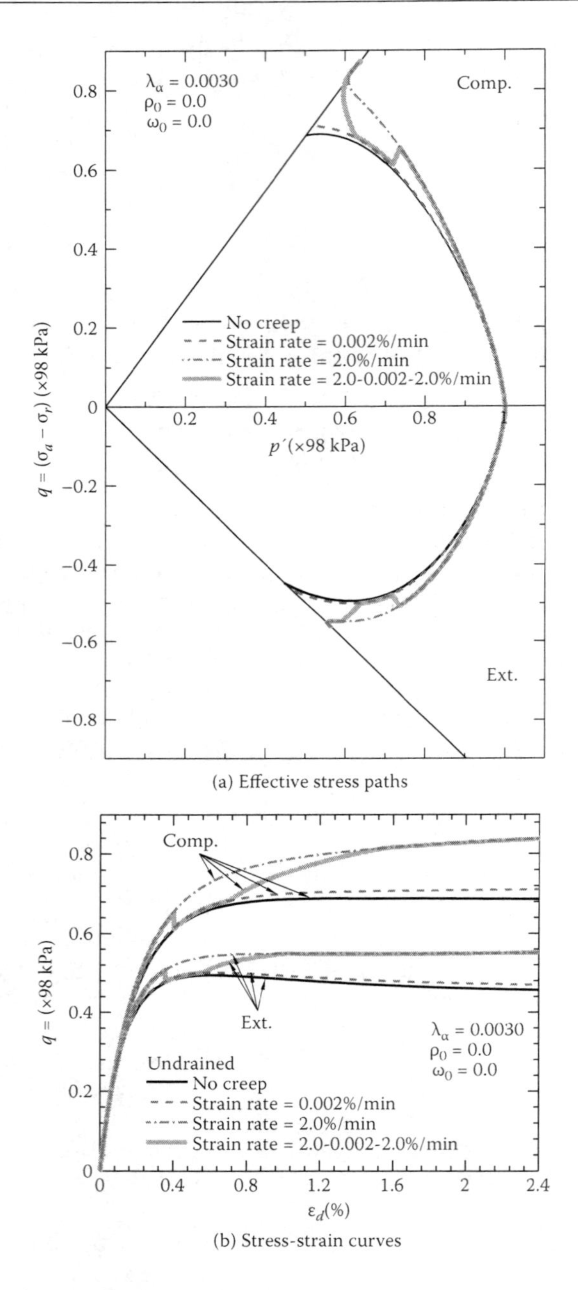

(a) Effective stress paths

(b) Stress-strain curves

Figure 6.46 Simulation of undrained triaxial compression and extension tests on a normally consolidated clay subjected to different strain rates.

increasing strain rates. It is also seen that the calculated stress–strain state shifts between the corresponding stress–strain curves and effective stress paths when the strain rate changes during shear—a phenomenon known as "isotache."

Figure 6.47 shows the calculated results of undrained creep on a normally consolidated clay under triaxial compression conditions. The clay is sheared up to a certain deviatoric stress condition with the strain rate of $\dot{\varepsilon}_a$ = 2.0%/min, and then the deviatoric stress is kept constant. Diagram (a) shows the stress paths during shear and creep condition, diagram (b) shows the creep curves during creep, and diagram (c) presents the relation between strain rate and elapsed time during creep. It can be seen that the model represents the typical undrained creep behavior, including transient creep, stationary creep, and accelerating creep (Sekiguchi 1984).

Figure 6.48 shows the calculated results—(a) the effective stress paths, (b) the stress–strain curves—of undrained compression tests under different strain rates on the normally consolidated and overconsolidated clays with the same initial void ratio (e_0 = 0.83) and the same initial equivalent rate of void ratio change ($(-\dot{e})^p_{(equ)0}$ = 1 × 10^{-7}/min). In these figures, the results of the normally consolidated clay (p_0 = 98 kPa) are the same as those under triaxial compression tests in Figure 6.46. It can be seen that although the mean stress p in the fast test is smaller than that of the slow test at the same deviatoric stress q for the normally consolidated clay with negative dilatancy, the tendency for the overconsolidated clay with positive dilatancy is the opposite.

Figure 6.49 shows the observed results of undrained triaxial compression tests on remolded Fukakusa clay in normally consolidated states (N-1, N-3) and overconsolidated states (O-1, O-2) with different axial strain rates (Oka et al. 2003). The initial void ratios of these samples vary from 1.16 to 1.18. The differences of the strain rate effect on the soil behavior between the normally consolidated clays and the overconsolidated clays derived using the present model simulations correspond qualitatively with the observed test results in Figure 6.49. Asaoka, Nakano, and Noda (1997) also performed undrained triaxial compression tests on a remolded overconsolidated clay with different strain rates and reproduced the same strain rate effects on overconsolidated clays by performing a soil water coupled analysis using their elastoplastic model.

Simulations of undrained triaxial compression tests on a structured clay with different strain rates are shown in Figure 6.50. Here, the structured clay is sheared under undrained conditions with different strain rates from the conditions (A) and (C) in Figure 6.40(a). Diagrams (a) and (b) in Figure 6.50 show the effective stress paths and stress–strain curves, respectively. As mentioned when commenting on Figure 6.46, the curves represented by "No creep" are the results without time effect, and the thick curves indicate the results of the tests in which the strain rate changes from

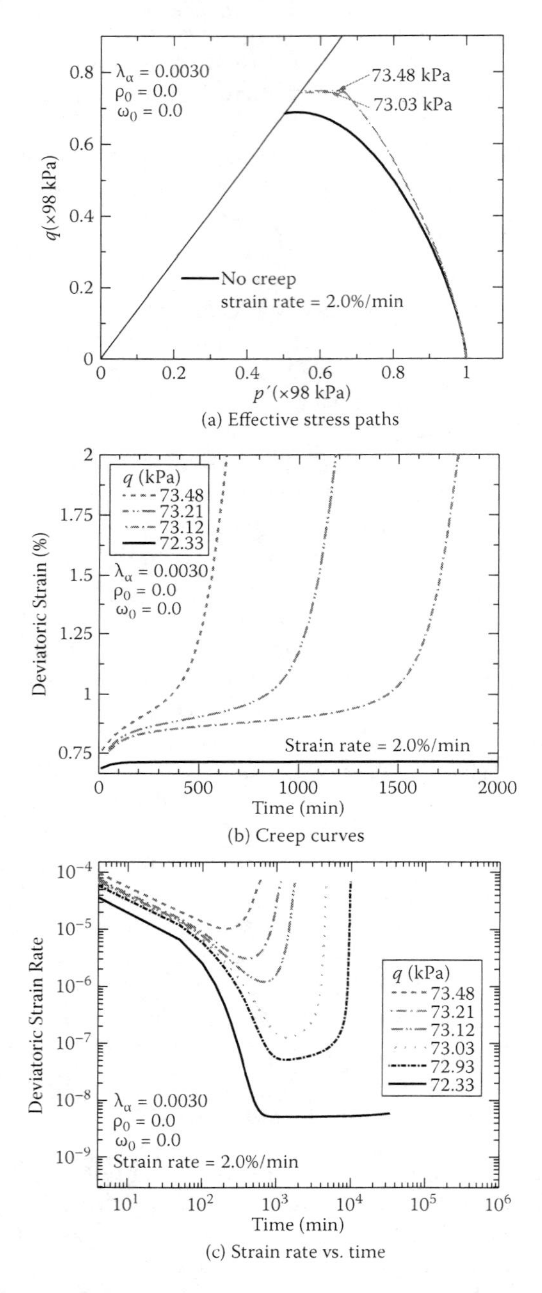

(a) Effective stress paths

(b) Creep curves

(c) Strain rate vs. time

Figure 6.47 Simulation of undrained creep tests after constant strain rate ($\dot{\varepsilon}_a$ =2.0%/min) triaxial compression tests on a normally consolidated clay.

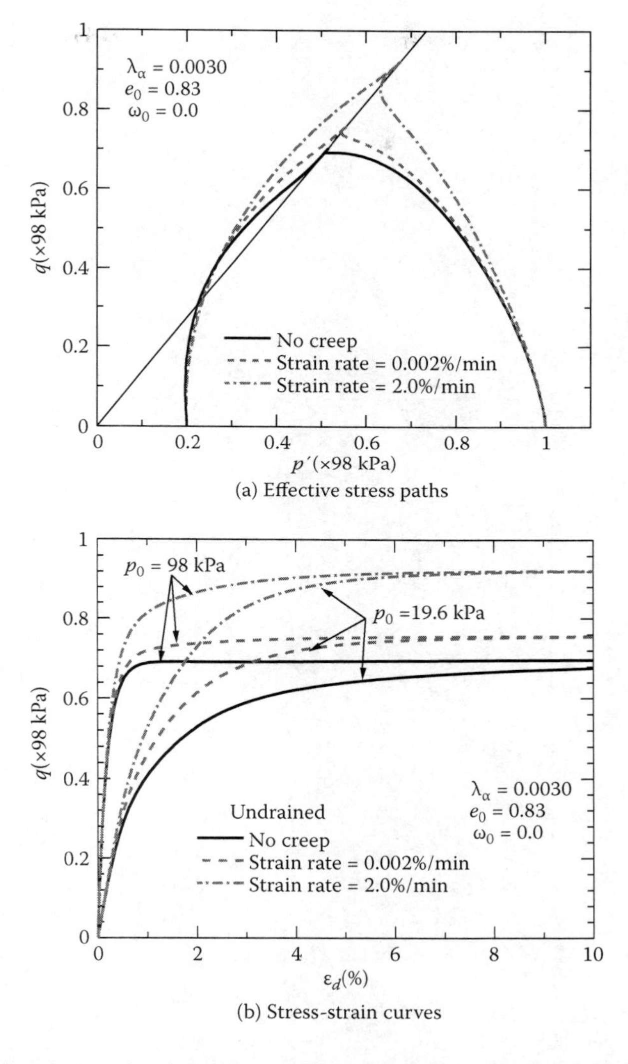

Figure 6.48 Simulation of undrained triaxial compression tests on normally consolidated and overconsolidated clays subjected to different strain rates.

2.0%/min to 0.002%/min and then 2.0%/min during shear. The phenomenon of isotache has been observed experimentally in structured soils as well as in nonstructured soils (Graham, Crooks, and Bell 1983). It can be seen that the present model describes observed time-dependent behavior of structured soils shown in Figure 6.51.

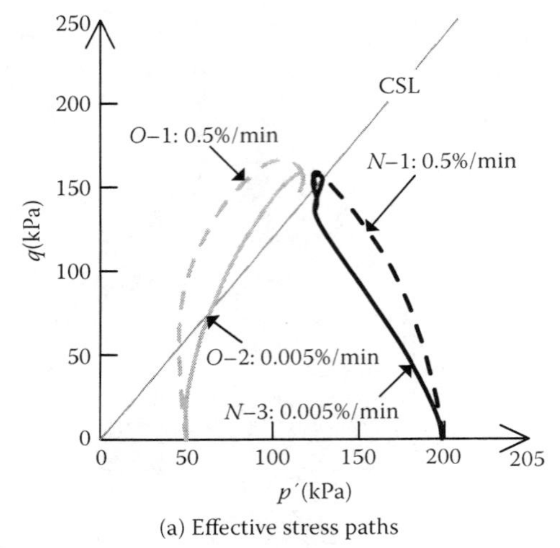

(a) Effective stress paths

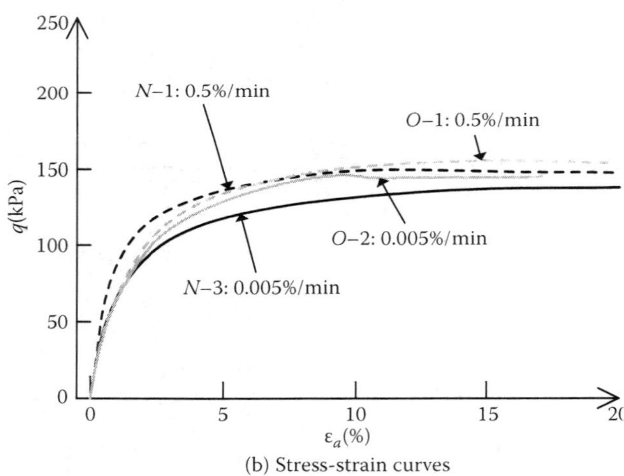

(b) Stress-strain curves

Figure 6.49 Observed results of undrained triaxial compression tests on normally consolidated and overconsolidated Fukakusa clays subjected to different strain rates. (Replotted from data in Oka, F. et al. 2003. *Soils and Foundations* 43 (4): 189–202.)

6.4.4 Modeling of some other features of soil behavior

In this section, the 1D models for describing temperature-dependent behavior and unsaturated soil behavior, which are explained in Section 3.10, are extended to 3D models.

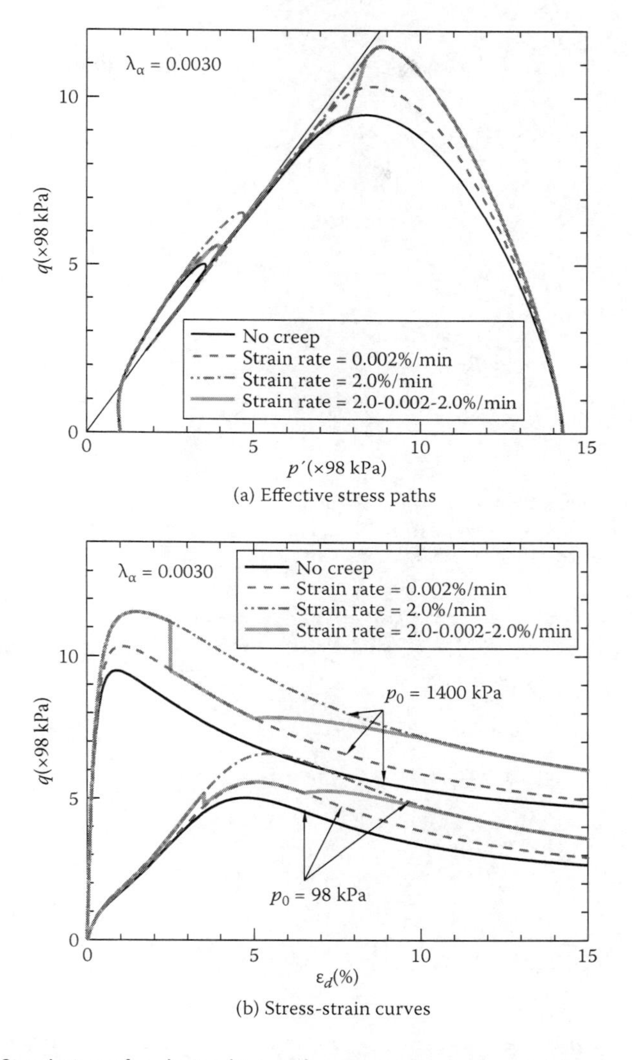

Figure 6.50 Simulation of undrained triaxial compression tests on a structured clay with different initial conditions and different strain rates.

6.4.4.1 Temperature-dependent behavior

In 3D stress conditions, the position of NCL on the e–$\ln t_{N1}$ relation is determined by the following equation as the function of temperature T in the same way as in eq. (3.54) in 1D stress conditions:

$$\psi - \psi_0 = \lambda_T (T - T_0) \tag{6.55}$$

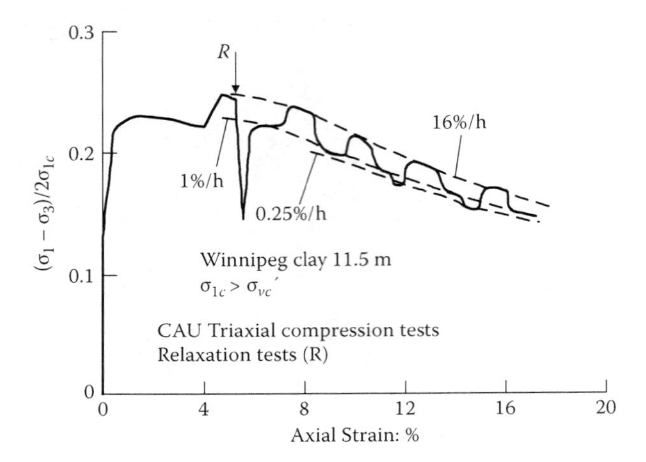

Figure 6.51 Observed result of undrained triaxial compression test on a natural clay with step-changed strain rates and relaxation. (Replotted from data in Graham, J. et al. 1983. *Geotechnique* 33 (3): 327–340.)

Referring to eqs. (3.58)–(3.60) in 1D stress conditions, the proportional constant Λ (see eq. 6.48) for the 3D model considering temperature-dependent behavior is expressed as

$$\Lambda = \frac{dF + d\psi^p}{(1+e_0)\left\{\frac{\partial F}{\partial t_{kk}} + \frac{G(\rho)}{t_N} + \frac{Q(\omega)}{t_N}\right\}}$$

$$= \frac{dF + (\lambda_T - \kappa_T)dT}{(1+e_0)\left\{\frac{\partial F}{\partial t_{kk}} + \frac{G(\rho)}{t_N} + \frac{Q(\omega)}{t_N}\right\}}$$

$$= \frac{dF + (\lambda_T - \kappa_T)dT}{h^p} \quad \left(\text{where} \quad dF = \frac{\partial F}{\partial \sigma_{kl}} d\sigma_{kl}\right) \tag{6.56}$$

Here, as described in eq. (3.56), κ_T is related to the coefficient α_T:

$$\kappa_T = -3\alpha_T(1+e_0) \tag{6.57}$$

Also, the elastic strain increment is given by the following equation using α_T in the temperature-dependent model:

$$d\varepsilon_{ij}^e = \frac{1+\nu_e}{E_e} d\sigma_{ij} - \frac{\nu_e}{E_e} d\sigma_{mm}\delta_{ij} - \alpha_T dT\delta_{ij} \quad \left(\text{where} \quad E_e = \frac{3(1-2\nu_e)(1+e_0)\,p}{\kappa}\right) \tag{6.58}$$

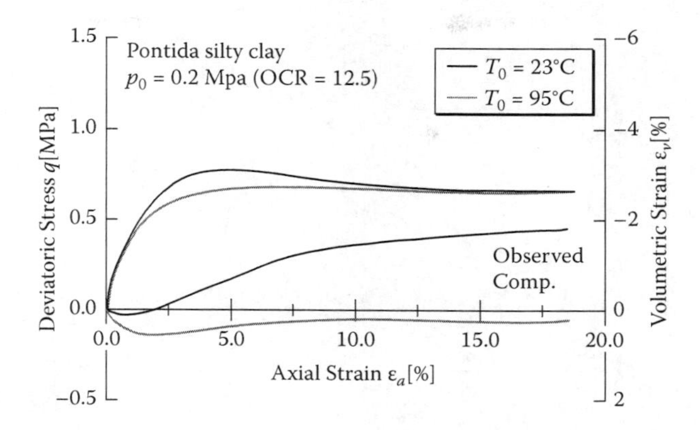

Figure 6.52 Observed results of drained triaxial compression tests on overconsolidated Pontida silty clay under different temperatures. (Replotted from data in Hueckel, T. and Baldi, G. 1990. *Proceedings of ASCE* 116 (GE12): 1778–1796.)

Figure 6.52 shows the observed stress–strain–dilatancy relation of an overconsolidated clay under different temperatures by Hueckel and Baldi (1990). It can be seen that though the stiffness, peak strength, and dilatancy become small with increasing temperature, the stress at residual condition (critical state) is independent of the temperature. Figure 6.53 shows the simulated results of Fujinomori clay, which are arranged with respect to the same relation as in Figure 6.52. In the calculations, the material parameters in Tables 5.2 and 6.1 (the same as those in Table 6.3 except for b) are used, and the additional material parameters concerned with the temperature effect are $\lambda_T = 0.003$, $\kappa_T = -0.0001$, and $T_0 = 20°C$, which are the same as the value used in the analyses in 1D conditions in Section 3.10.1. It can be

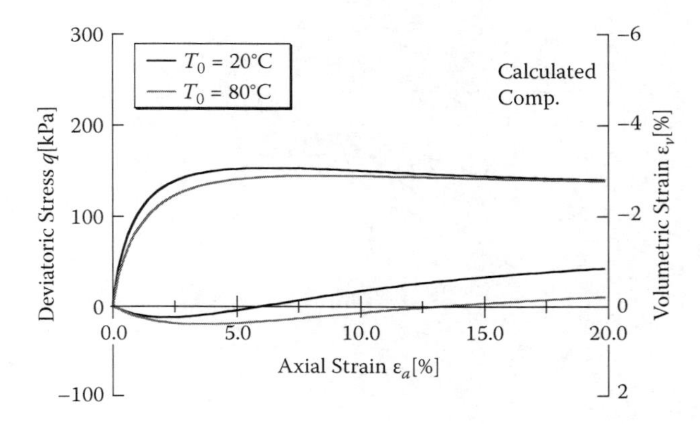

Figure 6.53 Simulation of drained triaxial compression tests on overconsolidated clay under different temperatures.

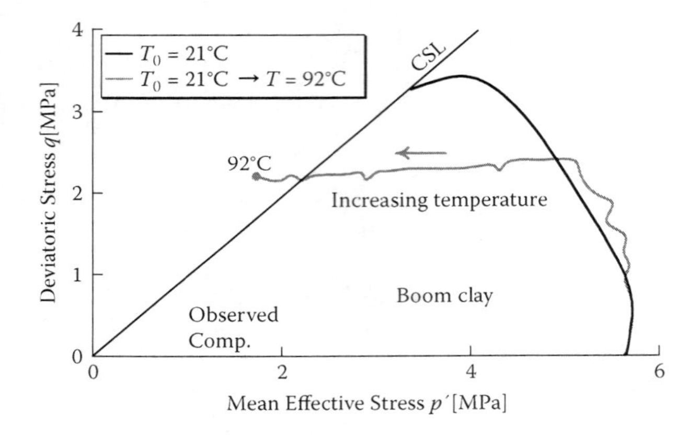

Figure 6.54 Observed effective stress paths of undrained triaxial compression tests on normally consolidated Boom clay under constant temperature and increasing temperature. (Replotted from data in Hueckel, T. and Pellegrini, R. 1991. *Soils and Foundations* 31 (3): 1–16.)

seen that the calculated results grasp well the observed temperature effects shown in Figure 6.52.

Figure 6.54 shows the observed effective stress paths in undrained heating tests on normally consolidated Boom clay (Hueckel and Pellegrini 1991). One test was carried out under constant temperature ($T = 21°C$) as usual. In the other test, the clay was heated from $T = 21°$ to $92°$ under creep condition after it was sheared up to a certain stress level at the temperature of $T = 21°$. Figure 6.55 shows the calculated results of Fujinomori clay.

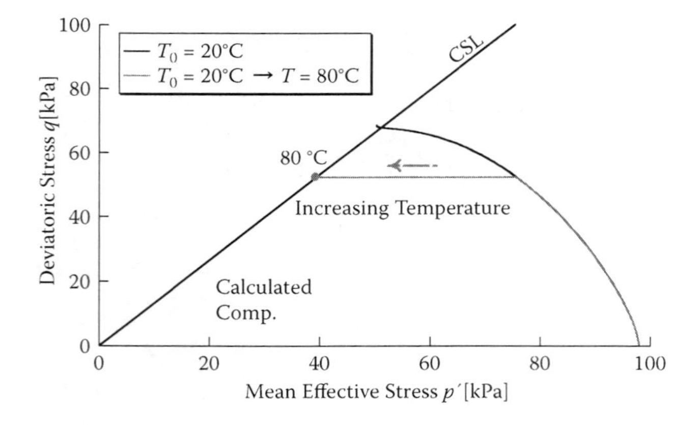

Figure 6.55 Simulation of effective stress paths of undrained triaxial compression temperature.

In the simulations, undrained shear tests were performed under $T = 21°$, and the heating was done from $T = 20°$ to $80°$. It can be seen from the observed and calculated results that a large deduction of effective mean stress occurs during heating in normally consolidated clay. This corresponds to the result that volume contraction due to heating occurs in heating of normally consolidated clay under drained conditions, as shown in Figures 3.38 and 3.39.

6.4.4.2 Unsaturated soil behavior

Bishop's effective stress in eq. (3.63) can be written by the following equation in 3D stress conditions:

$$\sigma''_{ij} = \sigma_{ij} - u_a\delta_{ij} + \chi s\delta_{ij} \cong \sigma_{ij} - u_a\delta_{ij} + S_r s\delta_{ij} = \sigma^{net}_{ij} + S_r s\delta_{ij}$$

$$(\text{where} \quad s = u_a - u_w \geq 0) \tag{6.59}$$

Sivakumar (1993) shows that the stress states of the critical state of saturated clay and unsaturated clays are uniquely not using σ^{net}, but using σ''_{ij} (see Figure 6.56). Also, it is experimentally shown by Kawai et al. (2000) that the stress–dilatancy relation of unsaturated soils, which is arranged using Bishop's effective stress σ''_{ij}, is independent of the degree of saturation.

Figure 6.57 shows the observed stress-dilatancy relation of saturated and unsaturated kaolin clay based on the t_{ij} concept, in which the data are arranged using Bishop's effective stress in eq. (6.59) (Kyokawa 2010). Here, the tests of saturated clay were carried out under constant mean principal stress, and the tests of unsaturated clay were carried out under

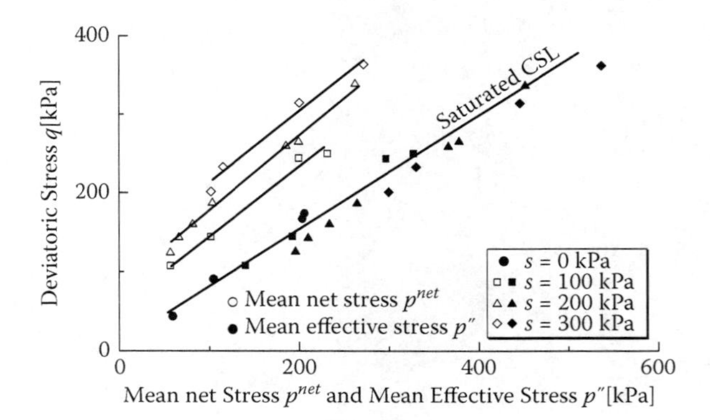

Figure 6.56 Stress condition at critical state of saturated and unsaturated soils. (Replotted from data in Sivakumar, V. 1993. PhD dissertation, University of Sheffield.)

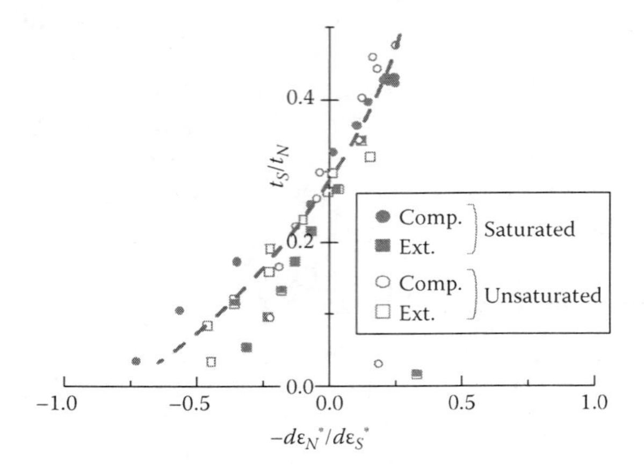

Figure 6.57 Observed stress–dilatancy relation based on t_{ij} concept on kaolin clay.

constant mean net stress and constant suction. The relation between stress ratio t_S/t_N and strain increment ration $d\varepsilon_N^*/d\varepsilon_S^*$ holds uniquely, regardless of the saturation and the magnitude of intermediate principal stress, though there is some scattering of data. It can be seen from these figures that when Bishop's effective stress in eq. (6.59) is employed, the concept of t_{ij} is considered to be effective for unsaturated soils as well as for saturated soils. Farias, Pinhiero, and Cordao Neto (2006) also proposed a constitutive model for unsaturated soil in general 3D stresses by referring to the Barcelona basic model (Alonso, Gens, and Josa 1990) and the t_{ij} concept.

The position of NCL on e–$\ln t_{N1}$ relation is determined by the following equation as the function of saturation S_r in the same way as eq. (3.66) in 1D stress conditions:

$$\psi-\psi_0=-l(1-S_r) \tag{6.60}$$

The proportional constant Λ (see eq. 6.48) for the 3D model for unsaturated soils is expressed as

$$\Lambda=\frac{dF+d\psi}{(1+e_0)\left\{\frac{\partial F}{\partial t_{kk}}+\frac{G(\rho)}{t_N}+\frac{Q(\omega)}{t_N}\right\}}$$

$$=\frac{dF+l dS_r}{(1+e_0)\left\{\frac{\partial F}{\partial t_{kk}}+\frac{G(\rho)}{t_N}+\frac{Q(\omega)}{t_N}\right\}}=\frac{dF+l dS_r}{h^p} \qquad \left(\text{where} \quad dF=\frac{\partial F}{\partial \sigma''_{kl}}d\sigma''_{kl}\right) \tag{6.61}$$

Here, dS_r is given by eq. (3.72) using SWCC as explained in the 1D model. More detailed formulation of this model is described in Kikumoto et al. (2010) and Kyokawa (2010).

To check the applicability of the present modeling of unsaturated soils, some simulations of unsaturated soils under various loading paths including soaking processes were carried out. At the initial states, two kinds of Fujinomori clays are assumed: one is in a normally consolidated state (loose sample: $e_0 = 0.83$) and another is in an overconsolidated state (dense sample: $e_0 = 0.68$). The initial degree of saturation ($S_r = 0.80$), the initial mean net stress ($p^{net} = 98$ kPa), and the initial suction ($s = 110$ kPa) are the

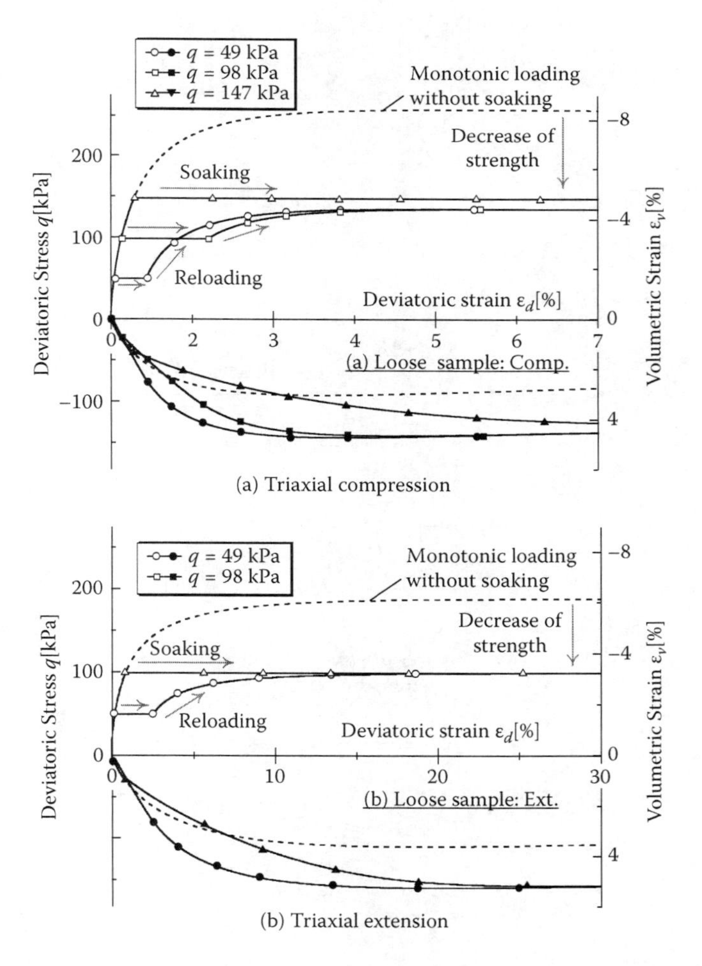

Figure 6.58 Simulation of drained triaxial tests of unsaturated clay with soaking (loose sample).

same for both samples. The material parameters in Tables 5.2 and 6.1 are used, and the additional material parameter is $l = 0.5$

These parameters are common with those used in the analyses in 1D conditions (Section 3.10.2). The clays are sheared up to a prescribed stress level (q = 49 kPa, 98 kPa, and 147 kPa) under constant mean stress (p^{net} = 98 kPa) and are soaked. The process of soaking is simulated by decreasing suction to zero under constant deviatoric stress. The clay that does not reach the failure state by the soaking is sheared again.

Figures 6.58(a) and (b) show the simulated stress–strain–dilatancy curves of triaxial compression and extension tests, including the soaking

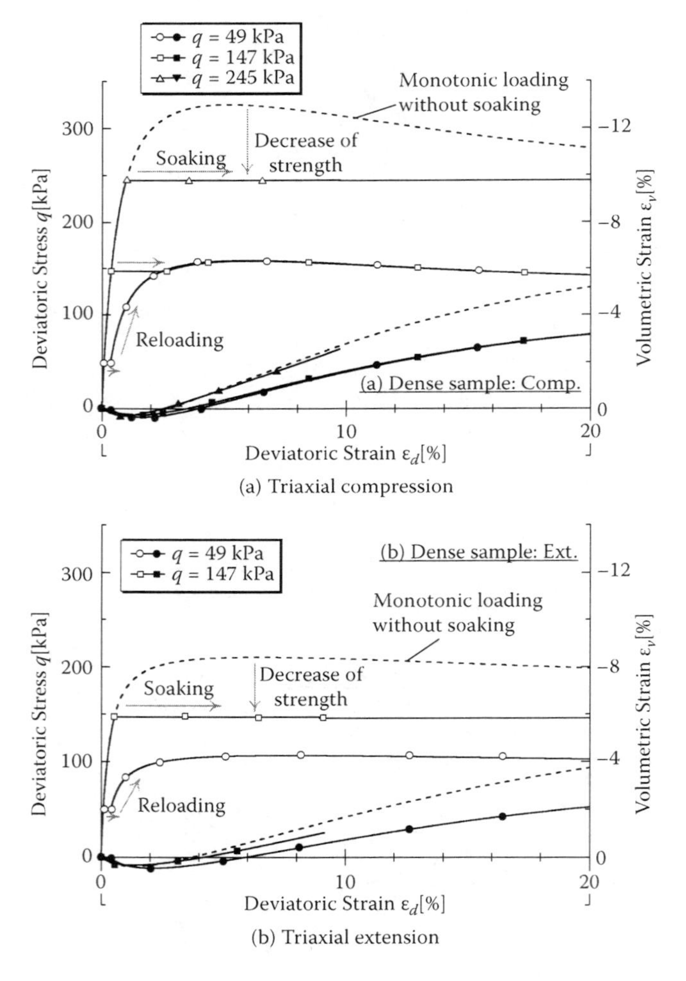

Figure 6.59 Simulation of drained triaxial tests of unsaturated clay with soaking (dense sample).

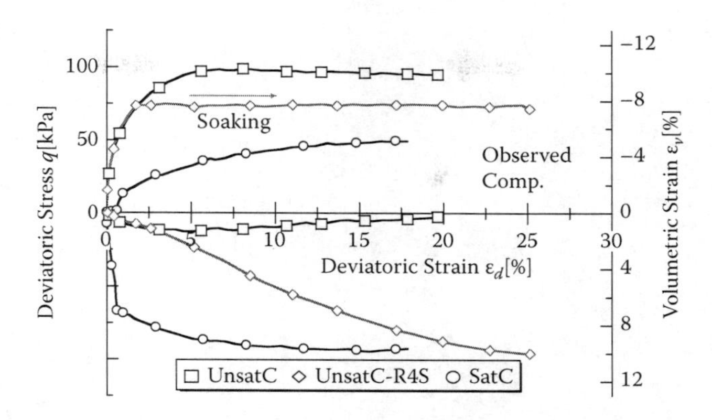

Figure 6.60 Observed results of triaxial tests of unsaturated kaolin clay with soaking.

process on loose sample. In these figures, the stress–strain–dilatancy curves without soaking are also drawn by broken lines. It can be seen that the stiffness and strength of unsaturated soil (broken curves) are much larger than those of soaked soils and are less contractive. Although during soaking the strains are not so large under small deviatoric stress (low stress ratio), the failure occurs by soaking in large deviatoric stress (high stress ratio). The simulation also shows that volume contraction occurs during the soaking process on the loose sample in the same way as in 1D conditions.

Figures 6.59(a) and (b) show the results of dense samples. Although general tendencies are the same as in the loose sand, volume expansion occurs in the soaking process in dense samples. It is possible to consider the influence of the intermediate principal stress on the deformation and strength characteristics of unsaturated soils because the model is formulated using the t_{ij} concept.

Figure 6.60 shows the observed results of triaxial tests on unsaturated kaolin clays. The initial condition is as follows: $e_0 \cong 1.5$ (loose sample), $p^{net} = 49$ kPa, and $s = 98$ kPa. Testing on unsaturated clay is carried out under constant net stress and constant suction. Here, UnsatC is the test without soaking, UnsatC-R4S is the test in which the clay is soaked at the stress ratio $\sigma_1/\sigma_3 = 4$, and SatC is the test in which the clay is first soaked (saturated) and then sheared. The observed behavior of unsaturated clay under shear loading including soaking corresponds to the simulation in Figure 6.58.

Chapter 7

Conclusions of Part I

In most geotechnical books and research papers about constitutive modeling of soils, the formulation of elastoplastic models is usually described under multidimensional conditions from the beginning—that is, how to determine yield surface, plastic potential function, hardening rule, flow rule, and others. In Part 1 of this book, the models were initially described under one-dimensional (1D) conditions for the sake of simplicity because such models can be formulated using one component of stress—the normal stress σ—and one component of strain—the normal strain ε (or the void ratio e). Then, it was shown that 1D models can be automatically extended to multidimensional ones by replacing the stress in the yield function of a 1D model by a scalar quantity defined by the stress invariants and assuming a flow rule. The summaries of each chapter in Part 1 follow.

In Chapter 2, some typical stress–strain behaviors of various materials and their modeling using elasticity and plasticity concepts were described under 1D conditions. Here, the meaning of fundamental concepts of elastoplastic theory, such as yield function, hardening parameter, consistency condition, and loading condition, were explained comprehensively. Next, it was shown that a 1D elastoplastic model can be extended to multidimensional stress conditions by assuming coaxiality and an appropriate flow rule.

In Chapter 3, after interpreting the well known 1D behavior of normally consolidated soils on the basis of conventional elastoplastic theory, the unified approaches to describe various features of 1D soil behavior were formulated based on the concepts of advanced elastoplasticity. These are applicable not only for normally consolidated soils, but also for overconsolidated soils and structured soils.

Firstly, a simple method to describe the behavior of overconsolidated soils was presented by introducing one state variable (ρ) related to density and by assuming a monotonic evolution rule for this variable (advanced model at stage I). This formulation is, in a sense, a 1D interpretation of the subloading surface concept proposed by Hashiguchi (1980).

Next, another state variable (ω), which represents the effect of bonding, was introduced. Assuming a monotonic evolution rule in the same way as that of the state variable (ρ), the behavior of structured soils such as naturally deposited clays was modeled (advanced model at stage II).

Since it is experimentally known that the normal consolidation line (NCL) in the e–lnσ relation shifts depending on effects such as strain rate, temperature, and suction (saturation), another state variable (ψ) that changes the position of the NCL was defined. Then, using this state variable, together with ρ and ω, a general method was proposed to consider the dependence of soil behavior on time, temperature, and degree of saturation in the constitutive model (advanced model at stage III).

As an application of the approach described, a unified, time-dependent model for normally consolidated, overconsolidated, and structured soils was presented. Time-dependent behavior was formulated without using the previous viscoplastic theories, such as nonstationary flow surface and overstress concepts. Rather, the state variable ψ was introduced and the subloading surface concept was utilized. Further, applications of the present methods to the modeling of other features such as temperature-dependent behavior and unsaturated soil behavior were presented.

The validity of these models (stage I and stage III) was confirmed by the simulations of various 1D tests in clays with different initial densities, bonding effects, and strain rates. Also, it was checked by soil-water coupled finite element analyses, where the computed results correspond to the well known consolidation features, including secondary consolidation, in oedometer tests for normally consolidated clays, overconsolidated remolded clays, and structured clays. Some simulations of 1D temperature-dependent behavior and suction-dependent behaviors of soils using the present modeling approach were also shown.

In Chapter 4, the framework of ordinary elastoplastic models formulated using stress invariants (p and q), such as the Cam clay model, was first presented. Then, it was explained that the Cam clay model is a three-dimensional (3D) extension of the well known linear relation between void ratio e and logσ for remolded normally consolidated soils. From this point of view, the formulations of the original and modified Cam clay models were presented. The applicability of the Cam clay as a 3D model was discussed based on test data of clay and sand. It was shown that models using the stress invariants p and q, such as the Cam clay model, cannot properly take into consideration the influence of intermediate principal stress on the deformation and strength of soils.

In Chapter 5, a simple and unified method to describe soil behavior under general 3D stress conditions was presented based on the t_{ij} concept. Using this concept, 1D models for normally consolidated soils (or the Cam clay model) were extended to 3D stress conditions in which the influence of the intermediate principal stress on soil deformation and strength is properly taken into consideration. The physical meaning of the t_{ij} concept was also discussed.

The modified stress t_{ij} is defined by the product of a_{ik} and σ_{kj}, where σ_{kj} is the Cauchy stress tensor and a_{ik} is the symmetric stress ratio tensor whose principal values are given by the direction cosines of the spatially mobilized plane (SMP). It was shown that, by formulating the yield function using the stress invariants of the t_{ij} tensor (t_N and t_S) instead of the ordinary stress invariants (p and q) and assuming the flow rule not in the Cauchy stress (σ_{ij}) space but rather in the modified stress (t_{ij}) space, any 1D model and any multidimensional model using p and q can be transformed to unified 3D models. Here, t_N and t_S are defined by the normal and in-plane components of t_{ij} to the SMP. The modified stress t_{ij} is, in a sense, a modified stress reflecting the induced anisotropy. The intrinsic problems of the model formulated using p and q and the advantages of the model based on the t_{ij} concept in numerical calculations were also explained.

The material parameters of the model based on the t_{ij} concept are the same as those of the Cam clay model. The validity of the model was confirmed by test data not only of the triaxial compression and extension tests but also of the true triaxial tests on normally consolidated clays.

In Chapter 6, the advanced 1D models (stages I–III), in which various features of soil behavior are taken into account, were extended to the 3D conditions, introducing the t_{ij} concept. In addition to the modeling of the features common to those in 1D conditions, the influence of stress path dependency on the direction of plastic flow, which is a feature of soil in multidimensional conditions, was also modeled. The validity of these models was confirmed by the comparison with various shear and consolidation tests on sand and clay and the simulations of various 3D soil features, including time-dependent behavior of normally consolidated clays and naturally deposited clays.

The influence of density was considered, introducing the state variable ρ in the same way as that in the 1D condition. Further, the influence of stress path on the direction of plastic flow is taken into consideration by dividing the plastic strain increment into two components: the component that satisfies the associated flow rule in t_{ij} space and the isotropic component under increasing mean stress. The model (advanced model at stage I) was formulated so as to coincide with the previous model, named the "subloading t_{ij} model" (Nakai and Hinokio 2004). The applicability of the model was confirmed by triaxial compression and extension tests on sand and clay with different densities, true triaxial tests on sand and clay, plane strain tests on clay, and torsional shear tests on clay.

The other soil features under 3D conditions were modeled in the same way as those in 1D conditions. The 3D model (advanced model at stage II) to consider the bonding effect corresponds to the previous "subloading t_{ij} model for structured soil" (Nakai 2007). The time-dependent model (advanced model at stage III) can describe not only the various strain rate effects under shear loading but also the typical creep behavior in the 3D condition (transient creep, stationary creep, and accelerating creep). It was shown that temperature-dependent behavior and unsaturated soil behavior also can be modeled by the advanced modeling in stage III. The advanced models in stages II and III were validated by simulations of various soil features in 3D conditions and their comparisons with the observed results reported in the literature.

Among the features of soil behavior described in Chapter 1, stress-induced anisotropy and cyclic loading have been usually modeled by employing the kinematic hardening rule and/or rotational hardening rule in most constitutive models. The author and collaborators also developed models using a rotational hardening rule in t_{ij} space, referring to the stress ratio tensor x_{ij} and others described in Table 5.1 (Nakai, Fujii, and Taki 1989; Nakai and Hoshikawa 1991; Nakai, Taki, and Funada 1993; Nakai and Muir Wood 1994; Chowdhury et al. 1999). However, as mentioned in Section 5.2, since the nature of the influence of the intermediate principal stress and the nature of stress-induced anisotropy seem to be the same, the author considers that the induced anisotropy should be described by the concept of the generalized modified stress t_{ij}. Therefore, modeling using rotational hardening is not included in this book. Rational modeling of induced anisotropy for geomaterials will be a future task.

Part 2

Numerical and physical modeling of geotechnical problems

Chapter 8

Introduction to numerical and physical modeling

8.1 INTRODUCTION TO PART 2

One of the main goals of constitutive modeling is its ultimate application in the analysis of practical geotechnical engineering problems with a high degree of confidence. On the other hand, most practical designs of earth structures, foundations, and countermeasures against earth disasters, among others, are based on elastic theory and/or rigid plastic theory, in which the deformation characteristics of geomaterials described in Part 1 are not considered. Following the development of the Cam clay model, nonlinear elastoplastic analyses have been carried out to solve boundary value problems. However, applications to practical design have been limited because most of the constitutive models used in the previous analysis cannot describe the soil behaviors comprehensively.

In Part 2 of this book, the constitutive model developed in Part 1 is used to analyze typical geotechnical problems, such as tunneling, open excavation, earth pressures, bearing capacity, and reinforced soils, to check its applicability to these applications and also investigate the associated deformation and failure mechanisms. Two-dimensional (2D) and three-dimensional (3D) finite element codes used in the analyses of these problems are called "FEMtij-2D" and "FEMtij-3D," respectively, and have been developed at the Nagoya Institute of Technology (NIT). Also, corresponding 2D and 3D physical model tests were carried out to validate the analyses.

Chapter 9 deals with various tunneling problems. Two-dimensional model tests of trapdoor problems and their corresponding numerical analyses were first carried out to clarify the fundamental mechanism of ground deformation and earth pressure development due to tunneling. Currently, shallow, large cross-sectional underground openings are needed, particularly in urban areas. Such large underground openings can only be realized by tunneling methods that involve multistage excavations of a number of smaller tunnels. To investigate the associated ground deformation and earth pressure development, both 2D model tests and numerical analyses of multiple trapdoor problems are performed in which the order of lowering each trapdoor is changed. Also, 3D model tests and numerical analyses are conducted in which a continuous excavation process in tunneling direction is simulated highlighting the 3D effect in excavation. In addition to the numerical and experimental studies on the trapdoor problem, circular tunneling excavation is considered. The influences of soil cover and existing building loads on ground behavior and earth pressure are discussed.

Chapter 10 relates to 2D numerical and experimental studies of earth pressure in various problems such as in retaining walls, braced open excavations, and bearing capacity of foundations. In particular, the earth pressure developed as a function of various wall deflection processes and modes and induced surface settlements is examined. In braced open excavations, not only the behavior of the ground and structure (wall and struts) during excavation in greenfield conditions (no nearby existing building loads) but also the influence of the existing building loads on these behaviors are investigated numerically and experimentally. As for bearing capacity problems, the behavior of foundations such as footing foundations, piled foundations, and piled raft foundations subjected to various kinds of loads (concentric load, eccentric load, and inclined load) is discussed.

Chapter 11 deals with reinforced soils. We examine 2D and 3D model tests and the corresponding numerical analyses of the caisson type foundation with reinforcements of steel bars under uplift loadings. Such types of foundations are already used to support electric transmission towers, but the reinforcing mechanism of steel bars is not clear. Based on both numerical and experimental studies, effective reinforcing methods for such types of foundations are presented. The other objective of this chapter is to clarify the mechanisms of the reinforcing earth method using geosynthetics to increase bearing capacity. For this purpose, 2D model tests and numerical analyses were carried out under different installation depths of reinforcement, different lengths of reinforcement, and different roughness of reinforcement.

Chapter 12 deals with the numerical simulation of localization and shear band development in plane strain tests and triaxial tests. Numerical simulations based on infinitesimal and finite deformation theories were carried out, assuming normally consolidated and overconsolidated Fujinomori clay described in Part 1 to investigate strain localization phenomenon in geomaterials under drained conditions.

To obtain reliable results from numerical analyses of soil–structure interaction problems, not only a comprehensive constitutive model of geomaterials, but also a proper description of the slip behavior at the interface between soil and structure is needed. In Section 8.2, a formulation of the elastoplastic joint element is presented, which can describe realistic slippage behavior at the interface.

Except for Chapter 12, 2D and 3D model tests and the corresponding numerical analyses are conducted. In every 2D model test, aluminum rods are used as an analogue material to represent the ground. In 3D model tests, alumina balls are used. The stress–strain–dilatancy curves of these materials obtained from shear tests and the material parameters used in the analyses are described in Section 8.3.

8.2 MODELING THE INTERFACE BEHAVIOR BETWEEN SOIL AND STRUCTURE

For simulating discontinuities between two layers, joint elements are usually used. In many soil–structure interface problems such as shallow or deep foundation, tunnel lining, braced excavation, or soil reinforcement, joint elements are adopted to simulate the interface of two different solids (Goodman, Taylor, and Brekke 1968). Although joint elements do not account for joint rotations explicitly, they are flat elements resembling a line without thickness. They have no resistance to tensile forces, but offer high resistance to compression in the normal direction to the interface.

Figure 8.1 shows the discontinuity in the boundary of two solids. The behavior on the discontinuous surface between the elements A and B is described by an elastoplastic joint element that can represent the frictional behavior; that is, the behavior of the discontinuous surface is uniform before reaching the ratio of tangential and normal force per unit length (p_s/p_n) to the coefficient of friction $\tan\delta$ (δ = friction angle of the interface). Once p_s/p_n reaches $\tan\delta$, slip will occur on the interface, keeping the value of (p_s/p_n) = $\tan\delta$. Therefore, the stiffness matrix of the joint element should be formulated to account for this phenomenon (see Figure 8.2).

The increment of stresses of the joint element (shear stress increment dp_s and normal stress increment dp_n) can be related to relative displacements of the joint element (increment of tangential component dw_s and normal component dw_n) by the following equation:

$$\begin{Bmatrix} dp_s \\ dp_n \end{Bmatrix} = [D_J]\begin{Bmatrix} dw_s \\ dw_n \end{Bmatrix} = \begin{bmatrix} k_{11} & k_{12} \\ k_{21} & k_{22} \end{bmatrix}\begin{Bmatrix} dw_s \\ dw_n \end{Bmatrix} \tag{8.1}$$

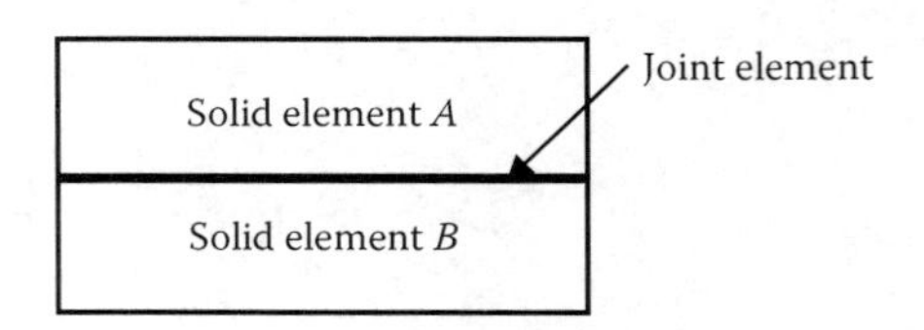

Figure 8.1 Interface between two solid elements.

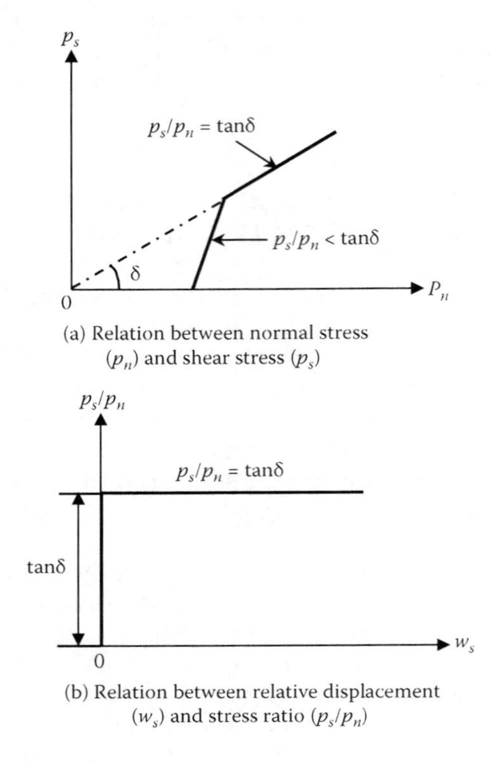

Figure 8.2 Ideal behavior at the interface.

As shown in Figure 8.2, the matrix $[D_J]$ is formulated so that slip does not occur when $p_s/p_n < \tan\delta$, but it occurs when $p_s/p_n = \tan\delta$ (Nakai 1985).

Case 1: $p_s/p_n < \tan\delta$

When $p_s/p_n < \tan\delta$, there is almost no relative displacement at the boundary surface both in tangential and normal directions. Therefore, the displacement of joint elements can be considered as infinitesimal elastic displacement. The formulation for this condition can be described as

$$dw_s = dw_s^e = \frac{1}{k_s} dp_s \tag{8.2}$$

$$dw_n = dw_n^e = \frac{1}{k_n} dp_n \tag{8.3}$$

Here, k_s is the unit stiffness along the joint element and k_n is the unit stiffness across, and large positive values should be assumed for k_s and k_n. In the equations for the formulation, superscripts e and p denote the elastic and plastic components, respectively. The matrix $[D_j]$ then can be given by

$$\begin{Bmatrix} dp_s \\ dp_n \end{Bmatrix} = \begin{bmatrix} k_s & 0 \\ 0 & k_n \end{bmatrix} \begin{Bmatrix} dw_s \\ dw_n \end{Bmatrix} = [D_J] \begin{Bmatrix} dw_s \\ dw_n \end{Bmatrix} \tag{8.4}$$

Case 2: $p_s/p_n = \tan\delta$

The normal displacement of the boundary is the same as in case 1—that is, eq. (8.3). However, plastic tangential displacement w_s^p occurs after p_s/p_n reaches $\tan\delta$, as shown in Figure 8.2. The relation of w_s^p and p_s/p_n is then given by the following form:

$$\frac{p_s}{p_n} = \xi w_s^p + \tan\delta \tag{8.5}$$

Here, ξ is a multiplier. When the value of ξ is very close to zero, $\tan\delta$ prevails in the right side of eq. (8.5). Then, the perfectly plastic condition shown in Figure 8.2 can approximately be expressed by making ξ a positive value close to zero. Differentiating and rearranging eq. (8.5), the increment of $w_s{}^p$ can be expressed by the following equation:

$$dw_s^p = \frac{1}{\xi} d\left(\frac{p_s}{p_n}\right) = \frac{1}{\xi} d\left(\frac{1}{p_n} dp_s - \frac{p_s}{p_n^2} dp_n\right) \tag{8.6}$$

The total increment of tangential relative displacement is expressed by the summation of the elastic component in eq. (8.2) and the plastic component in eq. (8.6):

$$dw_s = dw_s^e + dw_s^p = \left(\frac{1}{k_s} + \frac{1}{\xi p_n}\right) dp_s - \frac{p_s}{\xi p_n^2} dp_n \tag{8.7}$$

Combining eqs. (8.3) and (8.7), the relation between the relative displacement increment and the stress increment of the joint element is expressed as

$$\begin{Bmatrix} dw_s \\ dw_n \end{Bmatrix} = \begin{bmatrix} \dfrac{1}{k_s} + \dfrac{1}{\xi p_n} & -\dfrac{p_s}{\xi p_n^2} \\ 0 & \dfrac{1}{k_n} \end{bmatrix} \begin{Bmatrix} dp_s \\ dp_n \end{Bmatrix} \tag{8.8}$$

Therefore, the matrix $[D_J]$ of the joint element is expressed as the inverse matrix of eq. (8.8):

$$\begin{Bmatrix} dp_s \\ dp_n \end{Bmatrix} = \begin{bmatrix} \dfrac{\xi k_s p_n}{k_s + \xi p_n} & \dfrac{k_s k_n p_s}{k_s P_n + \xi p_n^2} \\ 0 & k_n \end{bmatrix} \begin{Bmatrix} dw_s \\ dw_n \end{Bmatrix} = [D_J] \begin{Bmatrix} dw_s \\ dw_n \end{Bmatrix} \tag{8.9}$$

When p_n becomes zero or reaches a negative value, eq. (8.4) is adopted and very small positive values of k_n and k_s are employed for describing separation at the interface. Also, when $dw_s \cdot p_s < 0$ (the direction of slippage is opposite to that of p_s), there is no slippage at the interface.

Figure 8.3 shows the calculated results of the elastoplastic joint element. Here, the parameters used in the joint element are as follows: $k_s = 2.0 \times 10^4 (\times 98$ kPa/ m), $k_n = 2.0 \times 10^4 (\times 98$ kPa/cm), $\xi = 0.01$, and $\delta = 20°$. Circle dots imply the case in which an increment of relative displacement is applied with increasing normal stress on the joint element ($dp_n/dw_s = 5000 \times 98$ kPa/ cm) from the initial condition of $p_n = 98$ kPa and $p_s = 0$ kPa; triangle dots imply the case in which an increment of relative displacement is applied with decreasing normal stress ($dp_n/dw_s = -5000 \times 98$ kPa/cm). It can be understood that the present elastoplastic joint element simulates the frictional behavior at interface under increasing and decreasing normal stress with adequate accuracy.

8.3 GEOMATERIALS USED IN 2D AND 3D MODEL TESTS AND THEIR MATERIAL PARAMETERS

The elastoplastic constitutive model for soils, named the subloading t_{ij} model (advanced model at stage I described in Section 6.2; Nakai and Hinokio 2004), is used in the finite element analyses in Part 2. As mentioned in Section 6.2, this model can describe properly the following typical characteristics of soils in spite of the use of a small number of parameters:

- Influence of intermediate principal stress on the deformation and strength of geomaterials

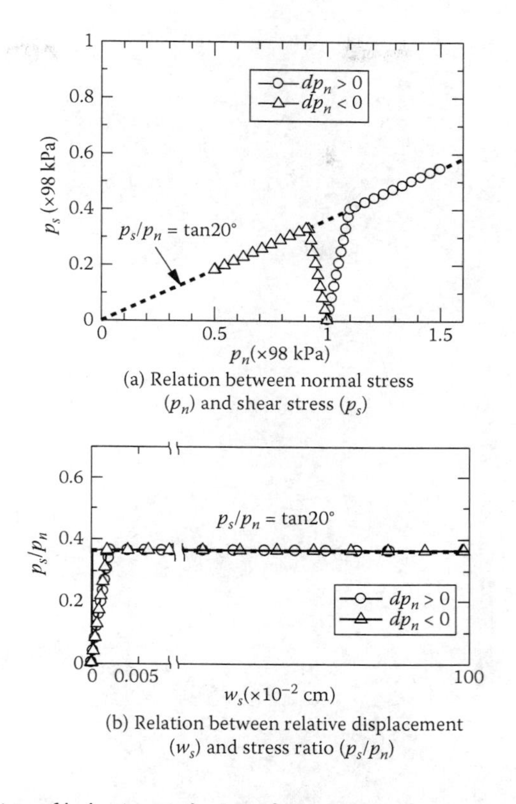

(a) Relation between normal stress (p_n) and shear stress (p_s)

(b) Relation between relative displacement (w_s) and stress ratio (p_s/p_n)

Figure 8.3 Simulation of behavior at the interface.

- Dependency of the direction of plastic flow on the stress path
- Negative and positive dilatancy during strain hardening
- Influence of density and/or confining pressure on the deformation and strength

8.3.1 2D model ground

The 2D model ground consists of a mix of 1.6 and 3.0 mm aluminum rods in the ratio of 3:2 by weight. The ground is prepared by piling up the stack of aluminum rods from the bottom. The unit weight of the model ground is 20.4 kN/m^3. Figure 8.4 shows (a) the aluminum rod mass and (b) the biaxial test apparatus to determine the material parameters for aluminum rod mass. Material parameters for the aluminum rod mass (2D ground) are shown in Table 8.1. The stress–strain–dilatancy curves for the biaxial tests and simulations with the model are shown in Figure 8.5. Diagram (a) show the results of

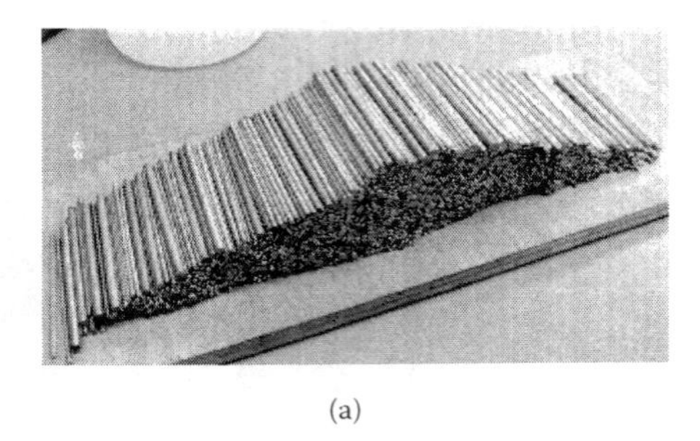

(a)

(b)

Figure 8.4 Material used in 2D model ground. (a) Aluminum rod mass; (b) biaxial test of aluminum rod mass.

Table 8.1 Values of material parameters for aluminum rod mass

λ	0.008
κ	0.004
$N(e_N$ at $p = 98$ kPa)	0.3
$R_{CS} = (\sigma_1/\sigma_3)_{CS(comp.)}$	1.8
ν_e	0.2
β	1.2
$a/(\lambda - \kappa)$	1300

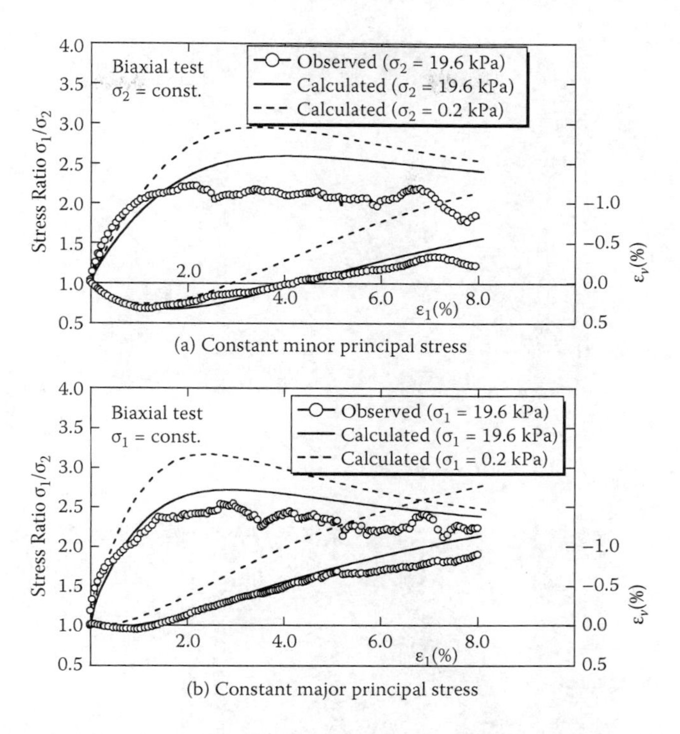

Figure 8.5 Test results of aluminum rod mass and calculated result using the material parameters in Table 8.1.

a constant minor principal test (σ_2 = const.) and diagram (b) shows the results of a constant major principal stress test (σ_1 = const.). Simulations for very low confining pressures (σ_1 or $\sigma_2 = 0.2$ kPa), representative of the stress level to be observed in the model tests, are also shown in the dotted curve in these figures.

The mass of aluminum rods behaves rather like a medium to dense granular soil, initially showing negative dilatancy followed by positive dilatancy. It is also seen that the dependency of stiffness, strength, and dilatancy on the confining pressure is described by the present constitutive model. In all analyses of 2D mode tests of tunneling, retaining wall, braced excavation, bearing capacity, and reinforced soils (described later), the common material parameters in Table 8.1 are used.

8.3.2 3D model ground

The 3D model ground is prepared in the following way: The mass of alumina balls (see Figure 8.6a) have diameters of 2.0 and 3.0 mm; it is mixed in the ratio of 1:1 in weight, filled in the funnel with a rubber hose (diameter

(a)

(b)

Figure 8.6 Material used in 3D model ground. (a) Mass of alumina balls; (b) triaxial compression test of mass of alumina balls.

Table 8.2 Values of material parameters for mass of alumina balls

λ	0.024
κ	0.014
$N(e_N$ at $p = 98$ kPa)	0.78
$R_{CS} = (\sigma_1/\sigma_3)_{CS(comp.)}$	2.0
ν_e	0.2
β	2.0
$a/(\lambda - \kappa)$	150

is 5 cm) attached to the end, and poured into the frame through the rubber hose. Here, the discharge opening of the hose is adjusted to be just above the ground surface so that alumina balls are deposited continuously without compaction. The unit weight of the mass of alumina balls prepared in such a way is 21.5–22.3 kN/m^3.

Table 8.2 shows the value of the material parameter of the mass of alumina balls (3D ground), which are determined by the triaxial tests (Figure 8.6b). Figure 8.7 shows the observed stress–strain–dilatancy

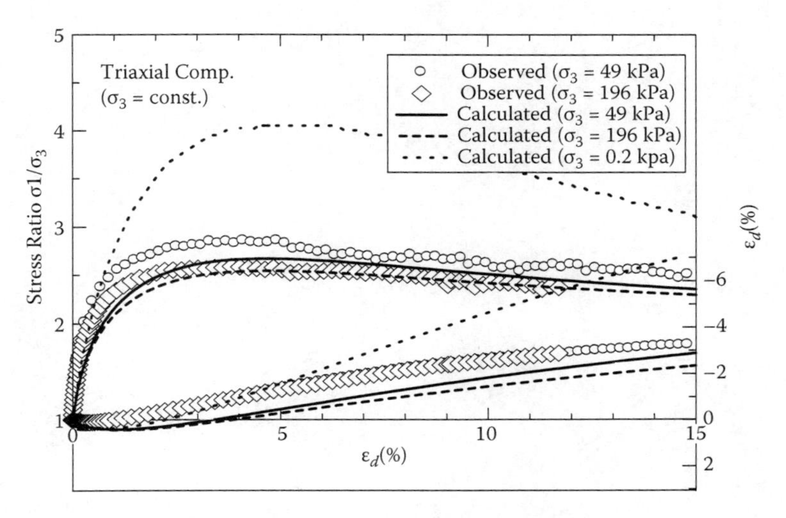

Figure 8.7 Test results of mass of alumina balls and calculated result using the material parameters in Table 8.2.

relation of triaxial compression tests (σ_3 = const.) and the corresponding calculated curves using the material parameters in Table 8.2, together with the calculated results under very low confining pressure. As shown in this figure, the behavior of the mass of alumina balls is also similar to that of the well known medium dense sand, in the same way as the aluminum rod mass. The 3D analyses in Section 11.1 are carried out using the material parameters in Table 8.2.

Chapter 9

Tunneling

9.1 TWO-DIMENSIONAL TRAPDOOR PROBLEMS

Two-dimensional (2D) model tests of tunnel excavation with corresponding numerical analyses were carried out to investigate fundamental mechanisms in ground behavior during tunneling. Numerical analyses were performed with the finite element method using the elastoplastic subloading t_{ij} model described in Section 6.2. Both model tests and numerical analyses were performed for two series of tunnel excavations. In the first series (series I), different tunnel depths were examined in order to investigate the influence of single tunnel excavation on surface settlement and earth pressure considering the ground as greenfield land (there is no structure on the ground surface above the tunnel).

Further, note that large underground openings in shallow ground necessitate multistaged excavations. Thus, in the second series (series II), the influence of construction sequence on surface settlement and earth pressure in tunneling with a large cross section was investigated for two different excavation sequences. In one case, the middle section was first excavated, followed by the removal of the two side sections. In the other case, the two side sections were excavated before excavation of the middle section. More detailed experimental and numerical results are described in Shahin, Nakai, Hinokio, Kurimoto, et al. (2004).

9.1.1 Outline of model tests

Trapdoor models have been used to investigate the mechanism of tunneling problems by many researchers (among them, Murayama and Matsuoka 1971; Adachi et al. 1994). Two-dimensional model tests are carried out to investigate the basic mechanism of the ground behavior and earth pressure for a tunnel excavation. The trapdoor model in Figure 9.1 is used. The model setup consists of 10 brass blocks (blocks A to J), each of which is 8 cm in width, set along the centerline of an iron table. These blocks can be moved upward and downward individually or simultaneously.

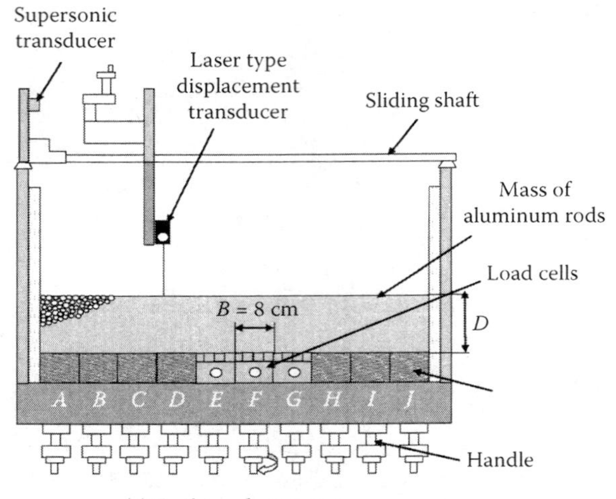

(a) Outline of apparatus

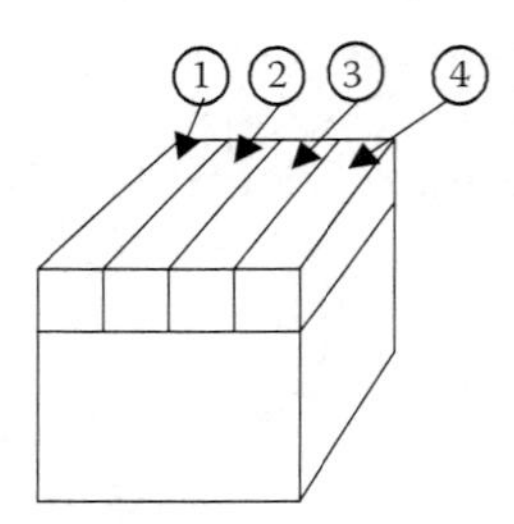

(b) Block with load cells

Figure 9.1 Trapdoor apparatus.

For measuring earth pressures, the blocks with load cells are placed just above the brass blocks by lowering them so that the top surfaces of all blocks, including the blocks with load cells, are in the same vertical elevation. The load cells can be placed at any of these blocks.

Each block with load cells is divided into four small parts to measure the earth pressure at four points in the block, which is shown in Figure 9.1(b). The downward movement of each block during the experiment is measured with an electronic transducer attached to the handle of a lowering block. Surface settlement is measured by using a laser type displacement transducer with accuracy of 0.01 mm. The position of the laser transducer is measured by using a supersonic wave transducer to get continuous readings of surface settlements. All data are recorded in a personal computer through a data logger. By taking photographs of the ground with a digital camera during the experiment, the deformation pattern inside the ground

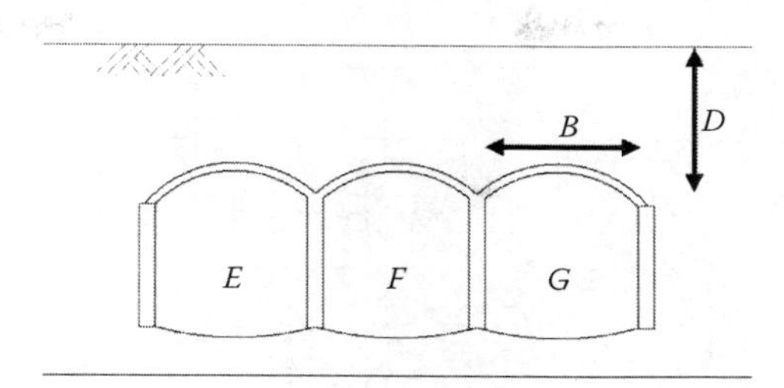

Figure 9.2 Shallow tunnel with large cross section.

can be visualized. In the model tests, aluminum rod mass is used as soil mass. The unit weight of the aluminum rod mass is 20.4 kN/m^3 at the experimental stress level. During preparation of the ground of each test, special attention is paid to getting a uniform ground.

Model tests can be performed with different depth-to-width ratios by varying the ground depth. In series I (single block excavation), model tests are conducted for four values of soil covers, $D/B = 0.5$, 1.0, 2.0, and 3.0, where D is the depth from the ground surface to the top of the tunnel and B (= 8 cm) is the width of the tunnel. Downward displacement of $d = 4$ mm is imposed to a single block in Figure 9.1(a).

In series II, two kinds of soil covers ($D/B = 0.5$ and 2.0) are employed. Shallow tunneling with large cross sections, as shown in Figure 9.2, is simulated by lowering three blocks, corresponding to the blocks E, F, and G of the trapdoor in Figure 9.1(a). In this series, excavation is carried out in two patterns, called case 1 and case 2. In case 1, the downward displacement of $d = 4$ mm is sequentially applied to block F, then to block E, and later to block G, in order. In case 2, the lowering sequence starts with block E, then block G, and finally block F. The intention of this series is to investigate the influence of the excavation sequences on surface settlement and earth pressure during tunnel excavation. The patterns of model tests in series I and II are tabulated in Table 9.1.

9.1.2 Outline of numerical analyses

Numerical analyses are conducted for the same scale of model tests considering plane strain conditions. Figure 9.3 shows the meshes for the ground in the case of $D/B = 2.0$. Isoparametric square elements of 1 cm a side are

Table 9.1 Patterns of excavations

Series	*Type of excavation*	*D/B*			
I	Single block excavation	0.5	1.0	2.0	3.0
II	Combination of three blocks	0.5		2.0	

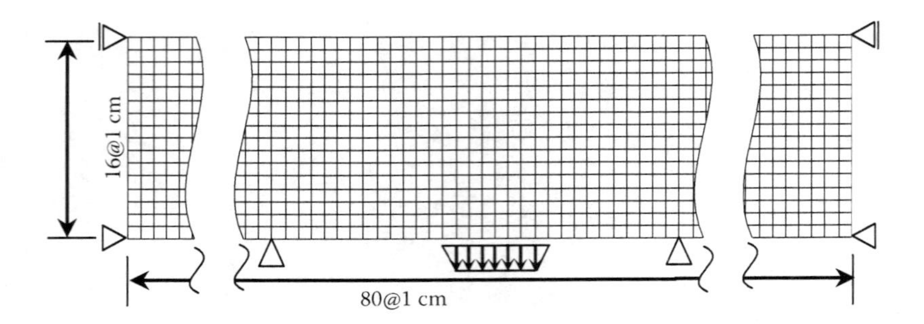

Figure 9.3 Mesh for finite element analyses: series I ($D/B = 2.0$).

used in the mesh of 2D finite element analyses. Both vertical faces of the mesh are free in the vertical direction, and the bottom face is kept fixed. To simulate the lowering of the blocks in the numerical analyses, vertical displacements are imposed at the nodal points, which correspond to the top of the lowering blocks in the model tests. Vertical displacements at the nodes are applied in increments of 0.002 mm. In series I, vertical displacements of 4 mm are imposed at the bottom nodes corresponding to block F, as shown in Figure 9.3. Figure 9.4 shows the simulation process of the excavation sequences for case 1 and case 2 in series II.

The initial stresses, corresponding to the geostatic (self-weight) condition, are assigned to the ground in all numerical analyses. This is accomplished by imposing body forces ($\gamma = 20.4$ kN/m^3) to all elements under one-dimensional (1D) consolidation conditions, starting from a negligible confining pressure ($p_0 = 9.8 \times 10^{-6}$ kPa) and an initial void ratio of $e = 0.35$.

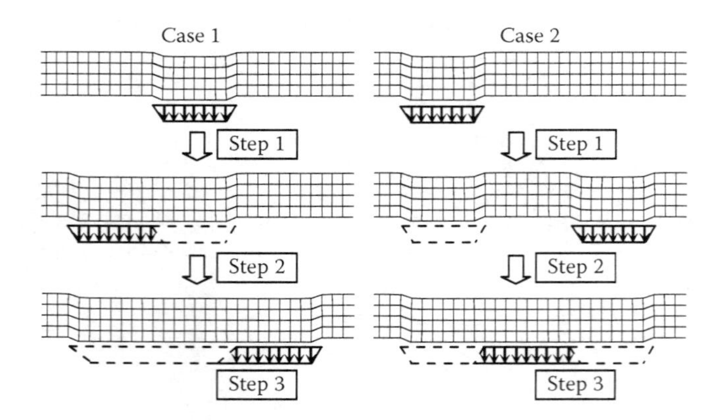

Figure 9.4 Simulation process in finite element analyses: series II ($D/B = 0.5$).

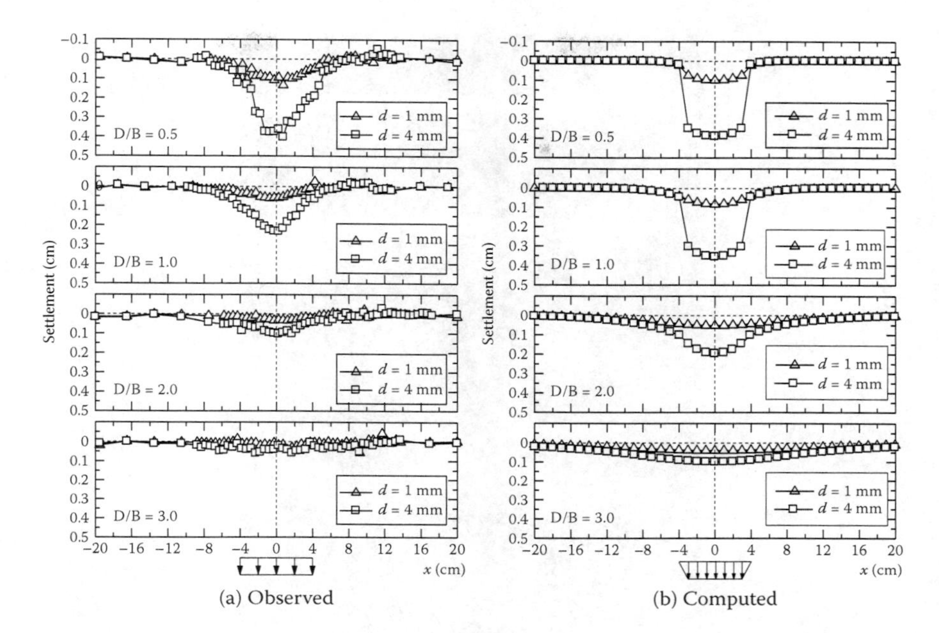

Figure 9.5 Profiles of surface settlement: series I.

9.1.3 Results and discussion

9.1.3.1 Series I (single block excavation)

Figure 9.5(a) shows the surface settlement troughs of the model tests for applied displacements of 1 and 4 mm in the case of $D/B = 0.5$, 1.0, 2.0, and 3.0. Figure 9.5(b) represents the computed surface settlements corresponding to the observed results in Figure 9.5(a). The prescribed displacement pattern of the lowering block and its extent are indicated at the bottom of each figure. From the model test results, it is seen that, for $D/B = 0.5$, the surface settlement at the center is almost the same as the imposed displacement of the lowering block, and that the range of influence of surface settlements is rather small.

On the other hand, surface settlements become smaller with increase of the tunnel depth, but they extend over a wider region. The results of numerical analyses in Figure 9.5(b) show the same tendency of model tests not only in shape but also in quantity. Figure 9.6 shows the movements of the model ground in the tests for different depths. These are the superimposed photos that were taken with a digital camera during each test. Two photos are superimposed in a figure; the first one was taken before lowering the block and the other was taken after lowering the block by 4 mm. The difference between the two photos is darkened using an image-processing software. It is noticed

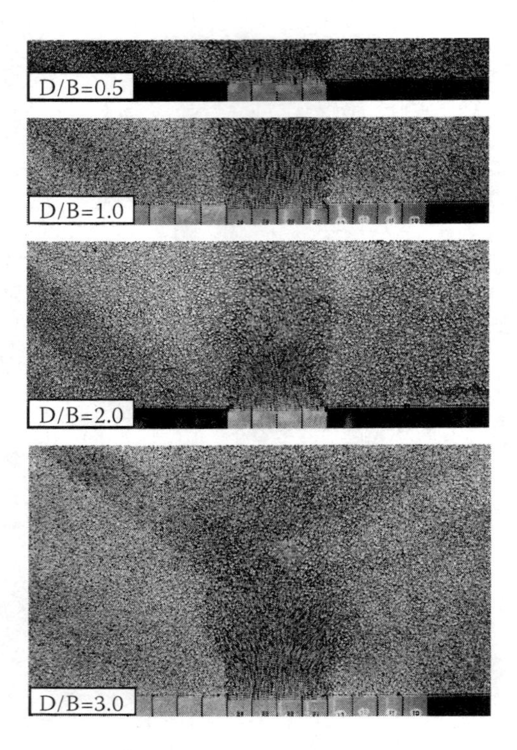

Figure 9.6 Deformed zones of the model ground: series I.

in this figure that the deformation zone spreads vertically from the top of the lowering block, and the range of this zone becomes wider for relatively deeper grounds. The movements of the ground in the deformation zone from the top of the lowering block to the ground surface are nearly uniform for $D/B = 0.5$ and 1.0. But, the ground is relatively less distorted near the surface compared to the top of the lowering block in the case of $D/B = 2.0$ and 3.0.

Figure 9.7 illustrates the displacement vectors computed in the numerical analyses. These figures are drawn with the same scale (shown at the bottom) for all ground depths. The shapes of the computed deformation zones for different ground depths are almost the same as the deformed zones in Figure 9.6.

Figures 9.8(a) and (b) show the distributions of the earth pressures observed in the model tests and computed in the numerical analyses, respectively. The earth pressures for different ground depths are plotted with the same scale of the actual earth pressure. The left vertical axis represents the earth pressures normalized by the initial vertical stress σ_{z0} (= γD) on the base level, where γ is the unit weight of ground materials and D is the depth of soil cover. The right vertical axis represents the actual values of earth pressure in

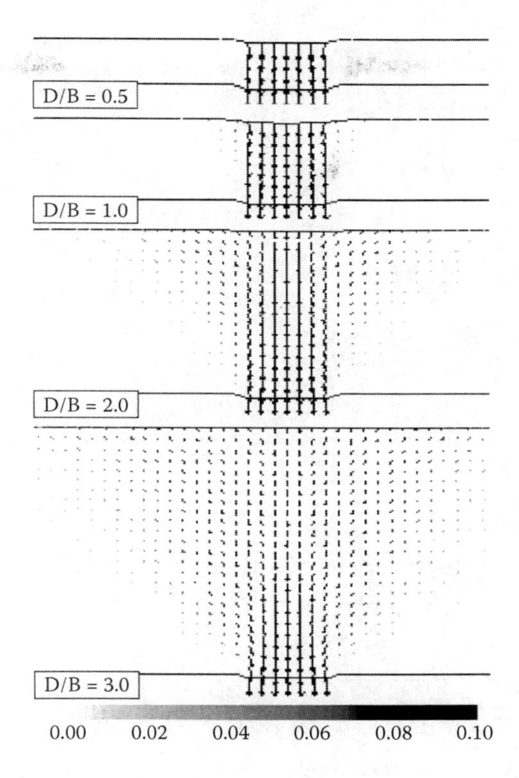

Figure 9.7 Displacement vectors of numerical ground: series I.

pascals. The dotted vertical line in these figures represents the center of the lowering block.

From both figures, for $D/B = 0.5$, almost no reduction of earth pressure at the excavation region is observed. But, for $D/B = 1.0$, the earth pressure on the lowering block decreases while increasing in the adjacent block due to an arching effect. However, the reduction of earth pressure at the central part of the lowering block is not so significant. In the case of $D/B = 2.0$ and 3.0, the arching effect is remarkable. The final earth pressures on the top of the lowering block for $D/B = 1.0$, 2.0, and 3.0 are almost the same. It is also noticed that the earth pressure on the lowering block decreases suddenly after applying only 0.05 mm displacement in this block. Further lowering the block, earth pressure on this block decreases gradually at a lower rate up to a certain limit, after which the earth pressure becomes almost constant. Sudden change in earth pressure is due to soil arching immediately after the ground is disturbed. Similarly to the surface settlements, earth pressures in the numerical analyses in Figure 9.8(b) show very

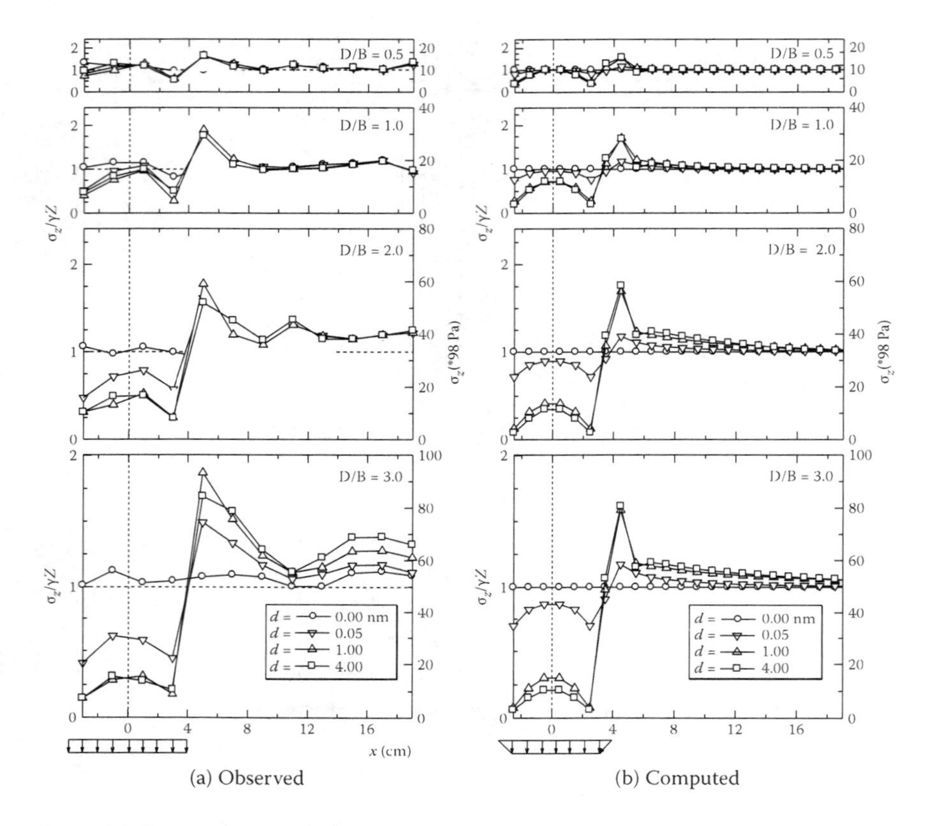

Figure 9.8 Earth pressure distributions on the trapdoor and basement: series I.

good quantitative and qualitative agreement with the results observed in the model tests in Figure 9.8(a).

Figure 9.9 illustrates the change of earth pressure at the place of the lowering block with applied displacement for different soil covers. This figure confirms the sharp change of earth pressure during tunnel excavation. The results of numerical analyses are in good agreement with the results of model tests. Earth pressures at the position of the lowering block are almost the same after lowering the block in the case of D/B = 1.0, 2.0, and 3.0. Hence, it can be concluded that, in 2D condition, if the soil cover is equal to or greater than the tunnel width, the acting earth pressure at the place of excavation becomes almost constant and independent of soil cover. The observed and computed results appear to be in agreement with

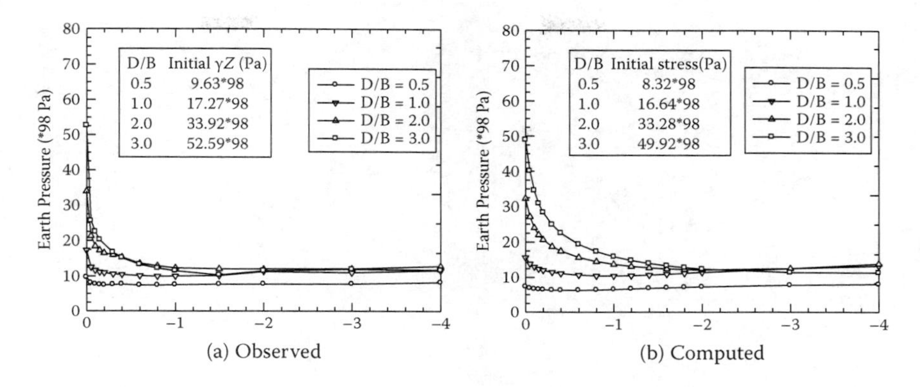

Figure 9.9 Earth pressure histories: series I.

the results of tunnel experiments performed by Murayama and Matsuoka (1971) and Adachi et al. (1994).

9.1.3.2 Series II (excavations with combination of three blocks)

The characteristics of surface settlement and earth pressure of a shallow tunnel with a large cross section are investigated by excavating three parallel tunnels in two different sequences. Figures 9.10(a) and (b) show the observed and computed surface settlements after lowering each block. Horizontal axes represent the distance from the center of the middle block. Legends indicate the amount of applied displacements. For $D/B = 2.0$, although the surface settlement at the center of the lowering block is about one-third of the imposed displacement after lowering one block, the final surface settlement at the center of the three blocks after lowering all three is almost equal to the imposed displacement. The surface settlement trough shows a smooth slope in the case of deeper soil cover ($D/B = 2.0$); however, it is rather steep for $D/B = 0.5$ because surface settlement occurs over a wider region for deeper excavations.

Final surface settlements in both case 1 and case 2 of series II are very close to each other. Although there is little difference between observed and computed results at the centerline of block F in the case of $D/B = 2.0$, the computed results explain well the characteristics of surface settlement troughs observed during the model tests at every step of excavation.

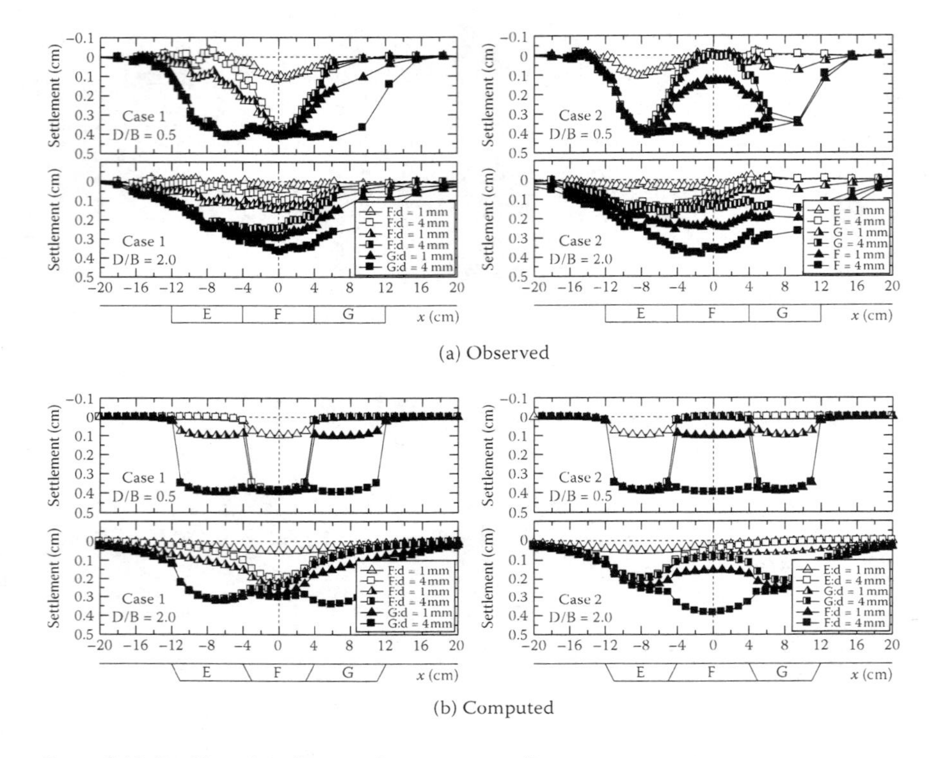

(a) Observed

(b) Computed

Figure 9.10 Profiles of surface settlement: series II.

Figure 9.11 shows the earth pressure distributions of case 1, and Figure 9.12 shows the same for case 2. Here, observed results in diagram (a) represent the earth pressures at blocks E, F, and G, since load cell is placed only at these positions. On the other hand, computed results in diagram (b) represent the earth pressures at all blocks, including blocks E to G. A gray frame is placed at the position corresponding to the values of observed results. In both figures, for all cases, computed results can simulate extremely well the observed results not only in shape but also in quantity. The observed and computed results reveal the following: Firstly, in the case of a very shallow tunnel ($D/B = 0.5$), lowering only one single block or more blocks in a sequence produces few changes in earth pressure distributions, despite some pressure decrease in the edges of lowering blocks and increase in the vicinity. In other words, in this case, much reduction of earth pressure due to arching cannot be expected.

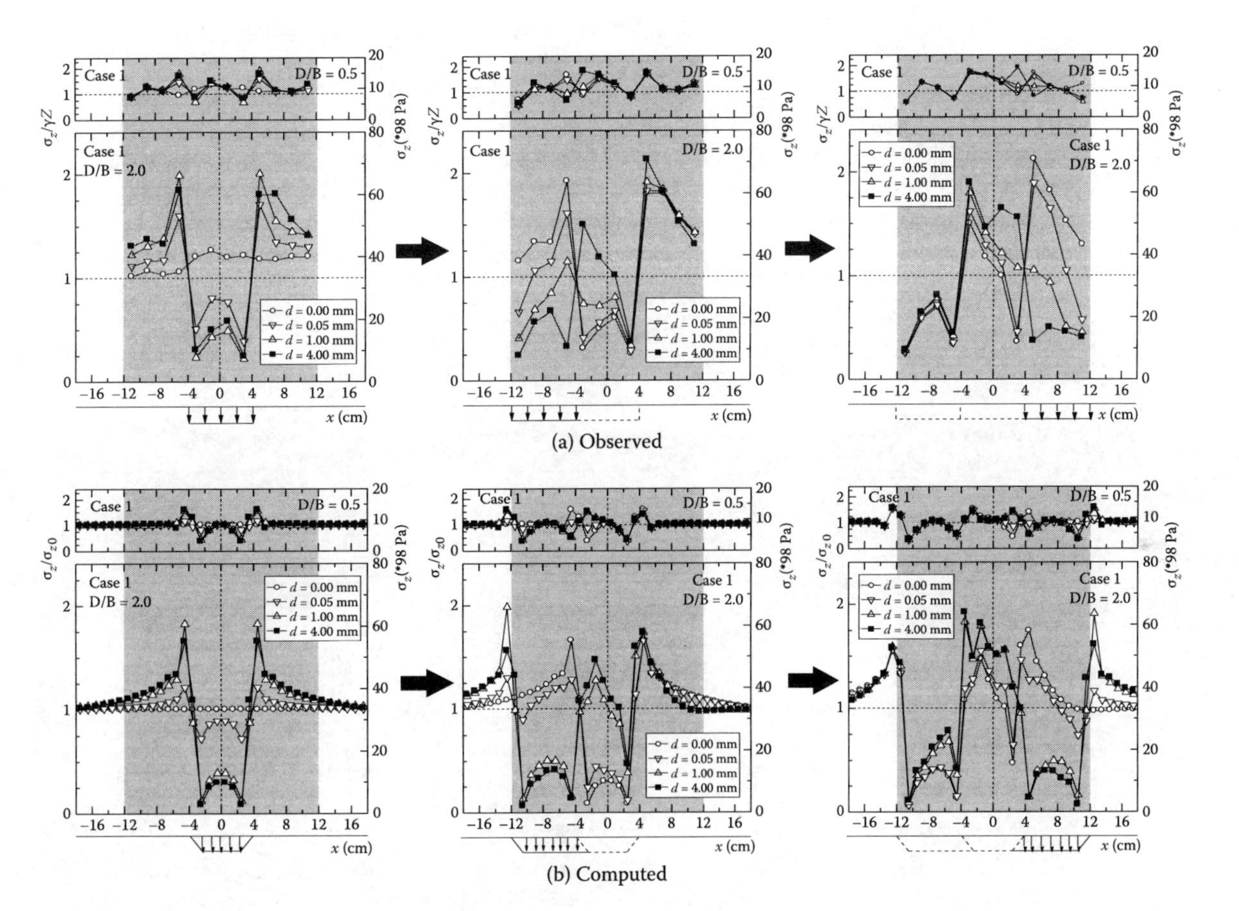

Figure 9.11 Earth pressure distributions—case 1 (sequence F-E-G): series II. (a) Observed; (b) computed.

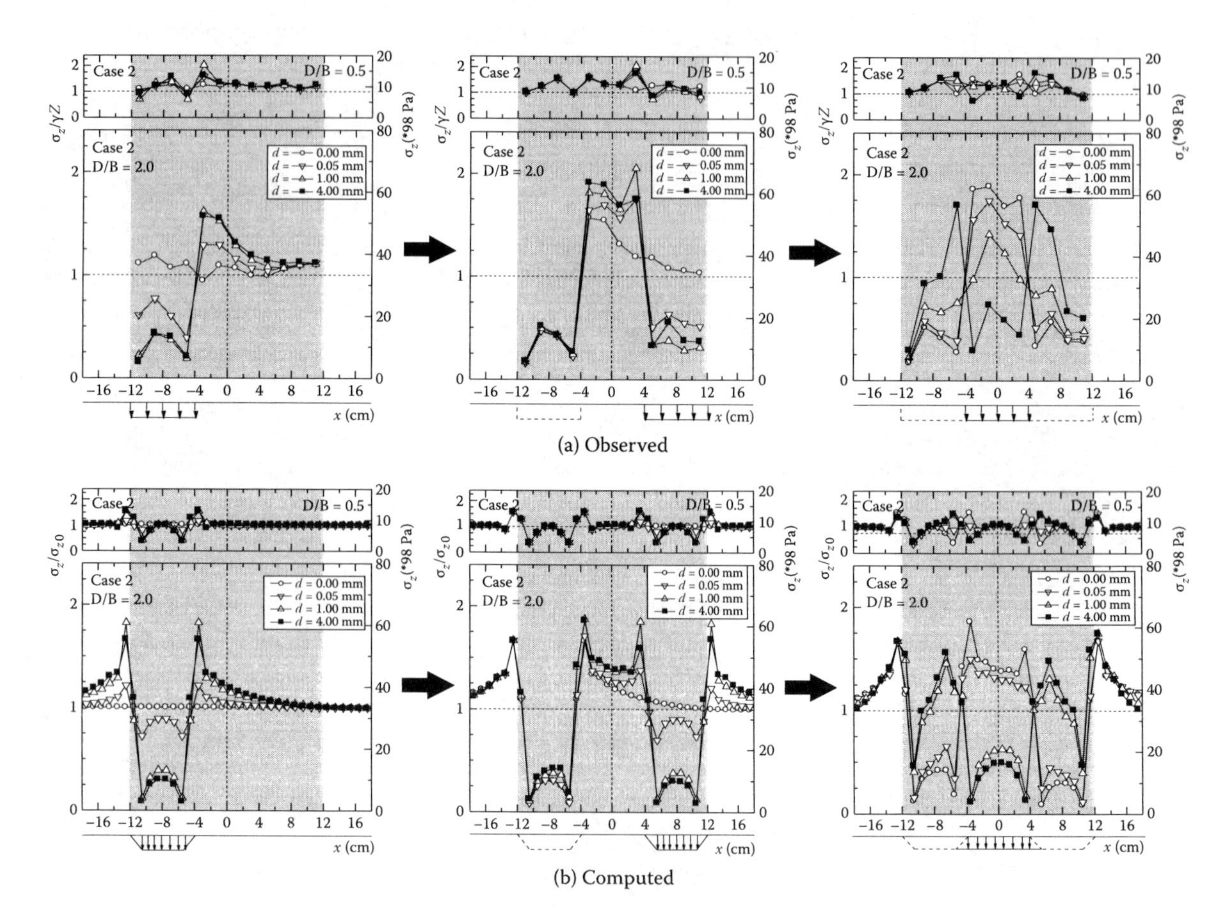

Figure 9.12 Earth pressure distributions—case 2 (sequence E-G-F): series II.

On the other hand, in the case of $D/B = 2.0$, an arching effect is significantly seen in each step of excavation (earth pressure decreases at the lowering block and increases at the adjacent blocks). When a single block is excavated (for $D/B = 2.0$) without any previous disturbance, an arching effect is fully exhibited for a small amount of imposed displacement. Once a region is disturbed, further arching still occurs due to the excavation of nearby blocks, but the rate of pressure change is lower than that for the undisturbed case. The final earth pressure distribution for case 1 and case 2 is very different. Therefore, deformation analyses considering the mechanical behavior of soil and the actual construction sequences can rationally predict the earth pressure during tunnel excavation.

9.2 THREE-DIMENSIONAL TRAPDOOR PROBLEMS

Three-dimensional model tests of trapdoor problems and the corresponding numerical analyses were carried out to investigate the influence of tunnel excavation processes on surface settlement and earth pressure. Then, for simulating real tunnel excavation in 3D conditions more precisely, an apparatus, called the pulling out tunnel apparatus, was developed. This apparatus can simulate the process of sequential tunnel excavation.

Experiments and analyses were conducted with various ground depths for simulating the influence of soil cover on tunnel excavations in the same way as the 2D trapdoor problems in Section 9.1. Surface settlements are measured at the transverse cross section of the ground. Since it is difficult to measure earth pressures at the top of the tunnel in the tests performed with the pulling out tunnel apparatus, earth pressures are measured adjacent to the tunnel cross section. More detailed experimental and numerical results on this topic are described in Shahin, Nakai, Hinokio, and Yamaguchi (2004).

9.2.1 Outline of model tests

Figure 9.13(a) shows the schematic diagram of the pulling out tunnel apparatus and Figure 9.13(b) indicates the process of excavation of this apparatus. Figure 9.14 is a picture of a model test using this apparatus. A wooden frame 80 cm in width and 80 cm in length in the direction of excavation

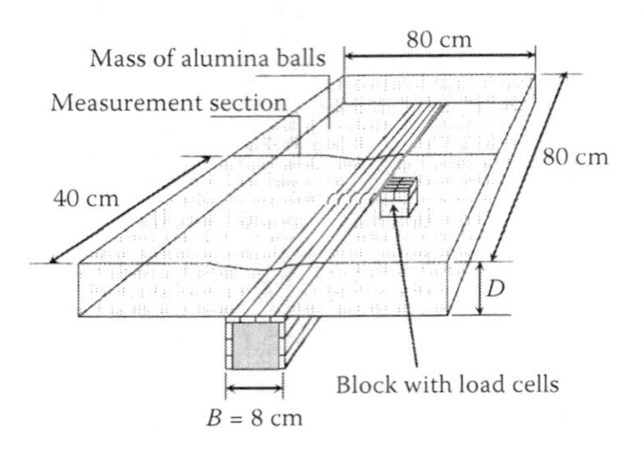

(a) Schematic diagram of apparatus

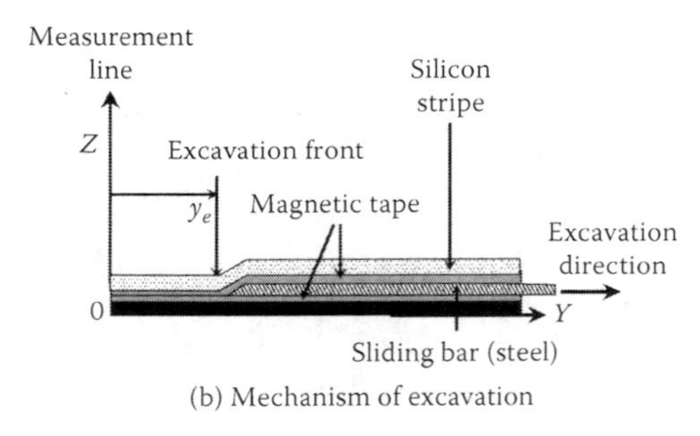

(b) Mechanism of excavation

Figure 9.13 Pulling out tunnel excavation apparatus.

Figure 9.14 Photo of model test with pulling out tunnel apparatus.

is set in the center on an iron table. There are four bars at the top and three bars at each side of the blocks of the tunneling apparatus, as shown in Figure 9.13(a). In the present study, only the top four sliding bars are pulled out to simulate tunnel excavation, as shown in Figure 9.14. The bars are 19.5 mm wide and 4 mm thick. To avoid friction between sliding the bars and soils, a silicon strip is placed just above the bars and a flexible magnetic tape is glued at the bottom of the silicon strip. When the bars are pulled out, the silicon strip fills the position of the bars, and the ground in the interface moves downward to fill the gap (4 mm), as shown in Figure 9.13(b). In this way, displacements equal to the thickness of the sliding bars are imposed. Since the sliding bars can be pulled out independently, the excavation sequence can be changed and model tests can be carried out with any desired excavation sequence.

In the pulling out apparatus, there is no device for measuring the earth pressure at the top of the sliding bars. Therefore, earth pressure is not measured at the position of tunnel excavation in the model tests. However, the block of load cells is placed adjacent to the tunnel for measuring earth pressures adjacent to the tunnel block, as shown in Figure 9.13(a). Surface settlement is measured at the transverse cross section of the ground (see Figure 9.13a), using the same system, consisting of a laser type displacement transducer and a supersonic wave transducer, as that in 2D trapdoor model tests.

In the same way as series I in the 2D trapdoor model tests, tests are conducted for four values of soil covers: $D/B = 0.5$, 1.0, 2.0, and 3.0, where D is the depth from the ground surface to the top of the tunnel and B is the width of the tunnel. In the 3D model tests, alumina balls are used as soil mass. The unit weight of the mass of alumina balls is 22.3 kN/m^3 at model stress level. The model ground of each test is carefully prepared in order to obtain a uniform ground.

9.2.2 Outline of numerical analyses

In the 3D finite element analyses, half of the ground is analyzed considering symmetry of the problem with respect to the direction of excavation. Isoparametric brick elements of different size are used in the mesh of 3D finite element analyses. Figure 9.15 shows the mesh for the ground with $D/B = 2.0$. Smooth boundary conditions are assumed in the lateral faces, while rough boundary conditions are assumed in the bottom face of the mesh. The downward displacement of $d = 4$ mm is imposed to the corresponding nodal points sequentially. The process of applying the displacements to the nodal points in the analyses is illustrated at the bottom of Figure 9.15. Surface settlements of the numerical analyses are measured at

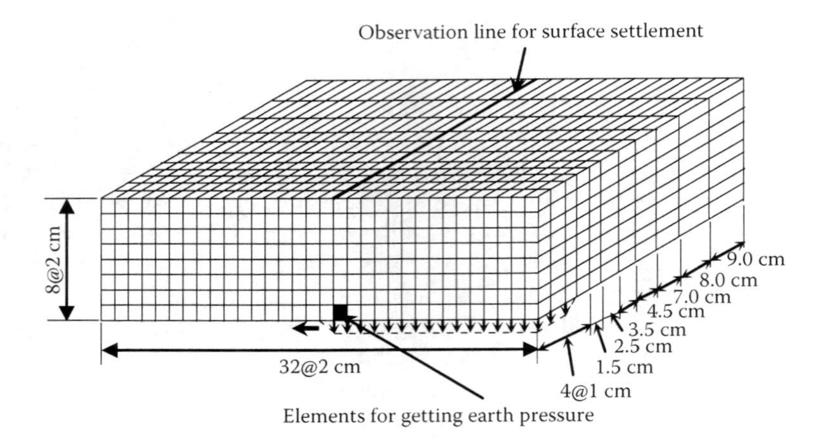

Figure 9.15 Mesh for 3D FF analyses ($D/B = 2.0$).

the place shown in this figure. Earth pressures in the numerical analyses are calculated along the elements shown at the bottom in Figure 9.15.

Figure 9.16 shows the results of triaxial tests on the mass of alumina balls used in 3D model tests and the results calculated using the material parameters in Table 8.1 in Chapter 8, which are the same as those for aluminum rod mass. It can be seen that the calculated results describe the observed stress–strain–dilatancy behavior of the mass of alumina balls

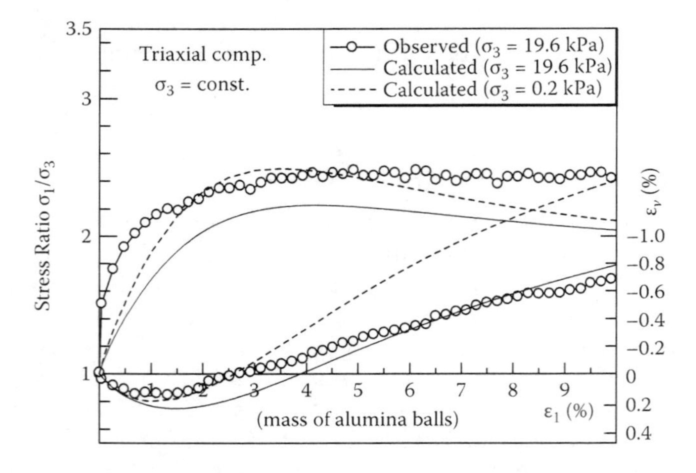

Figure 9.16 Test results of mass of alumina balls and calculated result using the material parameters in Table 8.1 in Chapter 8.

in the triaxial condition, though the strength is a little underestimated. Therefore, the same values of material parameters as those for the analyses of 2D model tests are employed in the present 3D analyses, in order to investigate the 3D effects on tunneling problems by comparing with the results of 2D conditions.

9.2.3 Results and discussion

Figure 9.17(a) represents the observed surface settlements of sequential excavations for $D/B = 0.5$, 1.0, 2.0, and 3.0. Figure 9.17(b) shows the computed surface settlements corresponding to the observed ones. Legends show the position of the excavation fronts, where value zero indicates that the excavation front is in the position of the observation section. In the legends, a negative sign indicates that the excavation front is behind the measuring section and a positive sign is used when the excavation front has passed the measuring section.

Figures 9.18(a) and (b) show observed and computed surface settlements at the center of the middle cross section of the tunnel, respectively. Abscissas represent the position of the excavation front from the middle cross section of the tunnel. Legends illustrate the depth-to-width ratio of the tunnel. For $D/B = 0.5$, surface settlement occurs when the excavation front is

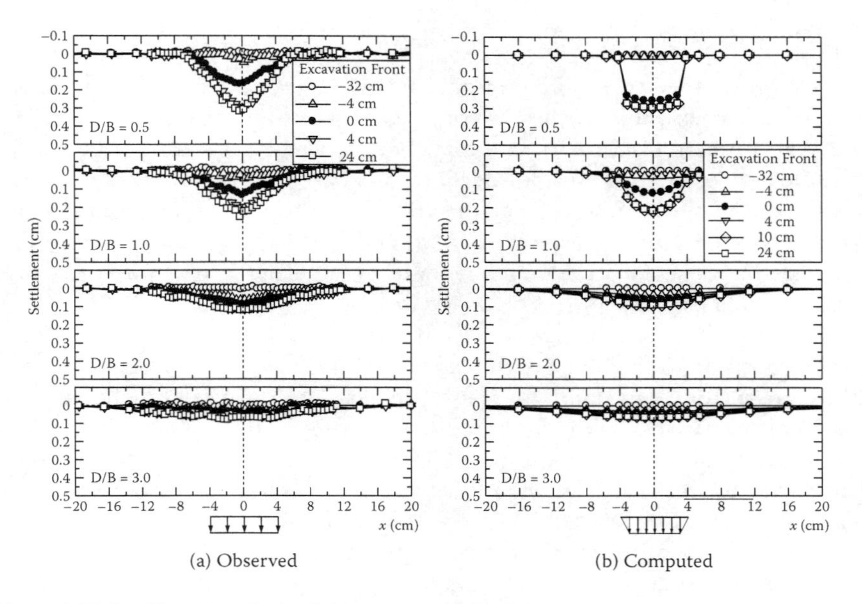

Figure 9.17 Profiles of surface settlement: series III.

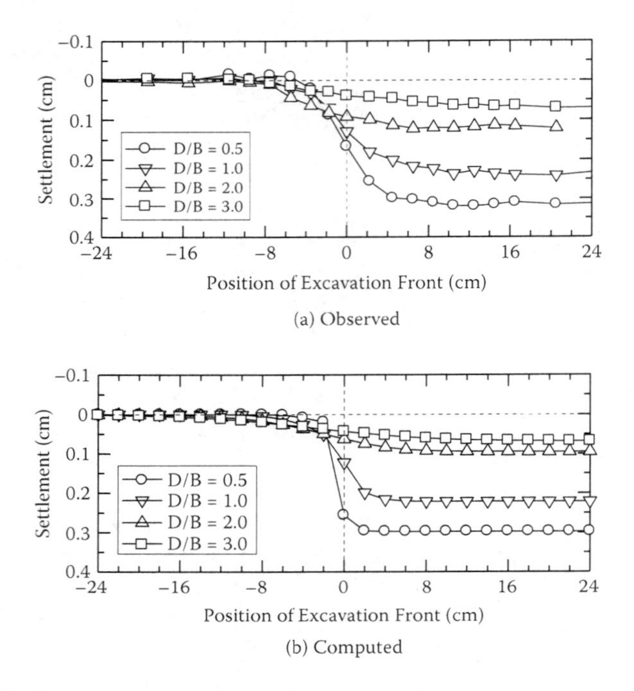

Figure 9.18 Settlement at the center of tunnel: series III.

very close to the measuring section, and the maximum surface settlement is noticed when the excavation front passes a certain distance beyond the measuring section. For further excavations, there is almost no influence at the measuring section. But, for deeper tunnel ground, the influence of excavation at the measuring section appears before the excavation front reaches that section. A similar response is obtained in the numerical analyses. The final shapes of surface settlement troughs in Figure 9.17 are similar to those of 2D conditions in Figure 9.5.

Figure 9.19 shows the history of earth pressure at a point that is 6.5 cm (at the position of the block with load cell in Figure 9.13a) from the center of the transverse cross section for the model tests and 6.75 cm for the numerical analyses, as shown in Figure 9.20. The point is located outside the tunnel block. The results of the numerical analyses were measured keeping similarity to the results of the model tests. The horizontal axis represents the longitudinal distance from the middle cross section of the tunnel. The left vertical axis represents earth pressure normalized by the product of the unit weight and depth of the ground for model tests and by the initial stress for numerical analyses. From these figures, it is revealed that earth pressure

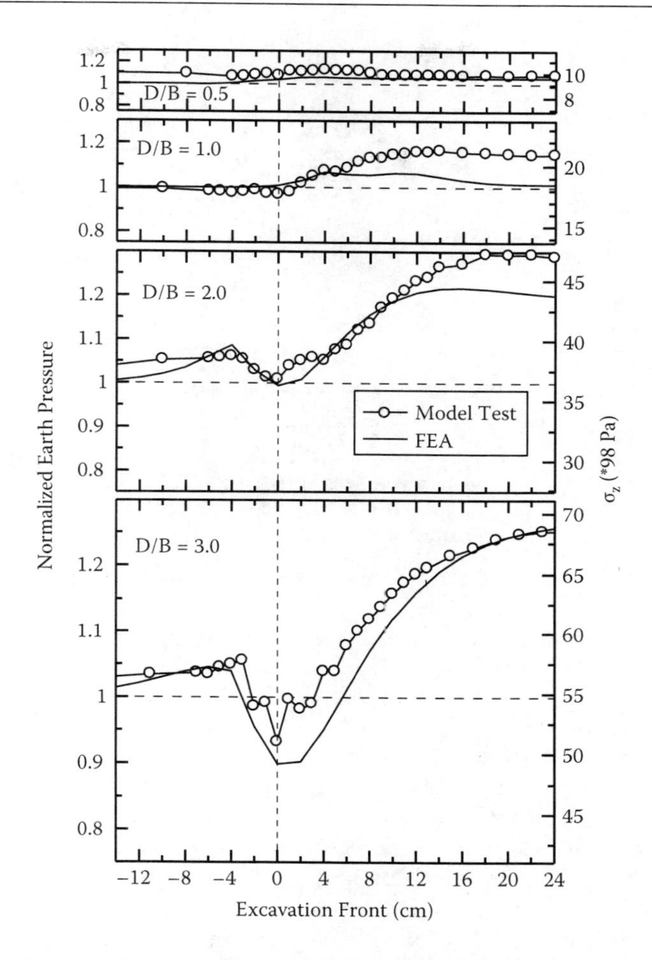

Figure 9.19 Histories of earth pressure at point outside tunnel cross section.

at the point increases before the excavation front reaches the same section. Since the region of arching is small for $D/B = 0.5$, significant reduction of earth pressure is not seen at this point.

However, for $D/B = 1.0$, 2.0, and 3.0, earth pressure of the point increases before the excavation front arrives at the measuring section and then decreases suddenly when this section is reached. For these depths of soil ground, the arching effect extends laterally up to the point where the load cell is located. Therefore, the region of arching increases with ground depth. Numerical analyses produce the same results as the model tests, both in shape and quantity.

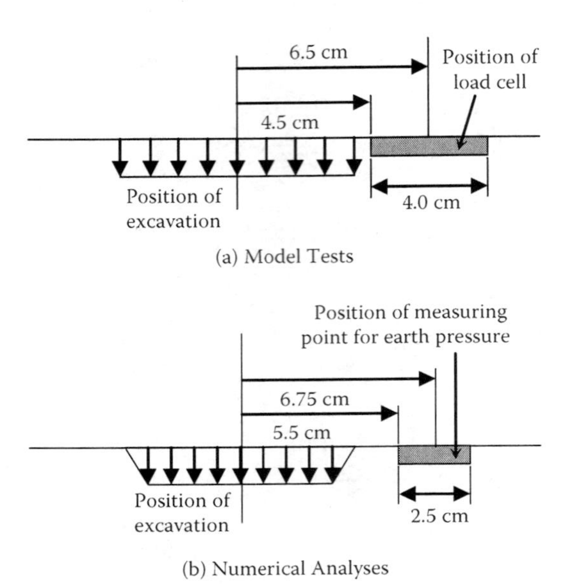

Figure 9.20 Position of the measuring point for earth pressure.

Figure 9.21 shows the computed earth pressure distributions along the transverse direction of the middle section. The legend indicates the position of excavation fronts in the same way as that in Figure 9.17. It is seen that earth pressure does not change so much when the excavation front is less than –4 cm, but it increases when the excavation front is –2 cm for all the values of soil cover. Earth pressure is nearly zero at the tunnel cross section when the excavation front is at the measuring section, and earth pressure increases gradually when the face goes away from this section. This is because, in 3D conditions, arching takes place at the measuring section in both transverse and longitudinal directions, even in the case of a very shallow tunnel (D/B = 0.5), and the advance of the excavation front from the measuring section breaks the formation of arching in the longitudinal direction.

Thus, earth pressure depends on the distance of the face of excavation. Although arching diminishes at the measuring section when the excavation front goes away in a certain distance, earth pressure of that position does not reach to its original (initial) value because of the redistribution of earth pressure in the measuring section. Since earth pressure becomes zero at the place of excavation, open face excavation such as New Austrian Tunnelling Method (NATM) is possible in practice.

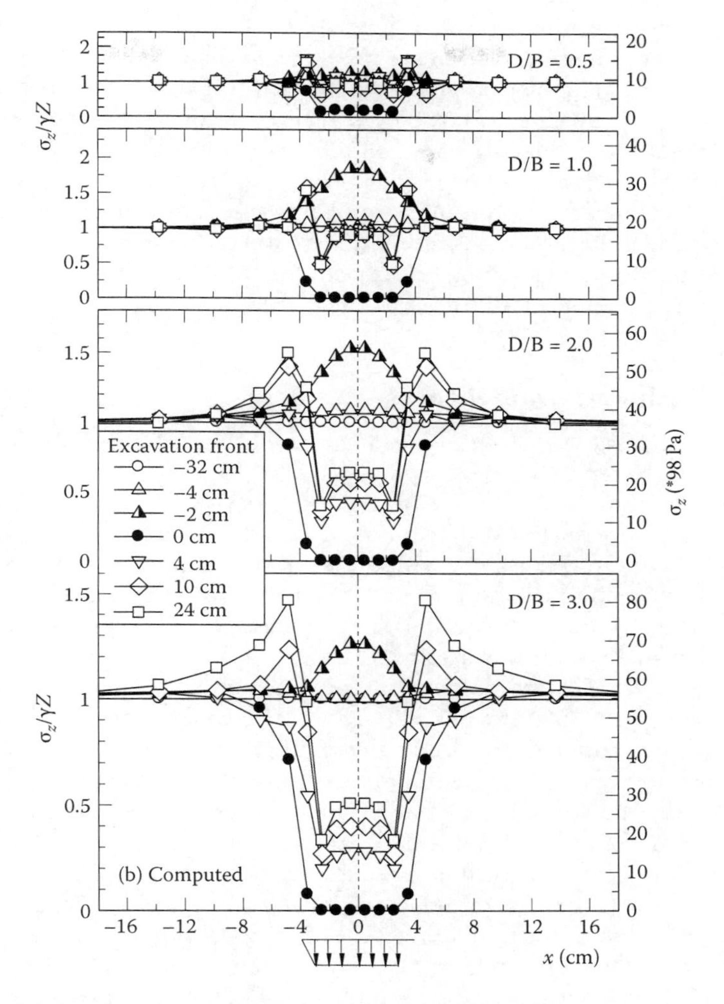

Figure 9.21 Computed earth pressure distributions: series III.

9.3 CIRCULAR TUNNELING

In the previous sections, the fundamental mechanisms of the ground behavior due to tunneling have been discussed based on the results of 2D and 3D trapdoor model tests and the corresponding numerical simulations. An apparatus was developed to model more realistic excavation of a tunnel in the laboratory. With this apparatus, 2D model tests are carried out to investigate the surface settlement and the earth pressure brought about by the circular tunneling. Particularly, the influence of excavation patterns and soil–structure interaction on the earth pressure and the ground movements during tunnel excavation are investigated.

Model tests and numerical analyses were performed for two series of tunnel excavations. In series I, not only the influence of the depth of circular tunnel but also the influence of the patterns of excavation on surface settlement and earth pressure considering the ground as greenfield are investigated. In series II, the effect of the interaction between the tunneling and existing nearby foundations are discussed. As the existing foundation, strip foundation and piled raft foundation are considered. More detailed experimental and numerical results on this topic are described in Shahin et al. (2011).

9.3.1 Outline of model tests

Figure 9.22(a) shows a schematic diagram of the 2D tunnel apparatus. Figure 9.22(b) represents a newly developed model tunnel with a circular cross section; it was constructed based on a device described by Adachi, Kimura, and Osada (1993). The tunneling device consists of a central shim (a shaft with a variable cross section) surrounded by 12 segments. One motor is attached to the shim to pull it out in a horizontal direction, thus shrinking the tunnel and causing the soil mass to converge to the tunnel center. The model tunnel is held by a vertical shaft, and it can be moved in a vertical direction with a second motor. Therefore, the device consists of two motors: One is used for shrinking the tunnel and the other can be used for moving the tunneling device in a vertical direction.

It is possible to make these motors work simultaneously and to control their speed. The reduction in tunnel diameter and the amount of radial shrinkage are obtained from a dial gauge reading. Twelve load cells are used to measure the earth pressures acting on the tunnel. The load cells are embedded in the blocks, which represent the tunnel lining. Including the load cell blocks, the total diameter of the model tunnel is 10.0 cm. The circular tunnel device is held in place at a certain height above the ground of an iron table over which the ground is filled, as will be explained later. Therefore, the earth pressure can be obtained at 12 points on the periphery of the tunnel at one time.

After setting the physical model, the excavation was simulated by controlling the two motors that move and shrink the tunnel device as explained before. The resulting surface settlement of the ground was measured using a laser type of displacement transducer with an accuracy of 0.01 mm, and its position in the horizontal direction was monitored with a supersonic wave transducer in the same way as the trapdoor model tests described in the previous sections. Photographs were taken during the experiments and used later as input data for the determination of ground movements with a program based on the technique of particle image velocimetry (PIV).

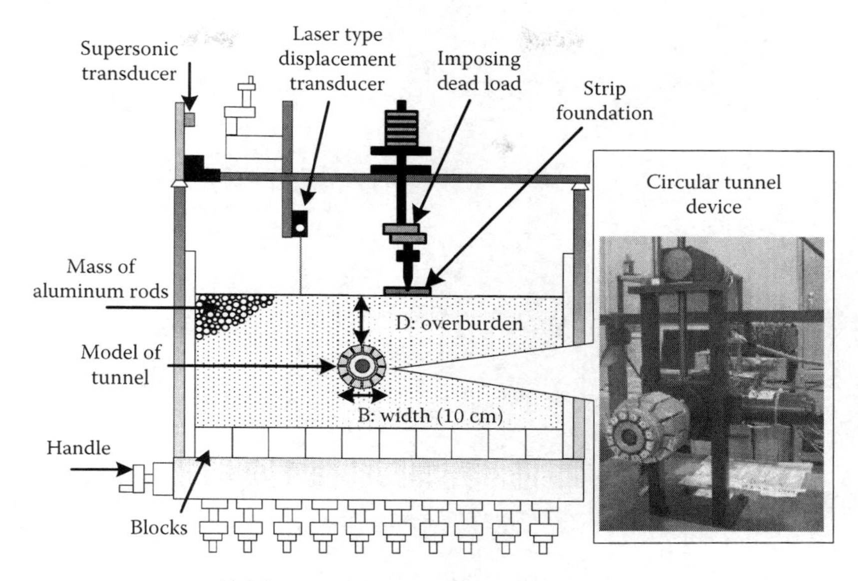

(a) Schematic diagram of whole tests apparatus

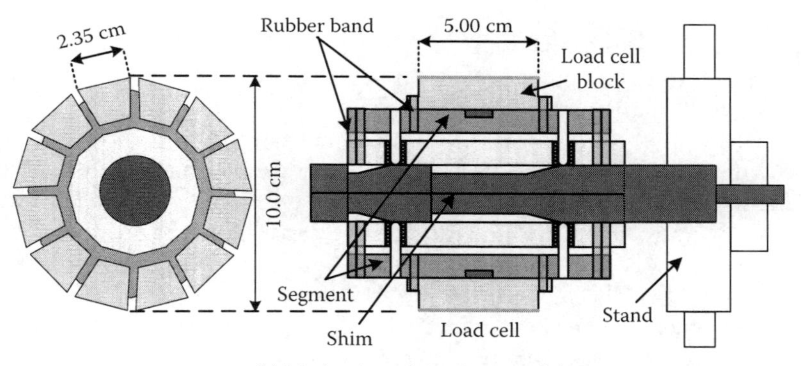

(b) Mechanism of circular tunnel device

Figure 9.22 Two-dimensional circular tunneling apparatus.

To simulate building loads, a strip footing and a piled raft were used to model the foundations. The strip footing was made of an iron plate of 8 cm in width and 1 cm in thickness. The cap of the piled raft was made of an aluminum plate of 8 cm in width and 2 cm in thickness. The piles were simulated using polyurethane walls because the model was 2D under plane strain conditions. Young's modulus of the pile material ($E = 1.06 \times 10^5$ kN/m^2) and the distance between two piles were chosen

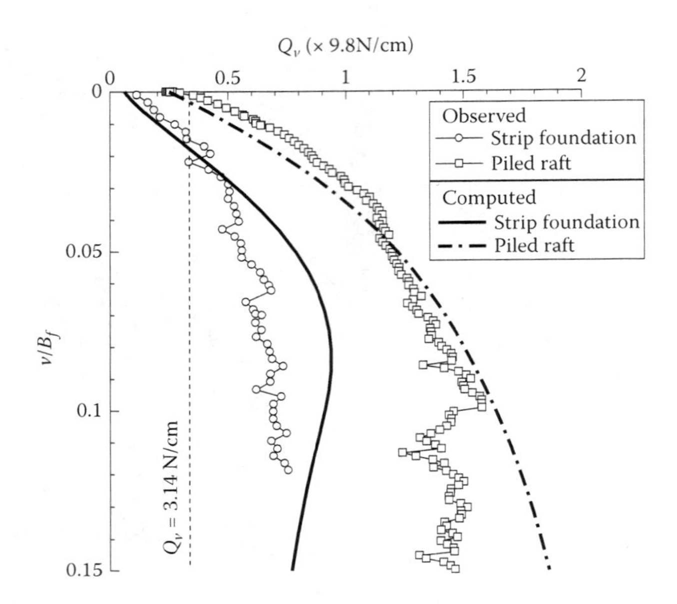

Figure 9.23 Observed and computed load-displacement curves for model ground.

to agree with the similarity ratio of 1:100 used throughout the model. For example, the 10 cm diameter model tunnel was intended to represent a real tunnel of 10 m diameter.

The thickness of the pile was 0.5 cm, the length of the pile (L_p) was 10 cm, and the distance between the front and rear piles was 5 cm. To impose the existing load, a constant value of dead load of Q_v = 0.32 (× 9.8 N/cm) was placed on the center of the foundation before performing the tunnel excavation. This load was kept fixed throughout the test and estimated from load tests performed in the model ground, as illustrated in Figure 9.23. In the load-displacement relation in Figure 9.23, the symbols denote the observed values, and the curves denote the computed results. The applied load Q_v = 0.32 (× 9.8 N/cm) was around one-third to one-half of the ultimate bearing capacity of the strip footing and much smaller than that of the piled raft.

In the model tests, aluminum rod mass was used as soil mass (unit weight is 20.4 kN/m^3) in the same way as the trapdoor models described earlier. The bottom of the iron table over which the mass of aluminum rods was placed had 10 moveable blocks used in the trapdoor tunnel experiments. The reason for using this type of base was to adjust the initial stress condition of the ground in such a way that the stress distribution would become similar to that of the ground without a tunnel (K_0 conditions for the mass of aluminum rods). This was achieved by adjusting the moveable blocks at the bottom of the apparatus until a

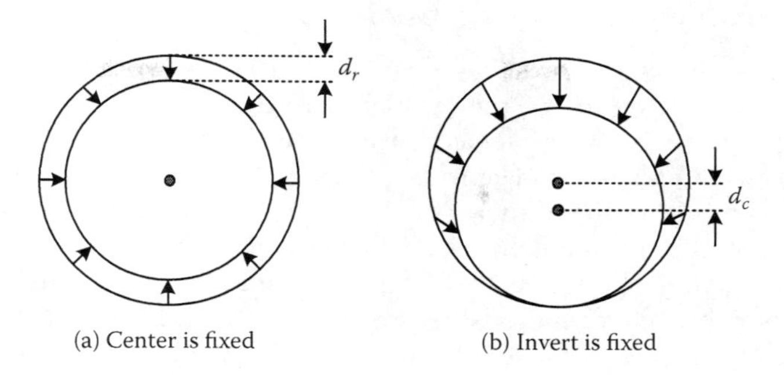

Figure 9.24 Schematic explanations of excavation patterns.

uniform vertical stress distribution, corresponding to the self-weight to the ground, was obtained. In the model test ground, the value of K_0 was about 0.7 and it was controlled by the moveable blocks at the bottom of the apparatus. Great care was taken to make a uniform ground and not to apply any undesired load on the ground.

Two types of excavation patterns were considered for the greenfield condition (series I). These patterns are illustrated in Figure 9.24, in which d_r represents the amount of shrinkage in the radial direction toward the center of the tunnel and d_c indicates the downward translation of the tunnel center. Pattern I corresponds to the excavation in which the center of the tunnel was kept fixed ($d_c = 0$) and the diameter of the tunnel was reduced, shrinking $d_r = 4$ mm all around the tunnel, as shown in Figure 9.24(a). Pattern II represents the excavation in which the invert was kept fixed (top drift excavation), as shown in Figure 9.24(b). This was achieved by lowering the tunnel device itself, $d_c = 4$ mm, during the application of shrinkage. Here, the same amount of shrinkage ($d_r = 4$ mm) was applied, but the overall excavation pattern resulted in a top drift of 8 mm and null invert movement.

In real tunnel excavations, however, the deformation mode is rather in between the previously mentioned excavation patterns. In both excavation patterns, the final volume loss around the tunnel was the same and was equal to 15.36%, in order to investigate the deformation behavior of the ground with such a huge volume loss. However, the data on different volume losses, starting from 0.20%, were recorded in the experiments. The model tests were conducted for three kinds of overburden ratios—namely, D/B = 1.0, 2.0, and 3.0, where D is the depth from the ground surface to the top of the tunnel and B (10 cm) is the diameter of the tunnel, as shown in Figure 9.22(a). For the ground with nearby existing building (series II), the excavation pattern in which the invert was fixed (pattern II) alone was employed, and the tests were performed for two kinds overburden ratios (D/B = 1.0 and 2.0).

9.3.2 Outline of numerical analyses

Figure 9.25 shows the mesh used in the finite element analyses for the tunnel excavation with the piled raft where the soil cover is $D/B = 1.0$. Isoparametric four-noded elements were used to represent the soil. The mesh was well refined with elements—1 cm in width in most regions—and even smaller around the tunnel excavation and below the foundations. The piles were modeled using hybrid elements (Zhang et al. 2000) consisting of elastic beam and solid elements. For the piles that had some thickness, the concept of hybrid element modeling was more realistic than either the modeling with beam element or elastic solid element in finite element analysis. In the case of using the beam element for the modeling pile, the volume of the pile body was being neglected.

On the other hand, if the pile was modeled with the elastic element alone, the bending effect would be neglected. In this research, the ratio of bending stiffness between the beam element and elastic solid element is 9:1, which produces the best fit curve of moment versus curvature for reinforced concrete pile and steel pipe pile. The frictional behavior (friction angle $\delta = 18°$) between the pile and the ground was simulated using elastoplastic joint elements (Nakai 1985), which is described in Chapter 8. The frictional angle, $\delta = 18°$, was obtained from sliding tests of the model pile (polyurethane plate) on the model ground (aluminum rod mass). Both vertical sides of the mesh were free in the vertical direction, and the bottom face was kept fixed. The analyses were carried out under plane strain conditions.

To simulate the tunnel excavation, horizontal and vertical displacements were applied to the nodes around the tunnel periphery. The analyses were carried out under the same conditions as those of the model tests. The

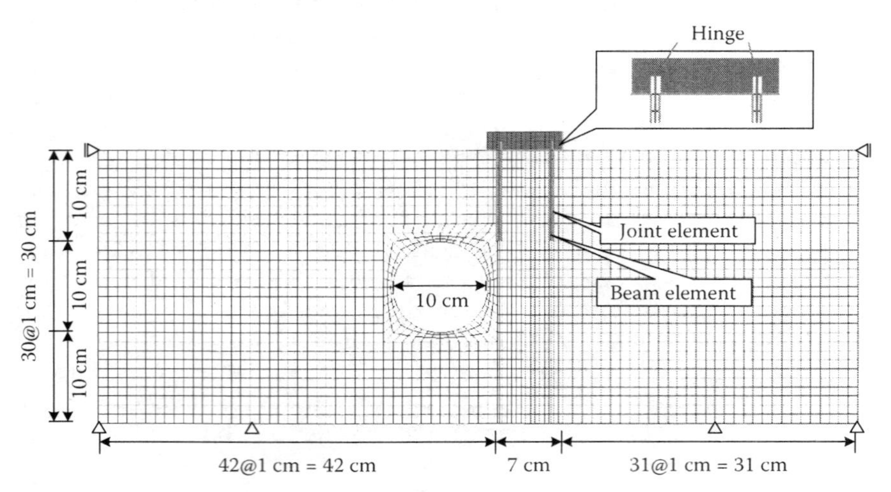

Figure 9.25 Mesh for finite element analyses ($D/B = 1.0$).

initial stress levels of the ground were calculated by applying the body forces due to the self-weight ($\gamma = 20.4$ kN/m^3), starting from a negligible confining pressure ($p_0 = 9.8 \times 10^{-6}$ kPa) and an initial void ratio of $e = 0.35$. After self-weight consolidation, the void ratio of the ground was 0.28 at the bottom and 0.30 at the top for $D/B = 1.0$. The value of K_0, derived from the simulation of the self-weight consolidation, was between 0.70 and 0.73; at the top it was 0.73 and at the bottom of the ground it was 0.70. In the case of the existing foundation, the ground was initially formed under geostatic conditions, and then a concentrated load was applied at the middle node of the foundation. The stresses, void ratios, and density parameters of the constitutive model at all integration points were stored and then used as the initial conditions of the ground before the tunnel excavation.

9.3.3 Results and Discussion

9.3.3.1 Series I (greenfield ground)

Figure 9.26(a) shows the observed surface settlement troughs for the tests with the fixed center excavation and fixed invert excavations for the amount of shrinkage $d_r = 4$ mm in the case of $D/B = 1.0$, 2.0, and 3.0. Figure 9.26(b) represents the computed results corresponding to the observed ones. The abscissa represents the distance from the center of the tunnel, while the vertical axis shows the amount of surface settlement. For both patterns of excavation, the maximum surface settlement occurred in the centerline above the tunnel crown. Surface settlements decreased with the increase in the tunnel depth, but they extended over a wider region. Similar results

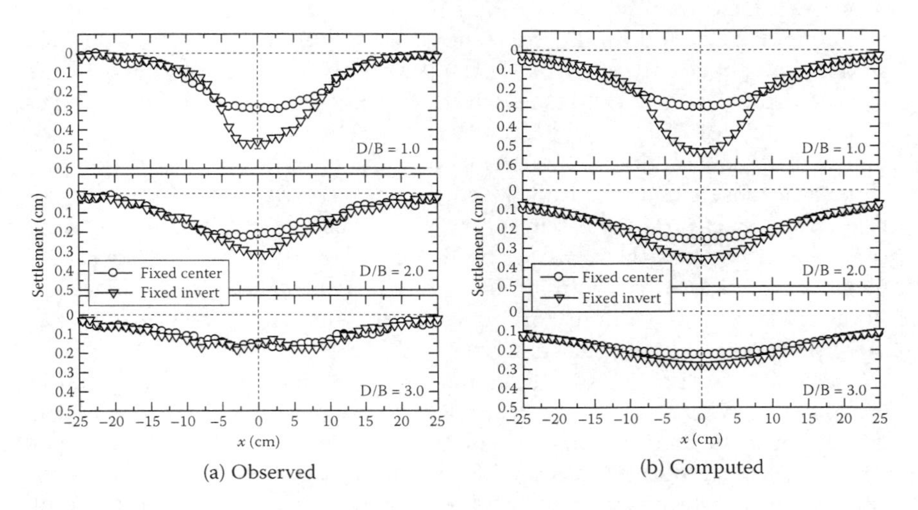

Figure 9.26 Profiles of surface settlement: greenfield.

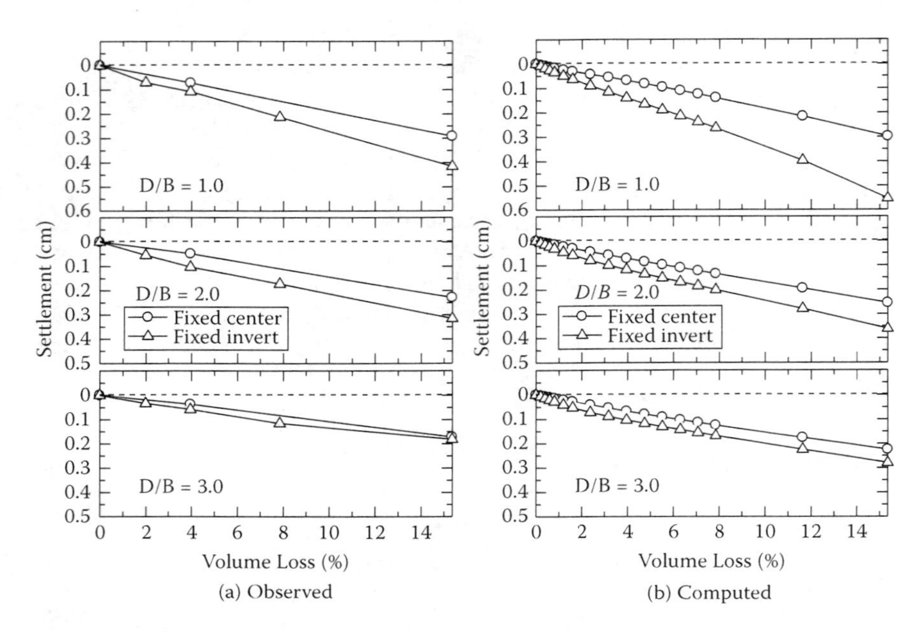

Figure 9.27 Observed surface settlement at the centerline: greenfield.

were observed in the results conducted with a trapdoor tunnel apparatus in Section 9.1. The shape of the surface settlement profiles was the same for all the soil covers in the case of fixed center excavation. For the same volume loss, the maximum surface settlement was larger in the case where the invert was fixed than when the center was fixed.

Figures 9.27(a) and (b) illustrate the observed and the computed surface settlements above the centerline of the tunnel against the volume loss during the tunnel excavation, respectively. It can be clearly seen that even for a volume loss of less than 2%, the two different excavation patterns produced different amounts of settlement in the case of shallow tunneling. This is because the applied displacement at the crown was 8 mm for the fixed invert excavation and 4 mm for the fixed center excavation, although the volume loss was the same for both excavation patterns. Surface settlement occurred locally for an applied displacement of 8 mm in the case of the fixed invert excavation; consequently, the shape of the surface settlement profile varied with the soil cover in this case.

The tendency of larger surface settlements for the fixed invert excavation was more significant up to $D/B = 2.0$. In the case of $D/B = 3.0$, however, the difference in surface settlement profiles between the two excavation patterns was less significant. It is revealed from these results that, for the same volume loss, surface settlement profiles vary with the excavation patterns in the case of shallow tunneling. Therefore, the surface settlement may not be properly estimated using the method of volume loss (Mair, Taylor, and Bracegirdle 1993) for very shallow tunneling.

The results of the numerical analyses show the same tendency as the model tests not only in shape but also in quantity. Some discrepancies may be observed in the computed values in the extreme boundaries of the model. This is because the lateral nodes were allowed to move freely in the vertical direction in the numerical analyses, when in fact there was some friction between the aluminum rods and the metal sides of the model. The agreement would have been improved if joint elements had also been used in this region.

The distributions of deviatoric strain in the model tests were obtained using the PIV technique. PIV was originally developed in the field of fluid mechanics (Adrain 1991). Two images (one was taken after the deposition of the initial ground and before the tunnel excavation, and the other was taken after shrinking the tunnel by $d_r = 4$ mm) were divided into a finite area; the average movement rate of the mass of aluminum rods for each area was extracted as nodal displacements. The strain for one grid was calculated from these displacements by using the shape functions and the B matrix that was used in finite element method to relate displacements and strains.

Figure 9.28 shows the distributions of deviatoric strain for the fixed center and the fixed invert excavations for $d_r = 4$ mm in soil cover $D/B = 1.0$. For the fixed center tunnel excavation, it is seen in the figure that the shear band of the ground developed from the tunnel invert, which covers the entire tunnel. For the fixed invert excavation, however, the shear band developed from the side of the tunnel. The shear band in this case was longer than that in the fixed center excavation. The different patterns of deviatoric strains, due to the different types of the tunnel excavations, led to the change in the ground behavior. The two different excavation patterns produced different kinds of deviatoric strain, although the volume loss was the same. The deviatoric strain of the numerical analyses showed very good agreement with the results of the model tests.

Figure 9.29 shows the observed and the computed earth pressure distributions for $D/B = 1.0$. The plots are drawn in the 12 axes corresponding to the radial direction of the 12 load cells toward the center of the model tunnel. The figures represent the value of earth pressure in pascals corresponding to the amount of applied displacement (amount of shrinkage). It is seen in the figure that the earth pressure decreased all around the tunnel for the fixed center excavation due to the stress relief. As the shear band developed around the entire tunnel (Figure 9.28), the surrounding ground became looser and reduced the stress levels in this area.

It is also noticed that the earth pressure decreased suddenly after applying a little shrinkage of the tunnel with a magnitude between 0.05 and 0.20 mm. The sharp change in earth pressure, with only a very

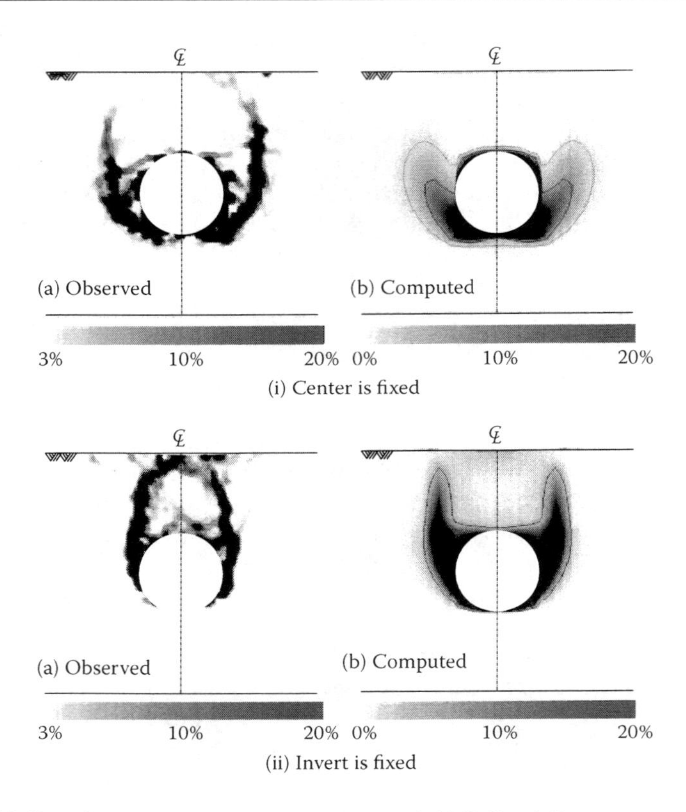

Figure 9.28 Distribution of deviatoric strain: greenfield (D/B = 1.0).

small shrinkage of the tunnel, confirmed what Peck (1969) described about soft-ground tunnels. Shrinking the tunnel even further, the earth pressure gradually decreased at a lower rate and it became almost constant after applying shrinkage to some extent. Sudden changes in earth pressure were due to the soil arching immediately after the ground was disturbed.

For the fixed invert excavation, the earth pressure distributions were different from those of the fixed center excavation. In this case, the earth pressure decreased all around the tunnel until d_r = 1 mm. With further shrinkage of the tunnel, the earth pressure increased in the bottom part of the tunnel, while it remained almost unchanged in the upper part of the tunnel. It can be said that the distribution of earth pressure is highly dependent on the excavation patterns.

Now, let us discuss the importance of the constitutive model used in the prediction of ground behavior by numerical simulations. Numerical analyses with a linear elastic theory have been carried out to compare the results

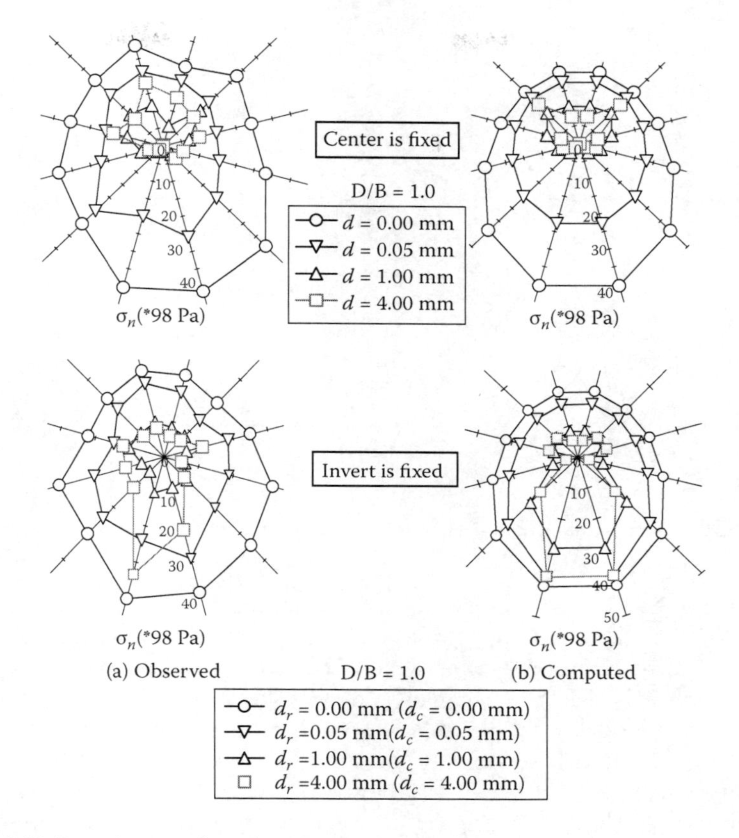

Figure 9.29 Distribution of earth pressure: greenfield.

with the elastoplastic analyses based on the present elastoplastic model. The comparisons here have been made by taking some typical results of the analyses. The Young's modulus for the elastic analyses has been chosen from the stress–strain relation (Figure 9.30) of the biaxial tests performed in the laboratory on the aluminum rod mass. The value of E = 5500 kPa is obtained from the figure, and the value of Poisson's ratio is assumed to be 0.33 for the aluminum rod mass.

Figure 9.31 shows the surface settlement profiles of the model tests, the elastoplastic analysis, and the elastic analysis for soil cover D/B = 1.0. It is seen that the elastic analysis produced a wider surface settlement profile compared to the observed one and that the maximum surface settlement was smaller as well. In this analysis, the displacement was applied to simulate the tunnel excavation; therefore, there was no influence of the magnitude of the Young's modulus on the shape of the settlement trough except for the value of Poisson's ratio.

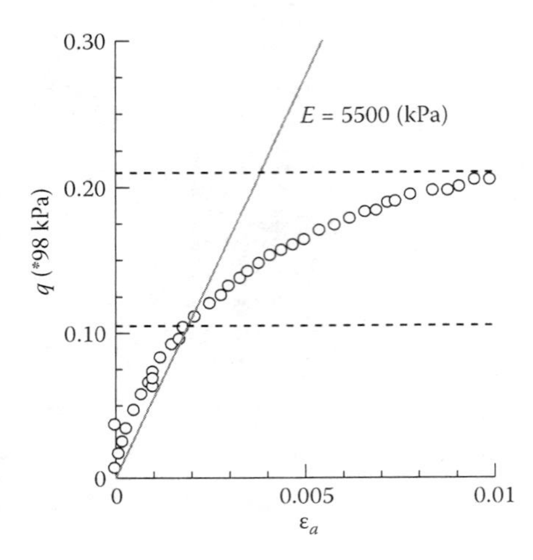

Figure 9.30 Stress–strain relation of aluminum rods' mass in biaxial test.

Figure 9.32 shows a comparison of the computed deviatoric strain distributions between the elastic analysis and the elastoplastic analysis. The results of the excavation pattern, where the center was kept fixed, are illustrated here. It is seen that the elastic analysis was not able to express the results of the model tests (Figure 9.28) of the tunnel excavation. In the elastic analysis, deviatoric strain was concentrated all around the tunnel.

To investigate the effect of the boundary, two types of analyses were carried out. In the first type, the distance between the bottom boundary and the invert of the tunnel was 10 cm. In the second type, the tunnel invert was 20 cm from the bottom boundary. In the elastic analyses, it is observed that the distributions of deviatoric strain depended on the distance from the tunnel invert to the bottom boundary. In contrast, both elastoplastic analyses produced the same shape of the shear band and the same distribution of deviatoric strain for the tunnel excavation, as observed in the model tests

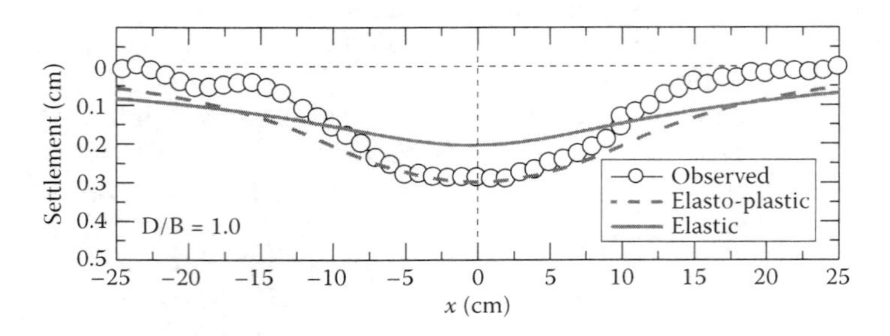

Figure 9.31 Comparisons of surface settlement profiles.

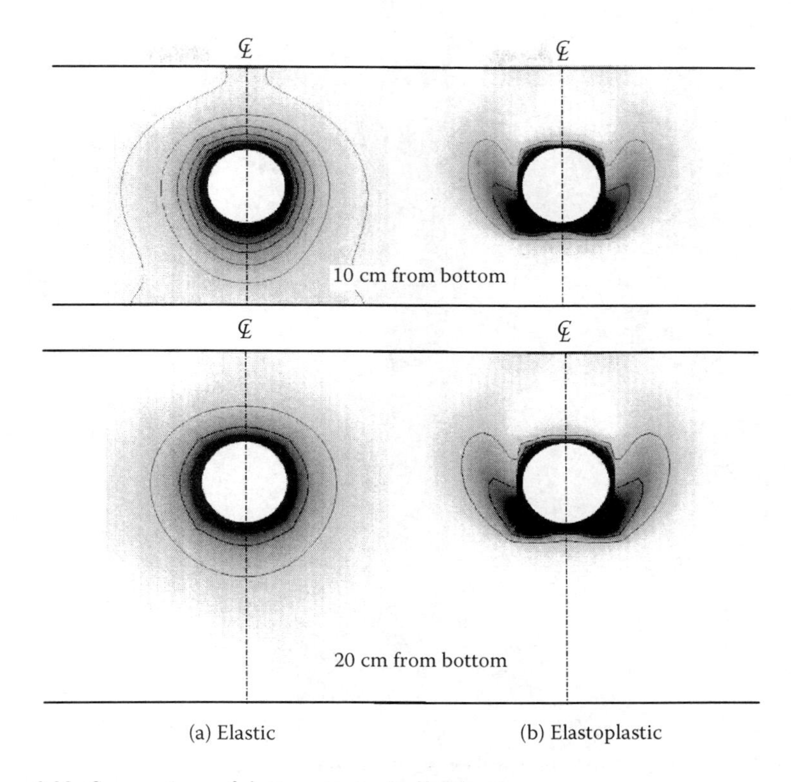

Figure 9.32 Comparison of deviatoric strain distributions.

(Figure 9.28). Therefore, it can be said that a proper elastoplastic constitutive model for soil is required to predict the ground deformation properly.

9.3.3.2 Series II (ground with nearby building load)

Figure 9.33 shows the observed and the computed surface settlement profiles for $D/B = 2.0$ in the case of a strip foundation at ground level for the fixed invert excavation. Figure 9.34 represents the same for the piled raft. For the sake of comparison, these figures also present the results of the greenfield condition for the shrinkage $d_r = 4$ mm, which is rendered with the solid line. In Figure 9.34, D_p denotes the vertical distance between the tip of the piles and the tunnel crown (see Figure 9.36 later). The position of the applied dead load is depicted at the tops of the figures.

It is seen in these figures that the maximum surface settlement occurred underneath the building load, and it was larger than that of the greenfield condition. Some tendency was obtained in 2D and 3D model tests using a trapdoor apparatus as well (Shahin, Nakai, Hinokio, Kurimoto, et al.

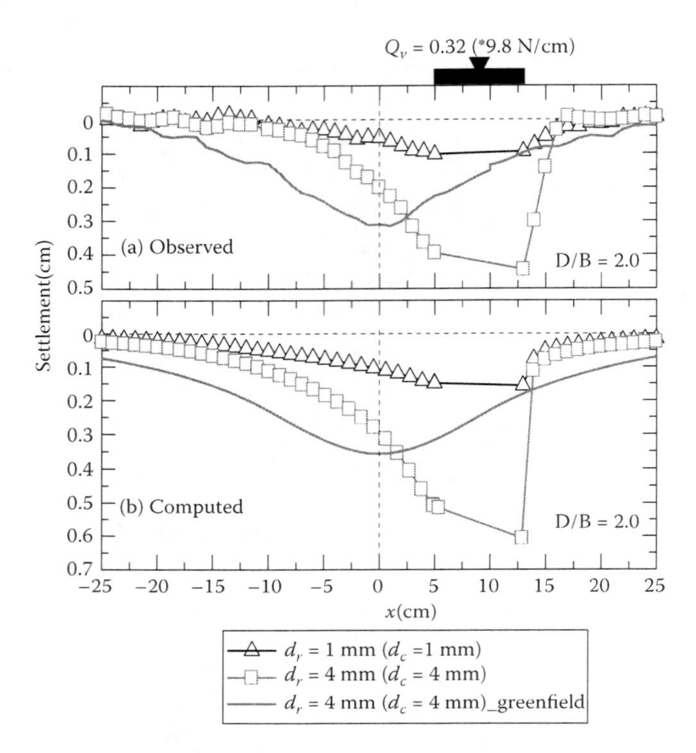

Figure 9.33 Surface settlement profiles: strip foundation.

2004; Sung et al. 2006). For the applied amount of building load, the strip foundation tilted in the opposite direction of the excavation. In contrast, the piled raft inclined toward the tunnel. The numerical simulations were able to explain the results of the model tests well for both foundations.

From these results, it can be said that the surface settlement in real field tunneling may not be maximum just above the tunnel axis when superstructures exist near the tunnel. It is also noticed that surface settlement troughs for tunnel excavations in grounds disturbed by existing buildings do not follow the usual pattern of a Gaussian distribution curve, as is generally observed for the greenfield condition.

Figures 9.35 and 9.36 illustrate the distribution of deviatoric strain when D/B = 1.0 and 2.0 for the strip foundation and the piled raft, respectively. It is demonstrated in the figures that the deformed zone spreads toward the foundation from the side of the tunnel, and the disturbed zone for the tunnel excavation spreads over the right edge of the strip foundation. The computed displacement vectors showed the same tendency of ground movements as the model tests. For the induced initial stress in the ground due to the building load, the development of the shear band was unsymmetrical on the left and

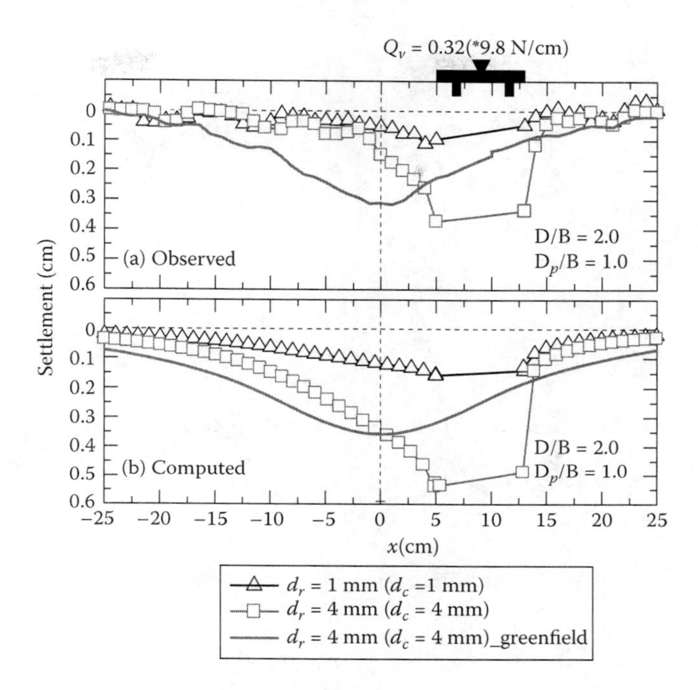

Figure 9.34 Surface settlement profiles: piled raft.

right sides of the tunnel. For $D/B = 1.0$, deviatoric strain levels were concentrated in the ground to the left and to the right of the strip foundation, while for $D/B = 2.0$, they were mainly concentrated on the right side of the foundation.

For the piled raft, a large deviatoric strain due to the tunnel excavation was concentrated around the tip of the farthest (rear) pile in soil covers $D/B = 2.0$ and $D_p/B = 1.0$; it was concentrated chiefly around the closest (front) pile in soil covers $D/B = 1.0$ and $D_p/B = 0.0$, where D_p is the vertical distance between the pile tip and the tunnel crown. Therefore, it can be said that the distance from the tip of the pile to the tunnel crown had a significant effect on the deformation mechanism. The intensity of the deviatoric strain on the left side of the foundations gradually decreased with the increase in soil cover.

The computed distributions of deviatoric strain for the numerical analyses showed a very good agreement with the results of the model tests. The shape and the development of the shear bands control the behavior of the foundations in shallow tunneling. This emphasizes the necessity of applying a proper constitutive model for soil in predicting the mechanism of ground deformation when superstructures exist in the vicinity of a tunnel excavation.

Figure 9.37 shows the earth pressure distributions for the strip foundation and the piled raft when $D/B = 1.0$ in the fixed invert excavation. The plots

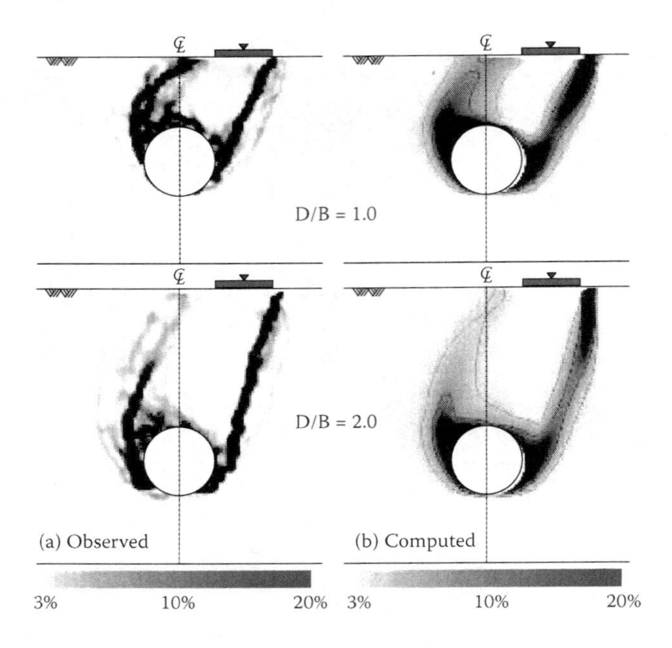

Figure 9.35 Distribution of deviatoric strain: strip foundation.

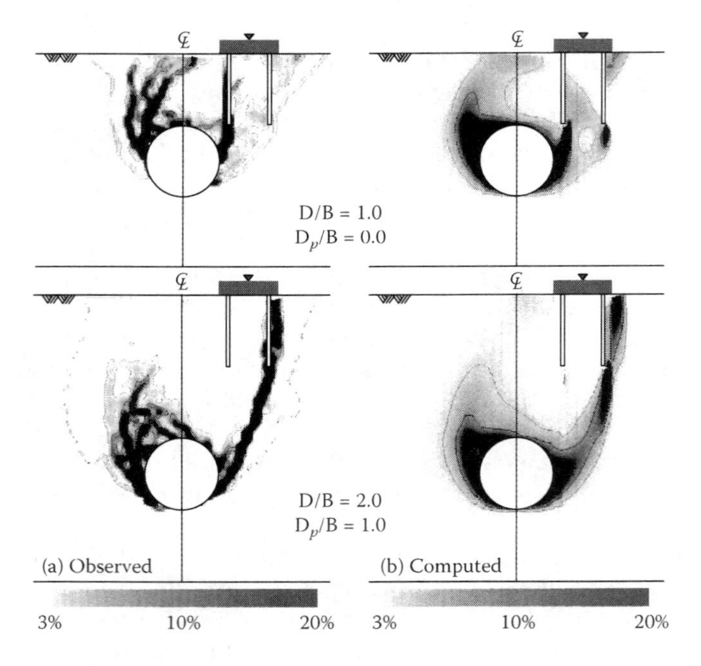

Figure 9.36 Distribution of deviatoric strain: piled raft.

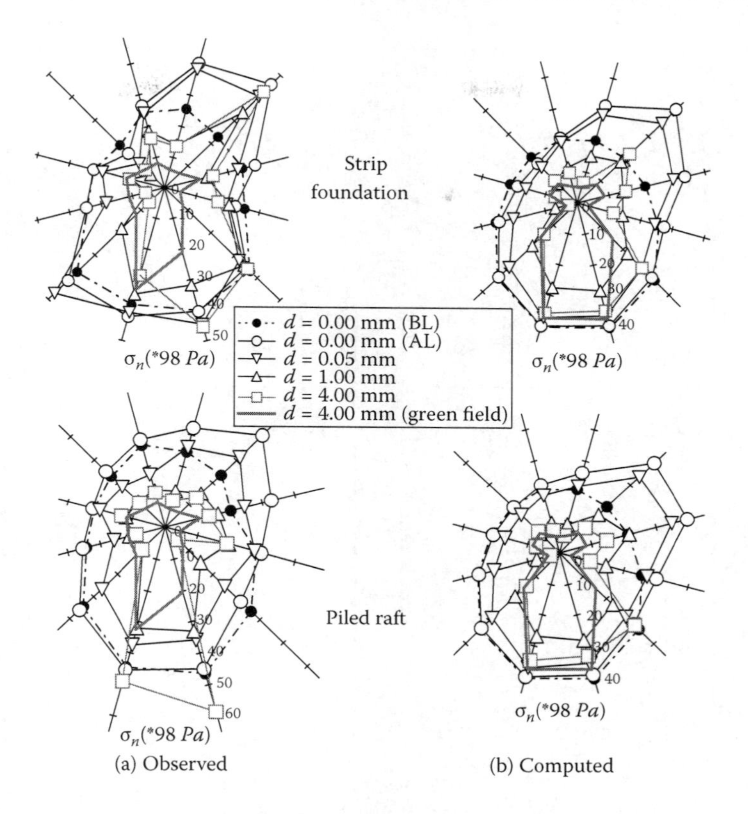

Figure 9.37 Distribution of earth pressure: *D*/*B* = 1.0.

are drawn using 12 axes corresponding to the radial direction of the 12 load cells toward the center of the model tunnel in the same way as in Figure 9.29. Here, the dotted curves with black circular marks represent the earth pressure levels before applying the building loads, the white circular marks represent the pressure levels after applying the building loads, and the solid line represents the earth pressure of the greenfield condition for $d_r = 4.0$ mm.

The earth pressure at the foundation side increased after applying the building loads. In the case of the strip foundation, the earth pressure decreased to some extent around the tunnel immediately after performing the tunnel excavation. However, with the advances of the tunnel excavation, the earth pressure at the foundation side increased again in this case. In contrast, the phenomenon of the increase in earth pressure after its reduction to some extent was not observed in the piled raft. For both foundations, an unsymmetrical earth pressure distribution was seen around the shallow tunneling, and the final earth pressure distributions were different from those of the greenfield condition.

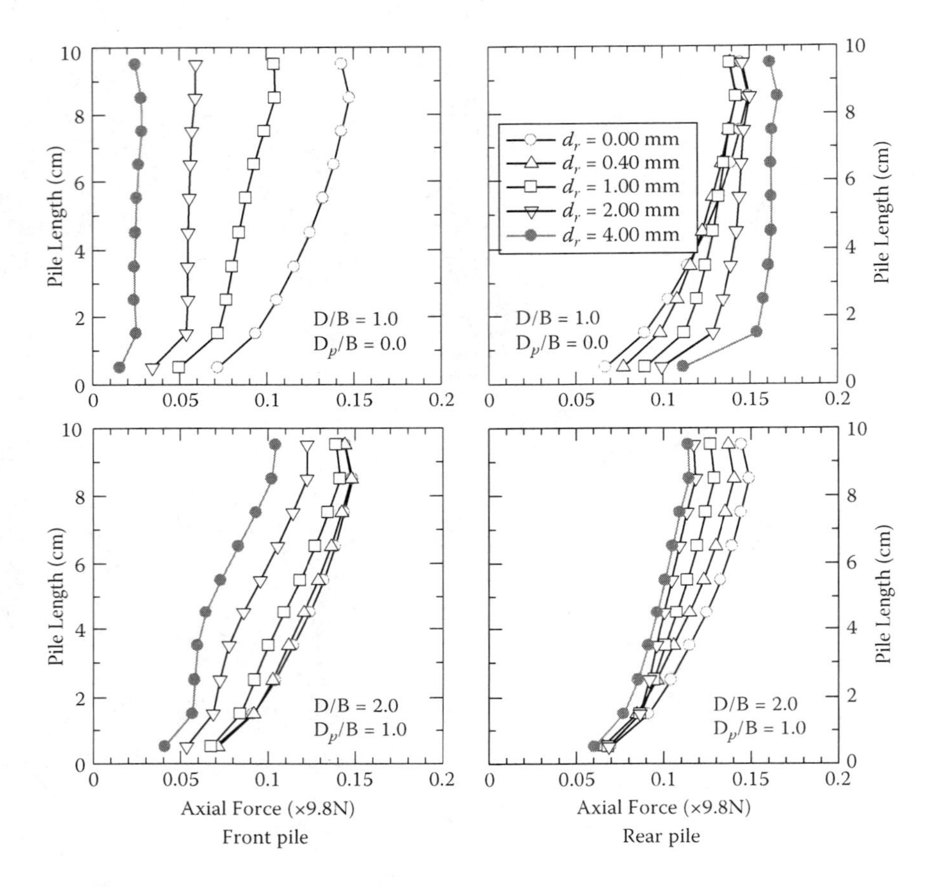

Figure 9.38 Computed distribution of axial force.

Therefore, the effect of the soil–structure interaction should be properly contemplated in the earth pressure computation around the tunnel lining. The simulations slightly underestimated the measured earth pressure levels around the tunnel invert. As a whole, the computed results predict well the earth pressure distribution of the model tests in the case of the soil–structure interaction problem as well.

Figure 9.38 shows the computed distributions of axial force at various stages of the tunnel excavation in the front and rear piles for $D/B = 1.0$ and 2.0. The vertical axis represents the length of the pile starting from the pile tip, and the compressive axial force is taken as positive in the figures. For soil cover $D/B = 1.0$, it is revealed that the axial force decreased remarkably in the front pile, while it increased in the rear pile. For soil cover $D/B = 2.0$, on the other hand, the axial force in both piles decreased due to the tunnel excavation.

Chapter 10

Earth pressure of retaining walls and bearing capacity of foundations

10.1 ACTIVE AND PASSIVE EARTH PRESSURE BEHIND RETAINING WALLS

Two-dimensional (2D) retaining wall model tests and the corresponding elastoplastic finite element analyses were carried out to investigate the influences of the wall deflection process and wall deflection mode on the earth pressures developed and the ground movements. In the first series (series I), the influence of wall deflection on the earth pressure behind the retaining wall and the ground movement in active and passive states is discussed. In the second series (series II), the influence of deflection mode of the wall is discussed in the active state. Although only 2D retaining wall problems are discussed here, a more detailed report on experimental and numerical results including 3D effects can be found in Iwata et al. (2007).

10.1.1 Outline of model tests

Figures 10.1 and 10.2 illustrate details of the 2D apparatus. The size of the frame is 500 mm wide and 320 mm high. Three moveable slide blocks (the height of each block is 80 mm) are placed on the right-hand side wall, as shown in the figure. One face of each slide block is divided into four small blocks (20 mm in height) with load cells, which are arranged vertically, to measure the earth pressure distribution on the slide block. Active and passive earth pressure conditions are produced by imposing horizontal movement to each slide block.

The 2D model ground consists of a mix of 1.6 and 3.0 mm aluminum rods in the ratio of 3:2 by weight in the same way as that in the 2D model tests in tunneling. The unit weight of the model ground is 20.4 kN/m^3. The initial horizontal earth pressure distribution in the 2D test is also shown by the broken line in each figure of the earth pressure distributions, which will be described later. The devices and the methods to measure the earth

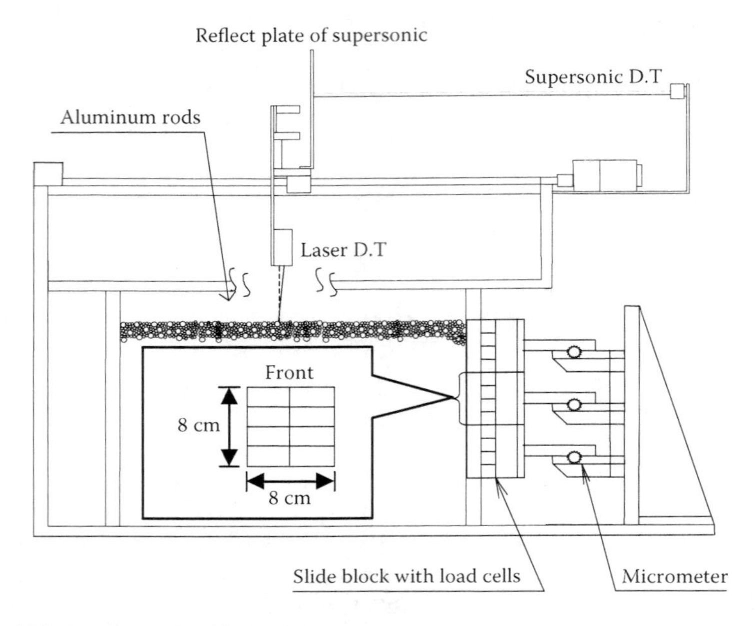

Figure 10.1 Apparatus for 2D model tests.

Figure 10.2 Whole view of 2D model test apparatus.

Table 10.1 Deflection process of wall in series I

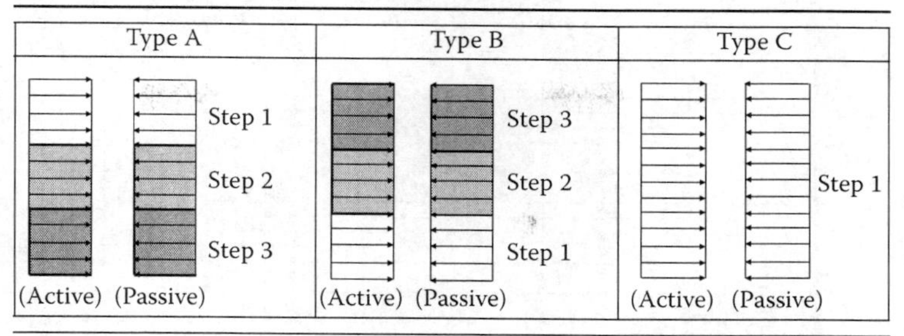

pressure and the surface settlements are the same as those used in the tunneling model tests. Also, the deformation patterns that develop in the ground are captured by a digital camera. Two series of the model tests are carried out.

In series I, in order to investigate the influence of the deflection process of the wall in active and passive states, the order of the blocks to apply the horizontal movement (d = 4 mm) is changed (see Table 10.1). In type A, horizontal displacement is applied one by one from the upper block to the lower block; in type B, horizontal displacement is applied one by one from the lower block to the upper block; and in type C, horizontal displacement is applied simultaneously to the three blocks. In every type of test, the final deflection mode of the wall is the same.

In series II, in order to investigate the influence of the deflection mode of the wall, four kinds of final deflection modes are employed, as shown in Table 10.2. In every test, the horizontal displacement is applied one by one from the upper block to the lower block. The applied displacement of each wall is described in the table. Here, the test of mode 1 is the same as that of type A at active state in series I.

Table 10.2 Deflection process of wall in series II

Mode 1	Mode 2	Mode 3	Mode 4
Step 1 (4mm)	Step 1 (4mm)	Step 1 (1mm)	Step 1 (1mm)
Step 2 (4mm)	Step 2 (2mm)	Step 2 (2mm)	Step 2 (2mm)
Step 3 (4mm)	Step 3 (1mm)	Step 3 (4mm)	Step 3 (1mm)

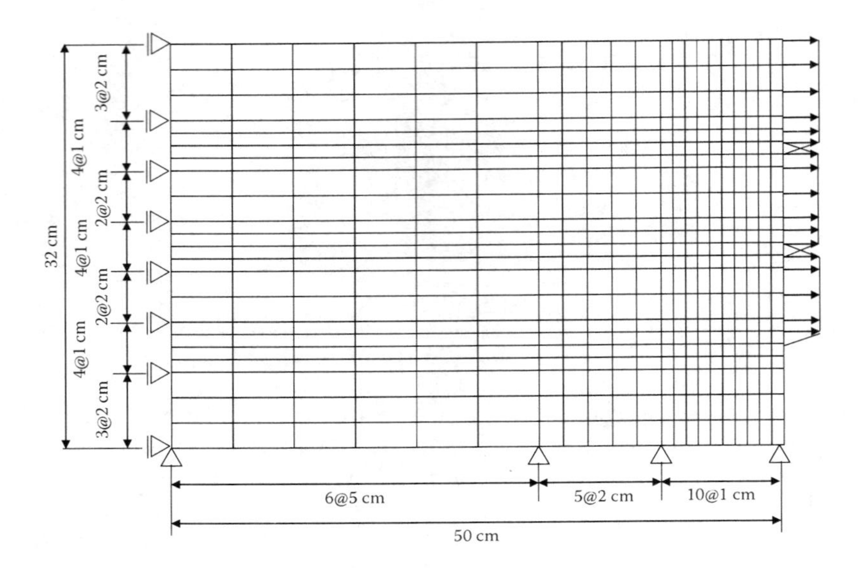

Figure 10.3 Finite element mesh for 2D analysis.

10.1.2 Outline of numerical analyses

Numerical analyses corresponding to the model tests are carried out at the same scale as that of model tests. Figure 10.3 shows the finite element mesh used in the analyses. In the analyses, smooth boundary conditions are assumed in the lateral faces, and the bottom of the mesh is assumed to be fixed boundaries. To simulate the movements of the slide blocks in the numerical analyses, horizontal displacements at the nodal points corresponding to the face of the moved block are applied. The values of the material parameters used in the analyses are the same as those in the 2D analyses of the model tests using the aluminum rod mass for tunneling problems (Table 8.1 in Chapter 8). The initial stresses of the ground are calculated by simulating the self-weight consolidation by applying body forces (unit weight).

10.1.3 Results and discussion

10.1.3.1 Series 1 (influence of deflection process of the wall)

Figures 10.4(a) and (b) show the observed and computed variations of lateral earth pressure coefficient K with wall displacement in active and passive states when the deflection process of the wall is type C (the three blocks move simultaneously). Here, K is defined as $K = P_h/P_{v0}$. (P_h is the horizontal total thrust against the moveable wall, and P_{v0} is the integration of vertical

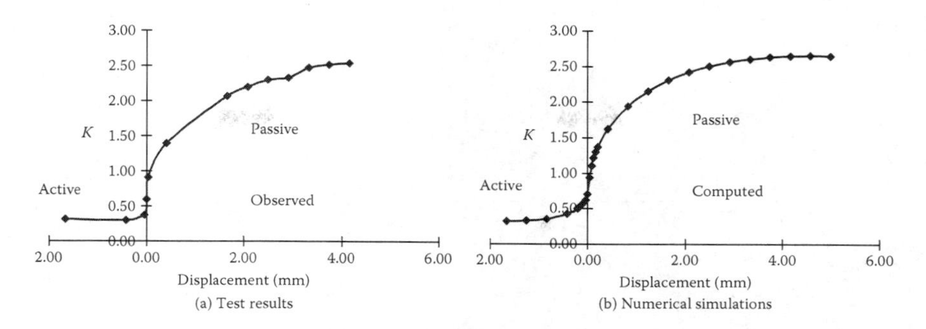

Figure 10.4 Variations of lateral earth pressure coefficient K with wall displacement in active and passive states.

stress along the moveable wall at rest.) It can be seen that there is good agreement between the observed results and computed ones. A small displacement of the wall is required only for developing a perfect active state, but a large displacement is required for developing a perfect passive state in the retaining wall problems, as described in the literature.

Figures 10.5(a) and (b) show the observed and computed profiles of the surface settlements for series I in the active state. In the figure, abscissa represents distance in horizontal direction from the wall, and the vertical axis represents surface settlements. Step 1, step 2, and step 3 imply the order of the slide block where the displacements are imposed. The computed results describe well the observed surface settlements qualitatively and quantitatively. It can be seen from the results of type A and type B in these figures that the maximum settlement does not increase significantly when the displacements are imposed at the middle and lower blocks. Also, there is not much difference in profiles of final surface settlements between type A and type B.

However, the profile of surface settlements for type C, in which displacement is applied to the three blocks simultaneously, is different from those of type A and type B: The surface settlements occur more widely, but the maximum settlements are smaller in type C. Therefore, it can be said that the region where subsidence occurs is overestimated and the maximum settlement is underestimated whenever the wall deflection process is ignored. Figures 10.6(a) and (b) show the observed and computed results in the passive state. It can be seen that surface heaving occurs more widely, but the maximum heave is smaller in type C than those in type A and type B.

Figures 10.7(a) and (b) show the observed and computed earth pressure distributions on the blocks for the three experimental cases in the active state. The vertical axis represents the depth of the wall and the horizontal one represents the earth pressure exerted on the block. The thick broken lines in these figures represent the earth pressure distribution at rest, and

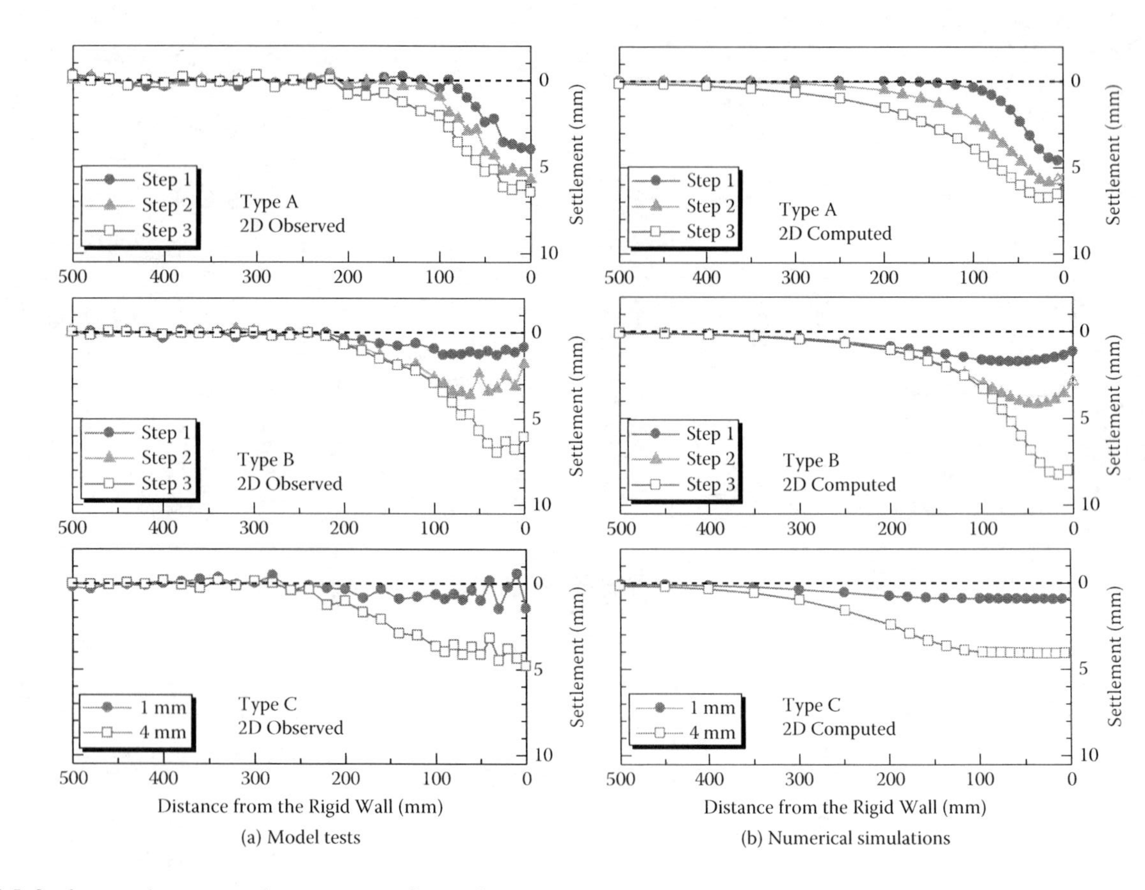

Figure 10.5 Surface settlements in the active state (series I).

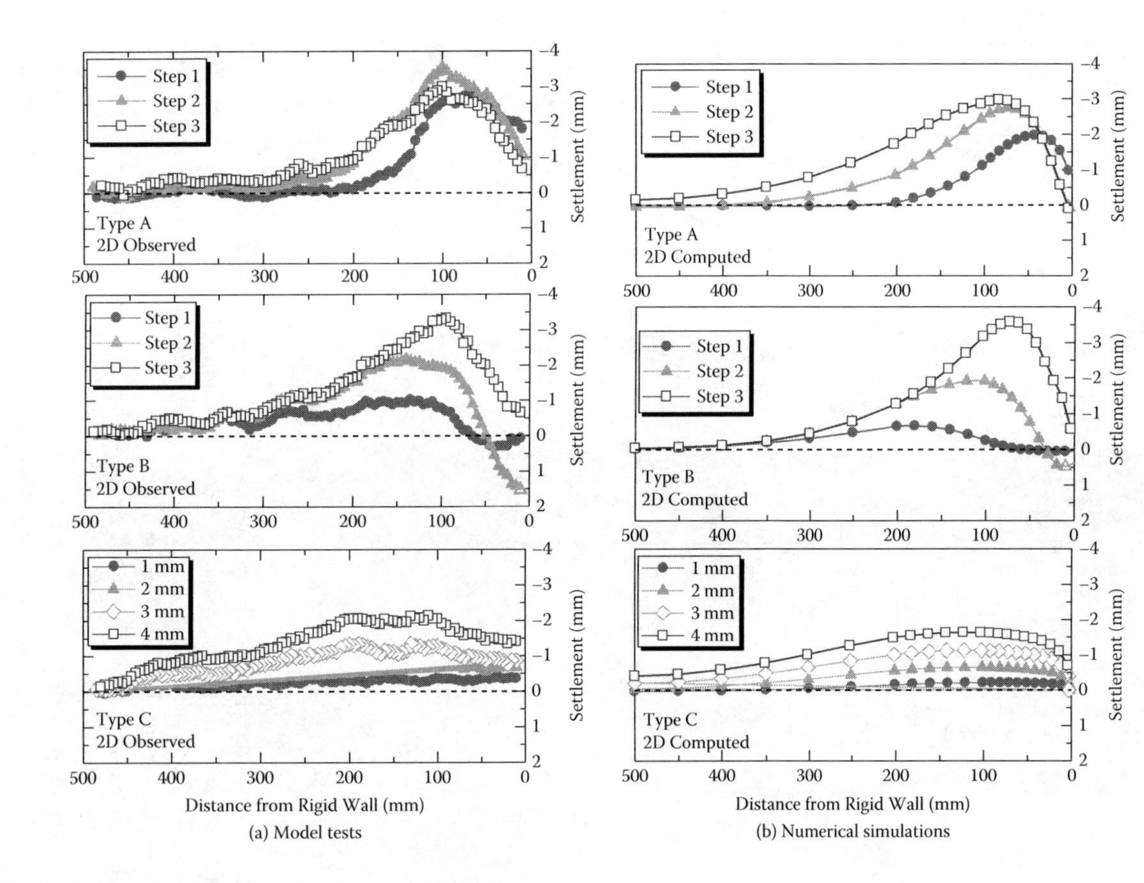

Figure 10.6 Surface settlements in the passive state (series I).

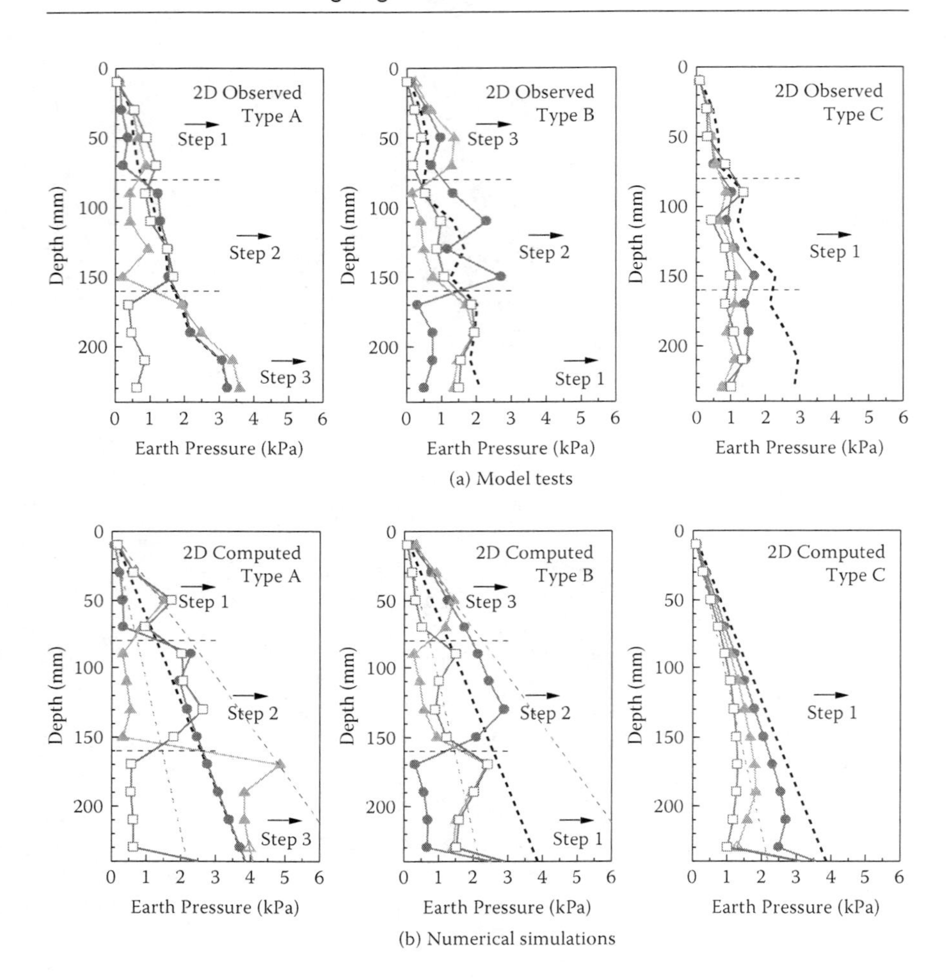

Figure 10.7 Earth pressure distributions on the wall in the active state (series I).

the thin broken lines in the numerical results correspond to the active and passive earth pressure distributions according to Rankine's theory. The final earth pressure distributions in these three types of wall deflection processes are summarized in Figure 10.8.

It can be seen from these figures that the magnitude of the earth pressure and the shape of earth pressure distribution in the active state are much influenced by the deflection process of the wall, even if the final deflection modes of the wall are the same. This is due to soil arching: Earth pressure on the moving block decreases and the earth pressure in the neighborhood increases in every step of the wall deflection processes. Further, it is also noticed from the results of type C that the earth

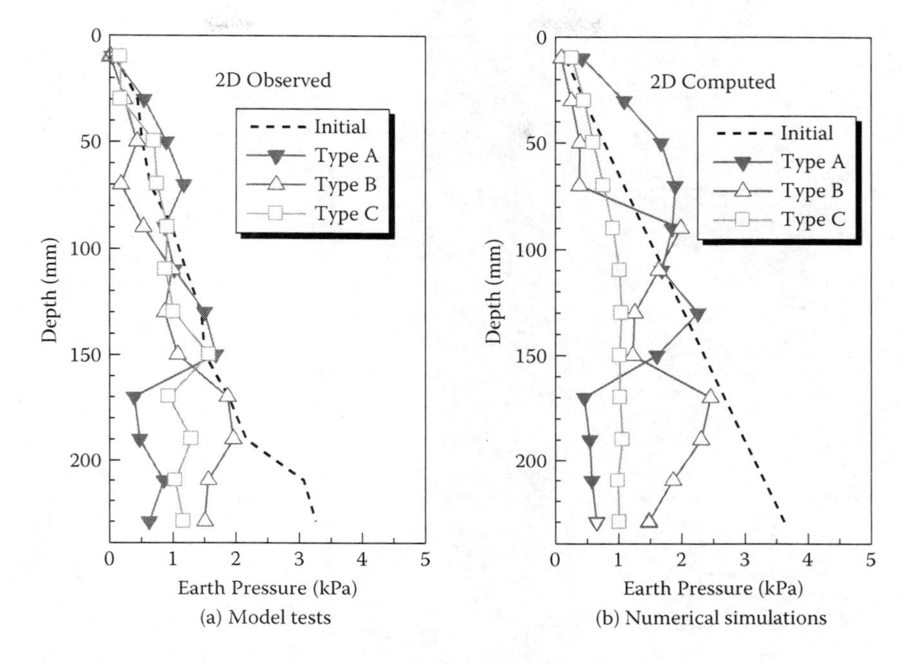

Figure 10.8 Comparison of final earth pressure distributions (series I).

pressure on the moving block decreases suddenly with a small deflection of the block that thereafter does not change much. The computed results describe well the differences in observed earth pressures between the three types of wall deflections.

Figures 10.9(a) and (b) show the observed and computed earth pressure distributions in the passive state. The earth pressure distributions are much influenced by the deflection process of the wall, in the same way as those in the active state. From the results in type A and type B, it can be noticed that when a block is pushed, the earth pressure on the block increases while the earth pressure in the vicinity of that block decreases by the arching effect.

Figures 10.10(a) and (b) show the observed and computed distributions of the deviatoric strains in the ground after a horizontal displacement of 4 mm is imposed on the three blocks in the active state. It can be seen that the zones where the deviatoric strains are relatively large develop in a narrower region closer to the wall in type A and type B than in type C. This is the reason why surface settlements in type A and type B occur more locally, as shown in Figure 10.5. The computed distributions correspond to the observed development of deviatoric strains for the three wall deflection processes.

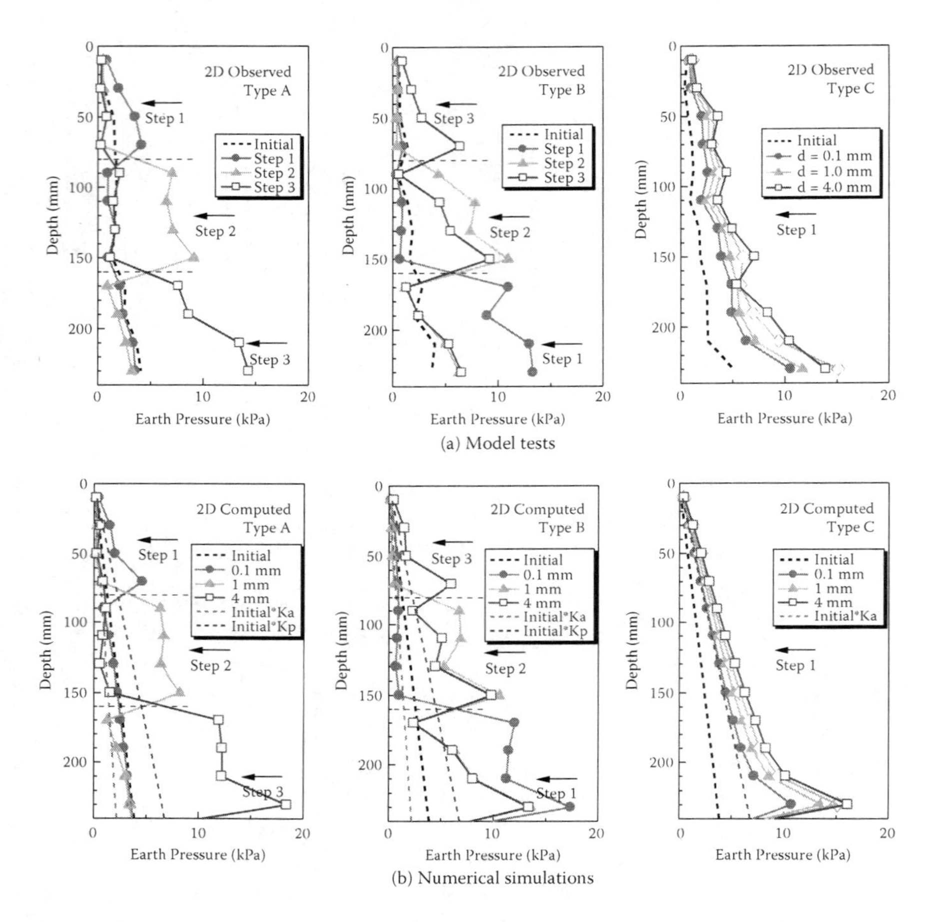

Figure 10.9 Earth pressure distributions on the wall in the passive state (series I).

10.1.3.2 Series II (influence of deflection mode of the wall)

Figure 10.11(a) shows the observed final profiles of the surface settlements for four kinds of wall deflection modes. The deflection modes imposed on the wall are indicated in Table 10.2. Figure 10.11(b) shows the corresponding computing profiles of the surface settlements. In every mode, the wall deflection is applied from upper to lower blocks. The computed results capture the observed settlements qualitatively and quantitatively, as for series I. It is seen from these figures that the surface settlement is not so large when the deflection of the upper part of the wall is small, even if the deflection of the lower part of the wall is large. Although the total amount of wall deflection in mode 2 and mode 3 is

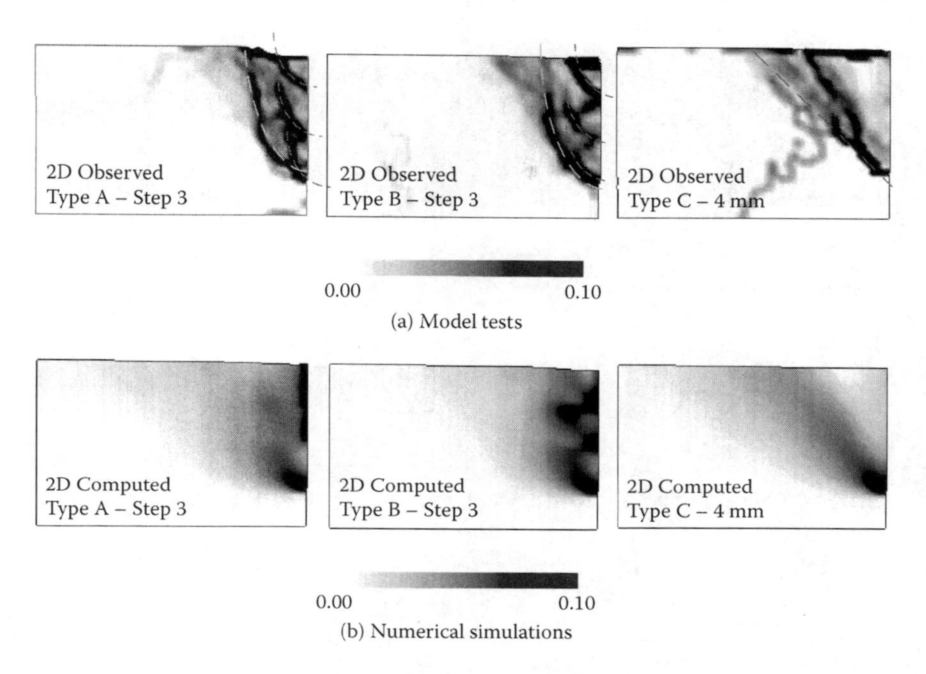

Figure 10.10 Distributions of the deviatoric strains in the ground at the active state (series I).

the same, the surface settlement in mode 2 is much larger than that in mode 3.

Figures 10.12(a) and (b) show the observed and computed earth pressure distributions in each step for series II, respectively. The computed results show good agreement with the observed ones. The final earth pressure distributions for these four types of wall modes are summarized in Figure 10.13. This figure shows that there is not much difference between these four types of deflection modes. In fact, it can be said from Figures 10.8 and 10.13 that the earth pressure distribution is not very much influenced by the deflection mode of the wall, but rather depends on the deflection process.

It has been stated in some textbooks that the shape of earth pressure distribution depends on the wall deflection mode (Terzaghi 1943; Tschebatarioff 1951; Yamaguchi 1969). This is true for the case when the wall deflects instantaneously. However, when the wall deflects gradually (progressively) from the upper part to the lower part (e.g., the open excavation with diaphragm wall), both the earth pressure distribution and its magnitude are not very much influenced by the deflection mode of the wall.

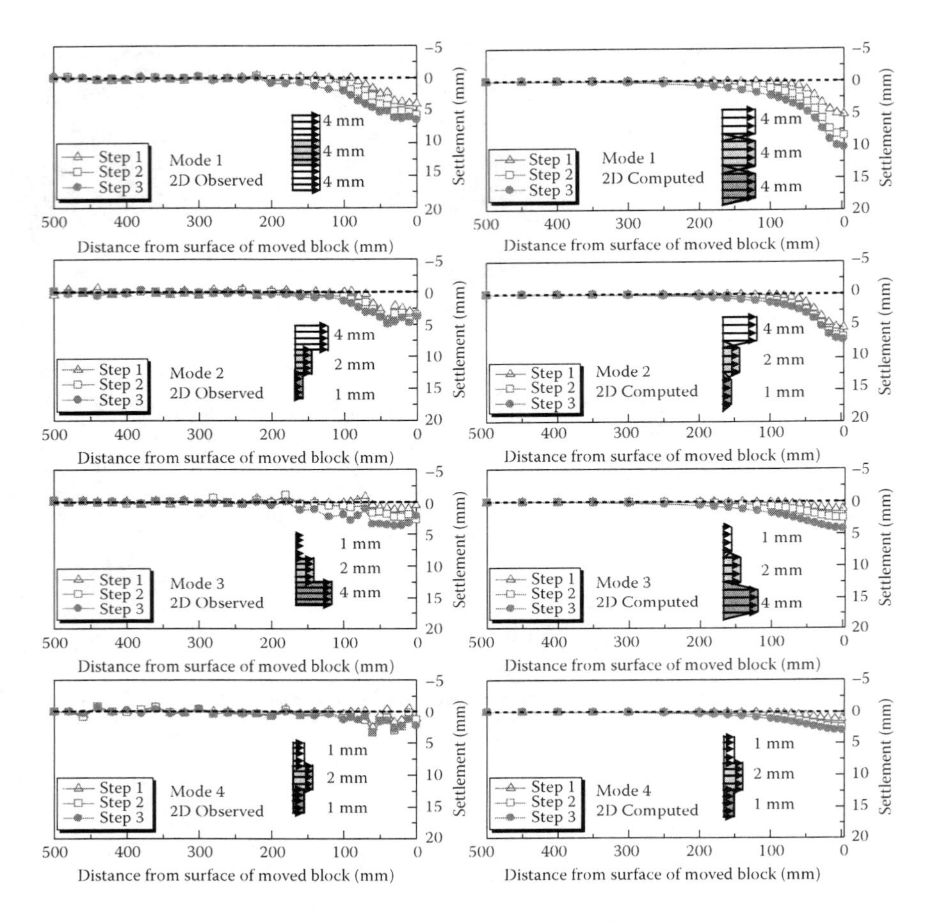

Figure 10.11 Surface settlements in the active state (series I).

10.2 BRACED OPEN EXCAVATION

As shown in the previous section, the earth pressure on the retaining wall and the ground movement in the backfill are very much influenced by the deflection process of the wall. On the other hand, in common design practice, both the earth pressure behind a retaining wall and its stability as a whole are evaluated within rigid plasticity theory such as Rankine's earth pressure theory. The wall deflection is usually calculated using the beam spring model with Rankine's earth pressure theory.

The resulting wall deflection then enters into an elastic finite element model that will then give an estimation of the ground deformations behind

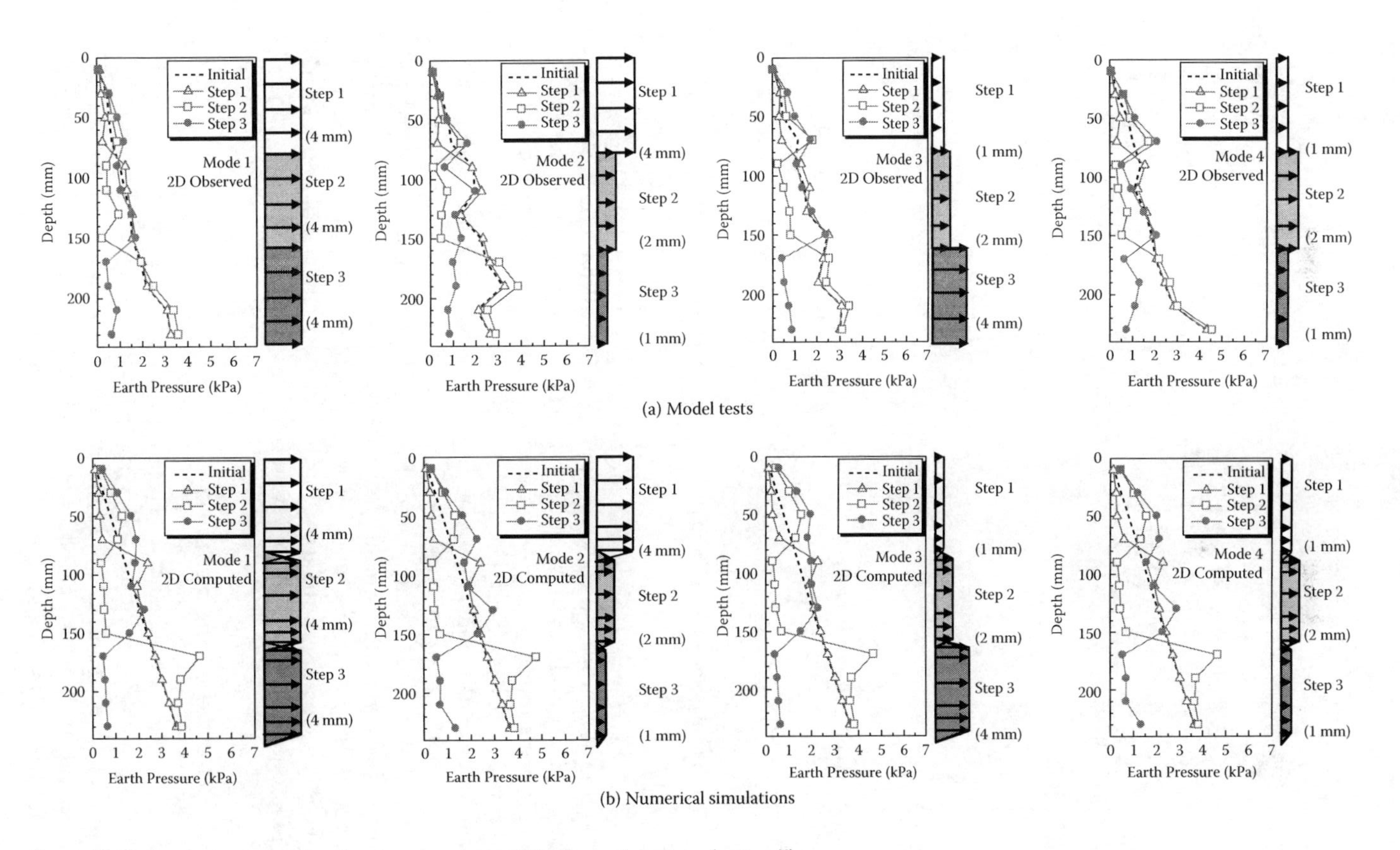

Figure 10.12 Earth pressure distributions on the wall in the active state (series II).

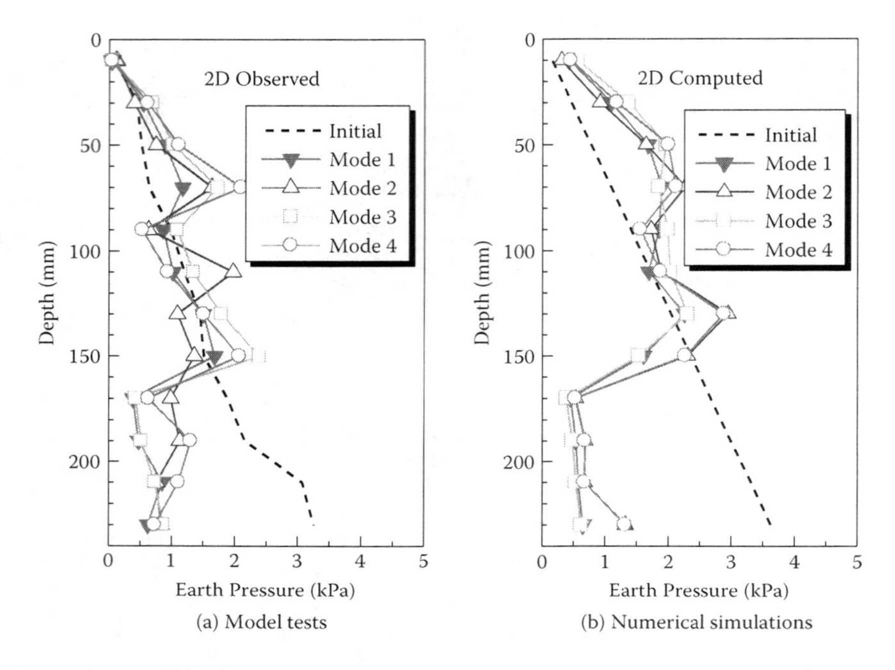

Figure 10.13 Comparison of final earth pressure distributions (series II).

the wall. Even other methods, in which the earth pressures, deflection of wall, and ground movements are obtained using different concepts, cannot consider properly the influence of the excavation sequence and other factors. It is also important to evaluate the influence excavation on nearby existing structures, particularly in dense urban areas.

Two-dimensional model tests and the corresponding numerical simulations have been conducted to investigate the deformation mechanism of ground, behaviors of existing structures, axial force in struts, and deformation of the retaining wall due to braced excavation. In the model tests and numerical simulations, four types of ground conditions are assumed: greenfield (no existing foundation), ground with strip foundation, ground with pile foundation, and ground with piled raft foundation.

10.2.1 Outline of model tests

Figure 10.14 shows a schematic diagram of the 2D apparatus. The size of the model ground is 68 cm in width and 45 mm in height. Aluminum rod mass is used as the model ground (unit weight of the mass is 20.4 kN/m^3) in the same way as in other 2D model tests. The retaining wall, 30 cm high × 6 cm wide × 0.5 mm thick, is a plate made of aluminum material ($EI = 0.88$

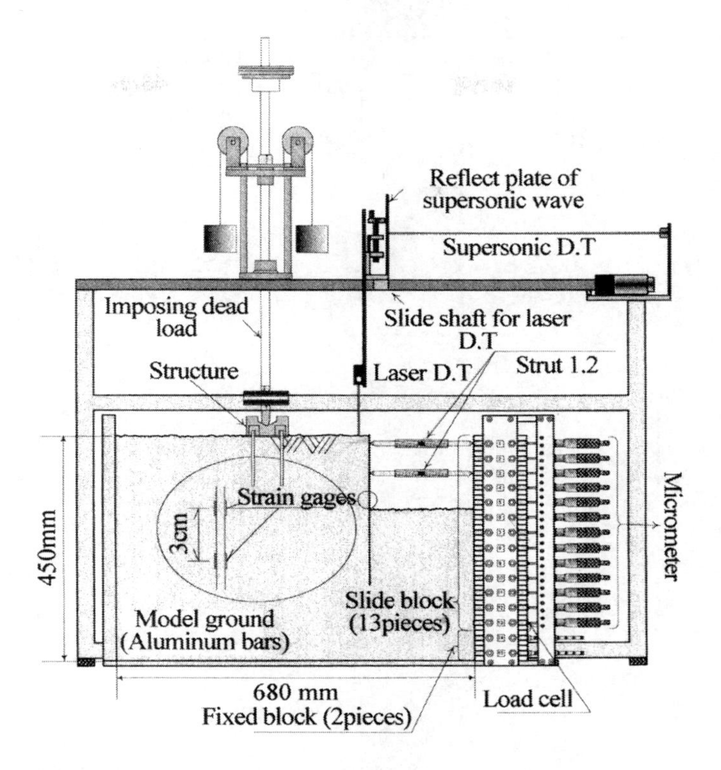

Figure 10.14 Schematic diagram of the apparatus.

N × m²/cm; $EA = 4.22$ kN/cm). It is set in the ground before excavation and located 20 cm from the right boundary of the ground. In the experiment, the model ground was excavated with a layer of 1.5 cm in thickness up to a depth of 18 cm in the ground. Two struts, located 1.5 and 7.5 cm from the ground surface, are set in the excavated ground when the excavation level reaches the depths of 3 and 9 cm, respectively. Both struts are made of aluminum consisting of springs between two aluminum rods. The stiffness of the upper strut is 3.64 kN/m/cm, and for the lower strut it is 3.04 kN/m/cm.

The dimension and stiffness of the wall and struts are chosen considering real ground conditions and assuming a scale of 1:100 between the model and prototype structures. The right side of the model ground is supported with 15 sliding blocks and two fixed blocks; each block contains a load cell that can measure axial force of the strut and horizontal earth pressure of the ground at the right-hand side. A laser type displacement transducer is used to measure surface settlement of the ground. By taking photographs with a digital camera and using the particle image velocimetry (PIV) technique, the ground movements and, consequently, the strain distribution in the ground are measured.

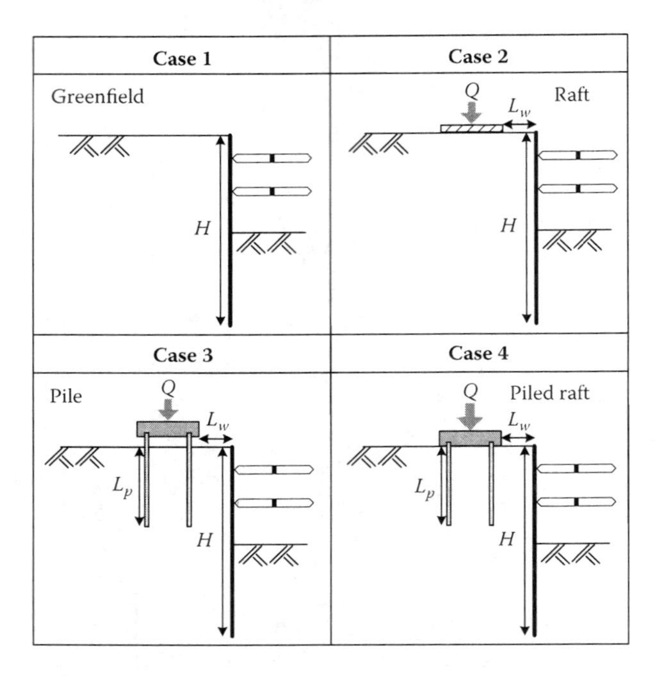

Figure 10.15 Patterns of experiments.

Figure 10.15 shows the experimental patterns considered in this research. Case 1 represents greenfield conditions where building load is not considered, while in cases 2–4, strip foundation, pile foundation, and piled raft are taken into account during braced excavation. The width of raft of the strip foundation is 8 cm, and 2D piles (polyurethane plates with thickness of 5 mm) are set up at the position of 1 cm from both edges of the raft (the distance between two piles is 5 cm) for pile foundation and piled raft foundation. In cases 2–4, the tests are carried out at the lateral distance $L_w = 4$ cm between the retaining wall and the edge of the raft of the foundation.

In the case of pile foundation and piled raft, pile length of 10 cm in the ground is considered. The dead load applied is about one-third of the maximum bearing capacity ground ($0.32 \times 9.8 = 3.14$ N/cm for case 2 and $0.60 \times 9.8 = 5.88$ N/cm for case 4). The size of the raft and the applied load for the strip foundation for cases 1–4 are the same as those used in the model tests in tunneling (see Figure 9.22). The dead load adopted in the case of the piled raft is larger than other foundation types to get significant deformation of the model ground, since the load-bearing capacity of the pile raft is larger.

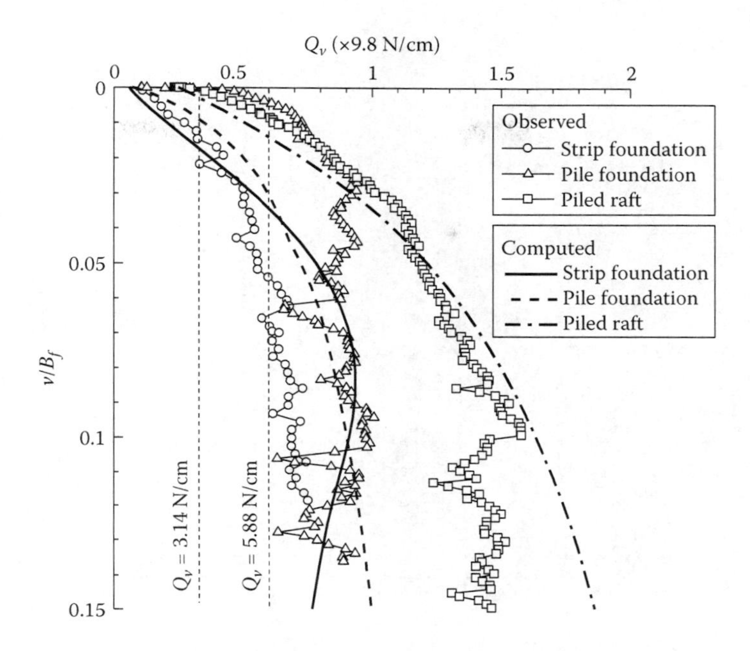

Figure 10.16 Load-displacement curves of three kinds of foundations.

Figure 10.16 shows the observed (symbols) and computed (curves) load-displacement curves for three kinds of foundations. The vertical dotted lines indicate the vertical loads applied to the foundations.

10.2.2 Outline of numerical analyses

Figure 10.17 shows a typical finite element mesh for the piled raft analyses. Isoparametric four-noded elements are used in the mesh for soil elements. Both vertical sides of the mesh are free in the vertical direction, and the bottom face is kept fixed. Analyses have been carried out at the same scale of the model tests considering plane strain drained conditions. The strip foundation is modeled as an elastic element, while piles and piled raft are modeled as a hybrid element (Zhang et al. 2000) consisting of elastic solid and beam elements. Elastoplastic joint elements (Nakai 1985) are used to model the interaction between the ground and the foundation, assuming an interface friction angle of $\delta = 18°$. The actions of struts onto the wall are modeled as elastic supports with axial stiffness.

The initial stresses, corresponding to the geostatic condition, are assigned to the ground in all numerical analyses by applying body forces due to the self-weight ($\gamma = 20.4$ kN/m^3). Before starting the excavation, the amount of

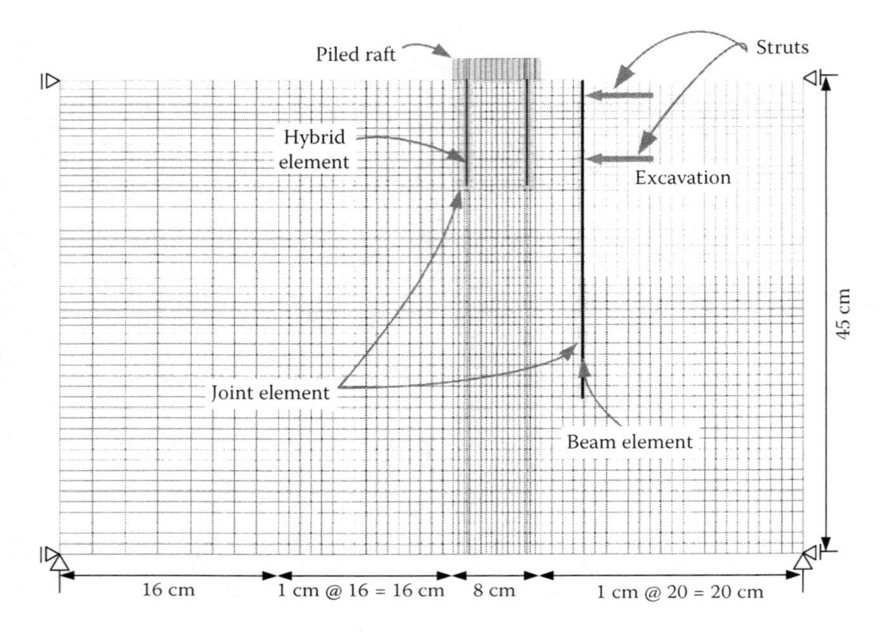

Figure 10.17 Finite element mesh for analysis (case 4).

dead load applied in the model tests is applied to the existing foundation as a nodal force in every simulation. The ground induced with the existing load is assumed as the initial ground condition for the braced excavation. The excavation is simulated by deactivating the soil elements, step by step, down to the required level.

10.2.3 Results and discussion

Figures 10.18(a) and (b) show the observed and computed horizontal displacements of the retaining wall for different cases: greenfield condition, with strip foundation, with piled foundation, and piled raft. The abscissas represent horizontal displacement and the vertical axes show the ground depth from the surface. The legend represents excavation depth (d). Here, the results at different excavation depths are shown to illustrate the deformation of the wall during various excavation stages. The arrows in the figures indicate the position of the upper and lower struts. It is seen in the figures that, for the case where nearby structures exist close to the wall, the wall deflects more compared to case 1, where building load is not considered (greenfield condition).

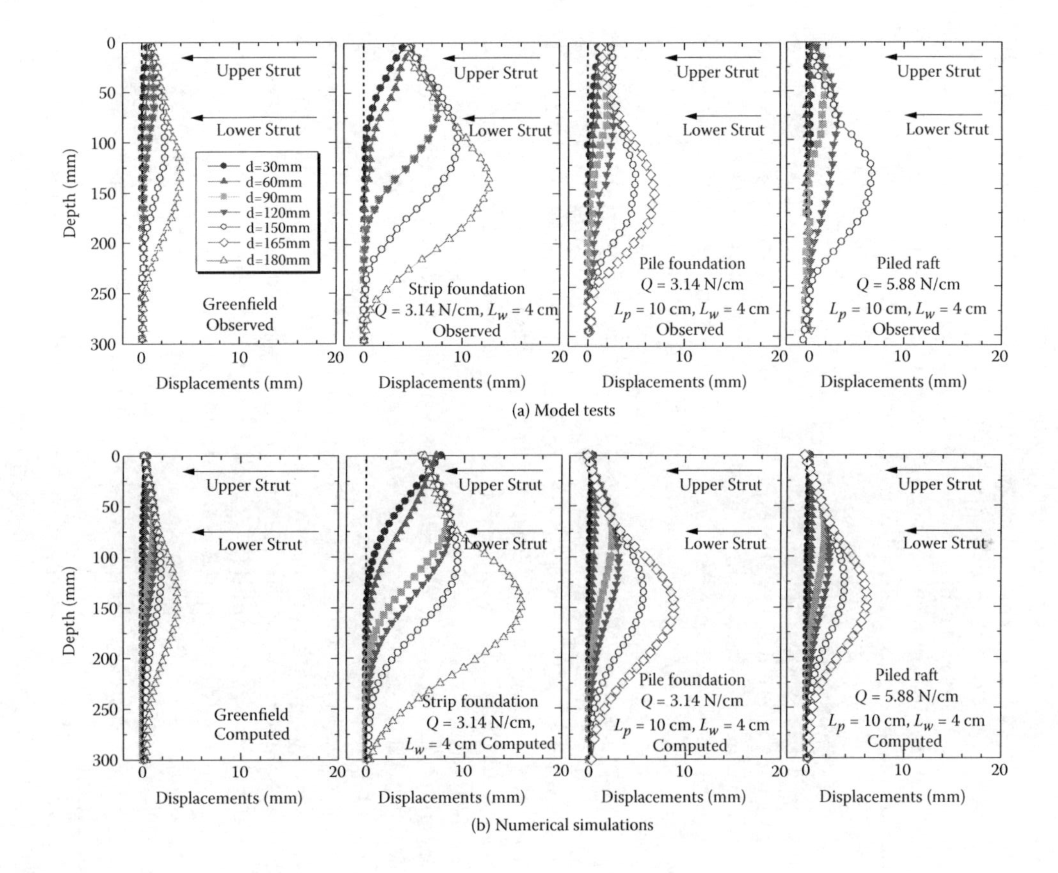

Figure 10.18 Horizontal displacements of retaining wall.

Comparing the results pertaining to the various foundation types, it is found that, for the strip foundation, the horizontal displacement is larger than that for the piled foundation and the piled raft. Although the magnitude of the applied load is larger in the piled raft, the deflection of the wall is smaller compared to the strip foundation, even for the shorter pile length (10 cm). As the excavation advances, the location where the maximum horizontal displacement occurs is deeper. In the cases of piled foundation and piled raft, if the excavation depth is shallower than the pile length displacement of the retaining wall is restrained because of the existing end-bearing capacity of the pile. However, once the excavation depth passes the pile tip, larger displacement of the wall occurs at the level below the tip of the pile. The results of the numerical analyses show the same tendency of the model tests both qualitatively and quantitatively.

Figure 10.19 shows the observed and computed surface settlement troughs for different cases corresponding to the deflection of the retaining wall. The abscissas represent the distance from the retaining wall and the vertical axes show surface settlement of the ground. It is seen in the figures that, for the nearby structures, the settlement of the ground is larger than that for case 1 (greenfield condition), and the maximum surface settlement occurs at the position of the foundation regardless of the foundation types when the structures exist within a certain distance from the wall. Comparing the results of the foundation types, it is found that the surface settlements at the position of the foundation for the strip foundation are larger than those for the pile foundation and the piled raft.

In contrast, the surface settlements adjacent to the wall for the pile foundation are larger than those for the strip foundation. Surface settlement for the piled raft with a larger amount of applied load is smaller than the other two foundation types. At the initial stage of the excavation (i.e., when the excavation depth is shallower), the strip foundation tilts toward the excavation, On the other hand, when the excavation depth is getting deeper, the foundation inclines to the opposite direction of the excavation, regardless of the foundation types. For the pile foundation and the piled raft, when the excavation depth passes the pile tip, the foundation rotates to the opposite direction of the excavation side. The finite element analyses can capture well the displacement and rotation of the foundation.

Figure 10.20 illustrates the distributions of the deviatoric strain in the model tests and the finite element analyses. The distributions of deviatoric strain of the model tests are obtained using the PIV technique in the same way as in other model tests. Here, deviatoric strains for two excavation depths are shown to demonstrate the sequence of the development of the deviatoric strain with the progress of excavation. It is seen that,

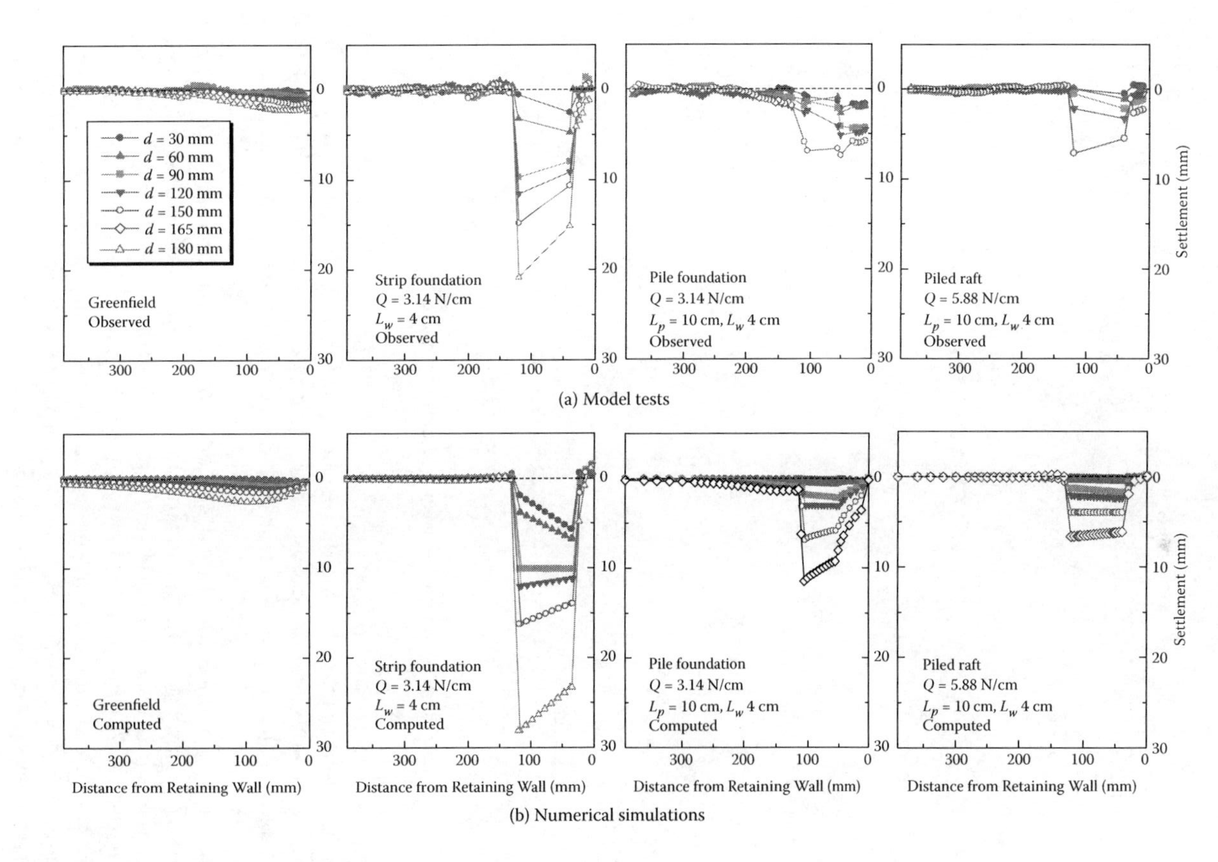

Figure 10.19 Surface settlement troughs.

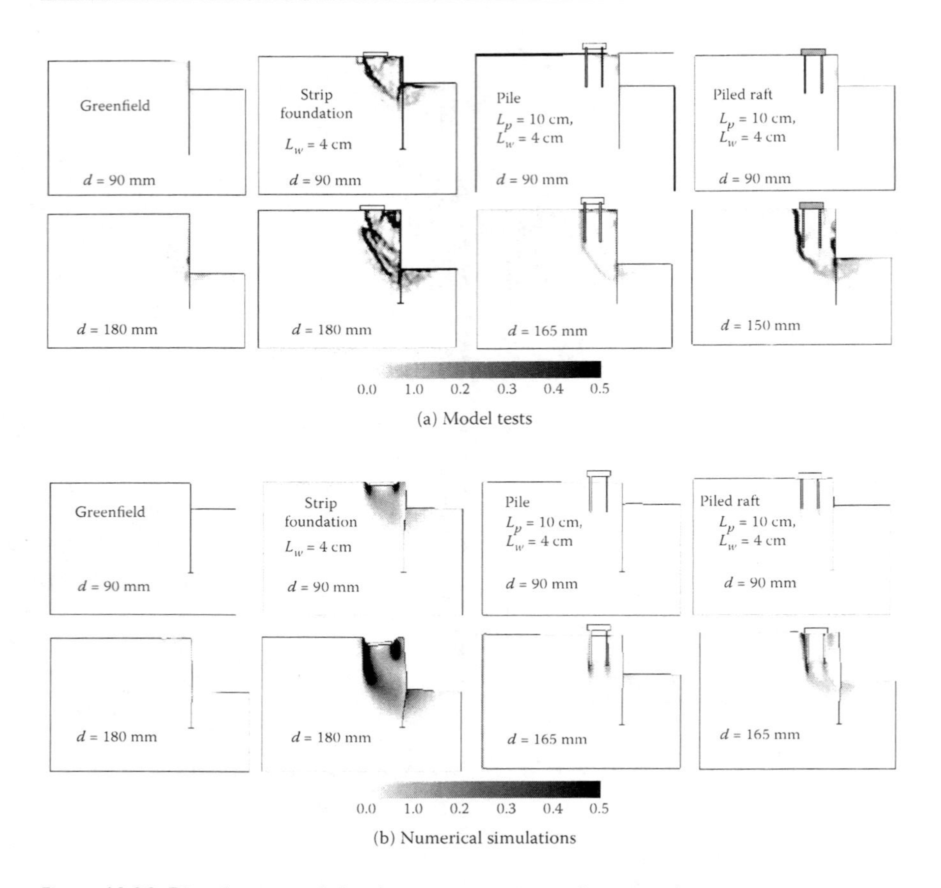

Figure 10.20 Distributions of the deviatoric strains in the ground.

for the existing building loads, a shear band develops from the excavation part and reaches both sides of the foundations when the excavation proceeds to some depth. A significant ground deformation is observed in the edges of the strip foundation and causes larger settlement, as seen in Figure 10.19.

In the pile foundation and the piled raft, a shear band also develops toward the pile tips with the advancement of the excavation. For the excavation depth deeper than the pile tip, concentration of deviatoric strain increases toward the pile tip that is farther from the wall. It is understood that these different developments of deviatoric strains are reflected in the difference of settlements and rotation for three types of foundations. The deviatoric strains of the numerical analyses show very good agreement with the results of the model tests.

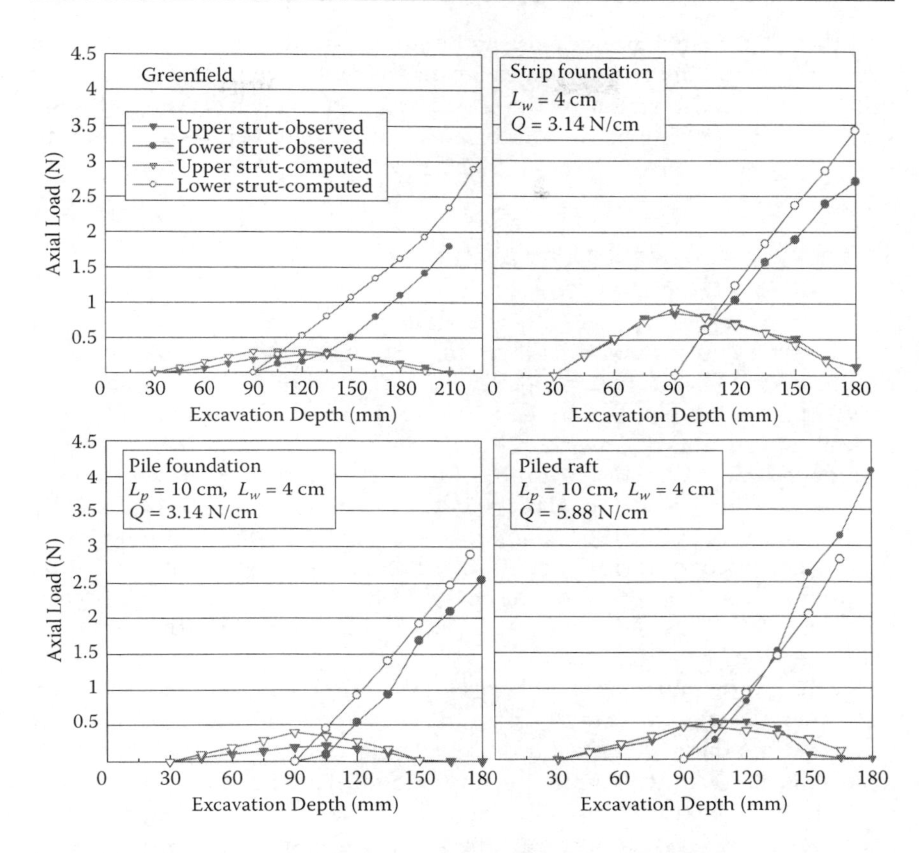

Figure 10.21 Variations of axial forces of struts with excavation depth.

Figure 10.21 illustrates the observed and computed changes of axial force of the upper and lower struts for different foundation types. The abscissas represent excavation depth, and the vertical axes show axial force in newtons. The figures show that, in the case of existing structures, axial forces for both upper and lower struts are larger than those for case 1, where building load is not considered. Comparing the results of the foundation types, it is found that, for the pile foundation and the piled raft, the maximum value of axial force of the lower strut and proportion of axial force with the upper strut are significantly larger than those for the strip foundation. The axial force of the upper strut becomes smaller after the lower strut is set up for all cases. The induced axial force of the lower strut is larger during the braced excavation. The numerical analyses capture well the observed axial force of the upper and lower struts in all conditions.

Although the influence of existing building loads on the behavior of the wall and ground in braced open excavation is described in this section, the influences of construction method, construction sequence, wall stiffness, wall friction, and others are discussed in other papers (Nakai, Kawano, et al. 1999; Nakai, Farias, et al. 2007).

10.3 STRIP FOUNDATION AND PILED RAFT FOUNDATION

The bearing capacity of soils is an important factor in many construction projects. Therefore, a proper prediction method of bearing capacity of soils is inevitably required for an accurate and economical design of foundations. Until now, bearing capacity has been mostly predicted using rigid plastic theories and/or empirical equations. The most common and widely used bearing capacity equation was proposed by Terzaghi (1943). Since then, many researchers have focused on developing an accurate empirical equation for bearing capacity of soils—for example, Hansen (1961), Meyerhof (1963), Vesic (1975), and Bolton and Lau (1993). However, these equations ignore many fundamental characteristics of soils.

Because of the inherent uncertainty of the bearing capacity analysis, the allowable bearing capacity is determined by reducing the applied stress from the foundation load using a large value of the safety factor, which makes the design of the foundation uneconomical in most cases. Even when estimating the bearing capacity using numerical analysis, a rigid, perfectly plastic analysis is used, although the behavior of soils is elastoplastic in nature. Considering these important aspects, bearing capacity of soils in 2D conditions is investigated with model tests and numerical analyses. Two types of foundations—strip foundation and piled raft—are used to check the load-displacement responses of different types of foundations. Changing from the location and the direction of loading on the foundation, the importance of loading conditions on the bearing capacity problem will be discussed.

10.3.1 Outline of model tests

Two-dimensional model tests of bearing capacity problems have been carried out using the apparatus in Figures 10.22 and 10.23. The figures show the case of piled raft foundation. Aluminum rod mass is used as the model ground in the same way as in other 2D model tests. The foundation is moveable in the horizontal direction with a slider. As shown in Figure 10.24(a), two types of foundations are employed in the model tests: strip foundation and piled raft foundation. The length of pile is $L_p = 12$ and 24 cm in the

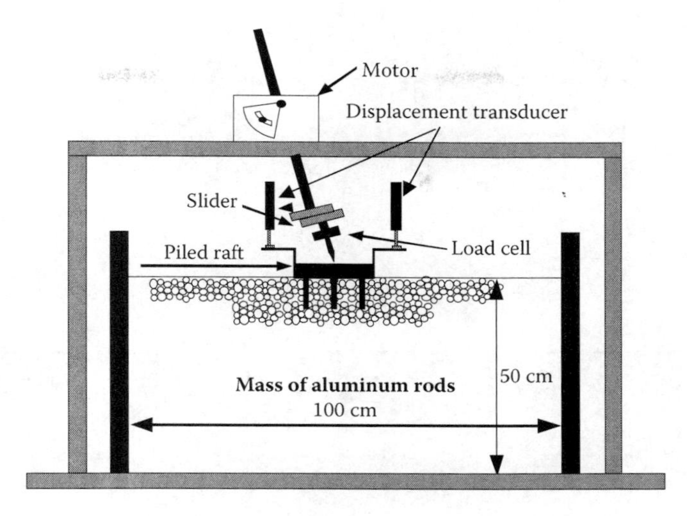

Figure 10.22 Apparatus for model tests of bearing capacity problems (piled raft foundation).

piled raft. Every foundation has a width of 12 cm and a height of 3 cm, and the width of the pile is 1 cm.

In piled raft foundations, stiff pile (elastic modulus: $E_p = 9.8 \times 10^9$ kPa) and flexible pile ($E_p = 8.8 \times 10^2$ kPa) are used to check the influence of stiffness on the behavior of bearing capacity of piled raft. In the case of flexible pile, the size and stiffness of the piled raft are obtained from the real field value considering the similarity ratio of 1:100 between the model and prototype structures. Three kinds of loading patterns are employed, as shown

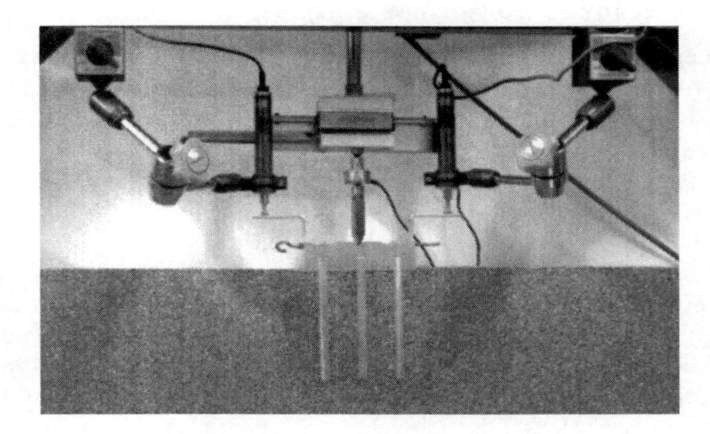

Figure 10.23 Piled raft foundation, measuring instruments, and model ground.

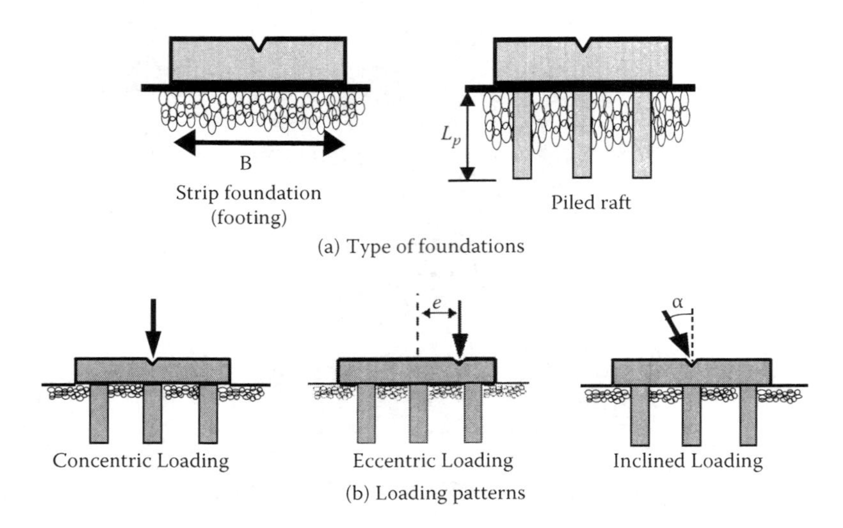

Figure 10.24 Explanation of foundations and loads.

in Figure 10.24(b): concentric loading, eccentric loading, and inclined loading. The load is applied at a height of $B/4 = 3$ cm from the bottom of the foundation in every test (B: width of the foundation). In the 2D model tests, the vertical and horizontal displacements (v and h) at the center of the foundation and its rotation angle (θ) are obtained from the vertical displacements of the two edges of the loading plate and the displacement of the slider. Digital photographs taken during the experiments recorded the ground movement under 2D conditions.

10.3.2 Outline of numerical analyses

For simulating 2D model tests, plane strain finite element analyses were carried out under drained conditions using FEMtij-2D. Figure 10.25 shows the mesh for a typical 2D analysis, where the bottom boundary is fixed and the lateral boundaries are free in the vertical direction. Isoperimetric four-noded quadrilateral elements are used to model the ground in the analysis.

The 2D model ground in every analysis is made by self-consolidating aluminum rod mass (unit weight = 20.4 kN/m^3, void ratio = 0.35 at a very low confining stress, $p = 9.8 \times 10^{-6}$ kPa). The strip foundation and the raft of the piled raft foundation are assumed to be an elastic material with large stiffness, and the frictional behavior between the foundation and the ground is simulated by an elastoplastic joint element. The friction angle between the foundation and the aluminum rod mass is assumed to be 15°. The analyses are carried out with the same conditions of the model tests.

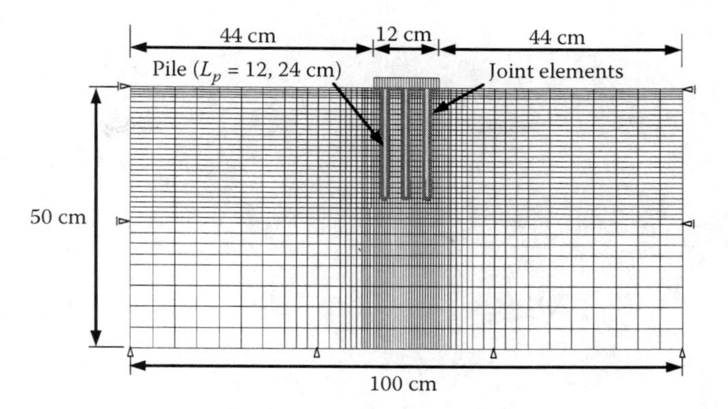

Figure 10.25 Finite element mesh for piled raft foundation.

After making initial ground, displacement is applied at a specific node on the strip foundation or raft according to the loading conditions.

10.3.3 Results and discussion

Figures 10.26–10.30 show the results of 2D model tests and corresponding numerical analyses under concentric loading, eccentric loading, and inclined loading. Diagram (a) in each figure refers to results of the foundations with short piles (L_p = 12 cm), and diagram (b) is the results of the foundations with long piles (L_p = 24 cm). Both diagrams show the results of footing (L_p = 0 cm) as well. In all figures, dots and curves denote the observed and computed results, respectively.

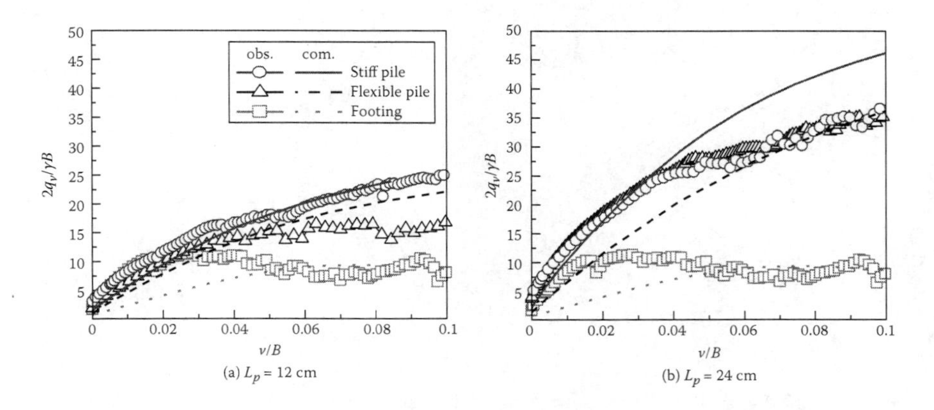

Figure 10.26 Vertical load versus vertical displacement under concentric loadings.

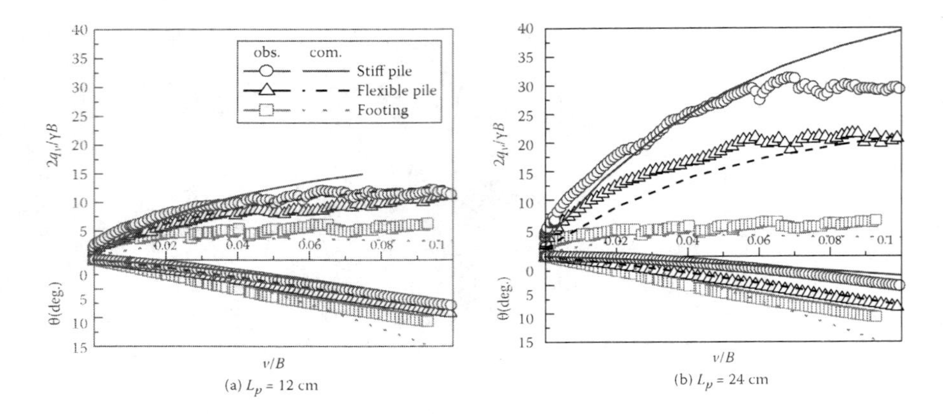

Figure 10.27 Vertical load and rotation angle versus vertical displacement under eccentric loadings.

Figure 10.26 shows the results between the vertical load and the vertical displacement of foundations with different types of foundations under concentric loading. The vertical axis indicates the load normalized with γB (γ is unit weight of soil and B is width of foundation) that corresponds to the coefficient of bearing capacity, and the abscissa indicates the displacement normalized with B. The observed load-displacement relations do not start from the origin because of the self-weight of

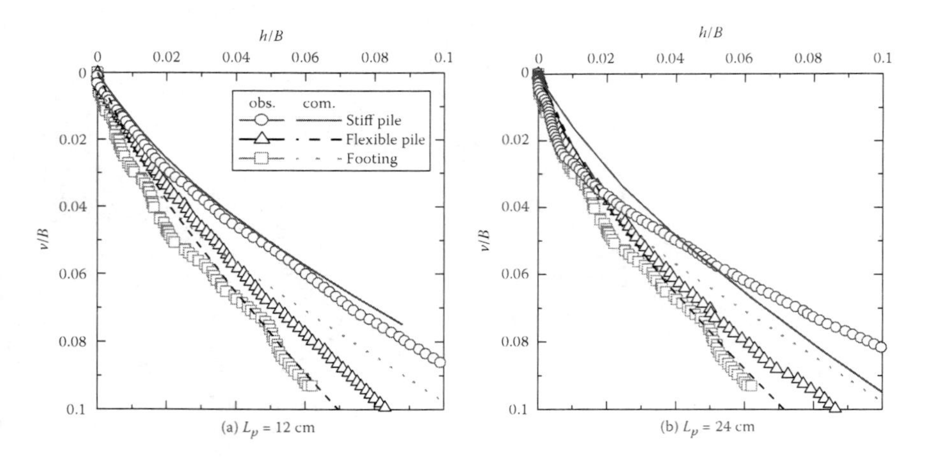

Figure 10.28 Vertical displacement versus horizontal displacement under eccentric loadings.

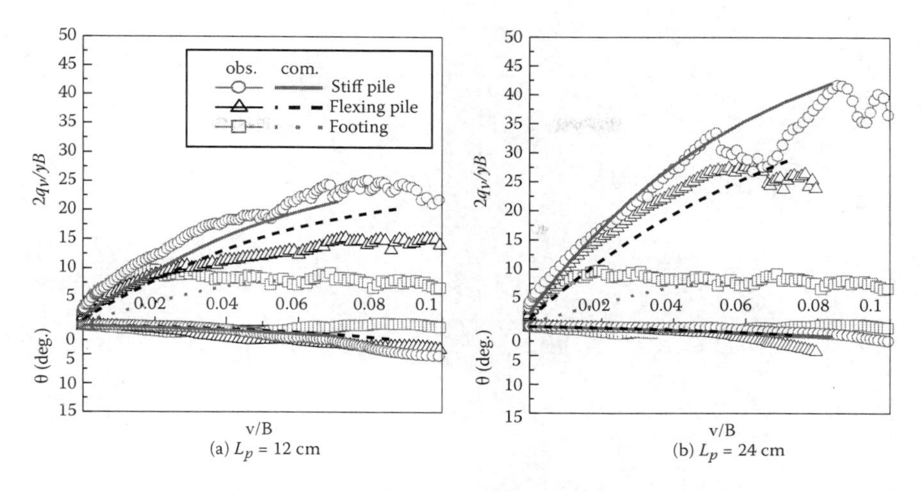

Figure 10.29 Vertical load and rotation angle versus vertical displacement under inclined loadings.

foundations. The computed results not only qualitatively but also quantitatively predict well the observed load-displacement curves of various kinds of foundations.

Figure 10.27 shows the observed and computed load-displacement relations under eccentric loading (eccentric ratio: $2e/B = 0.5$). In this figure, not only the normalized vertical load ($2q_v/\gamma B$) but also the angle of rotation (θ) of the foundation are plotted against the normalized vertical displacement (v/B). The angle in a clockwise direction is defined as positive. It is seen from the observed results that, though the bearing capacity and stiffness of the ground under eccentric loading are smaller than those under concentric loadings (except for the case of a foundation with stiff, long piles), the length of piles does not influence the rotation of foundation very much. The computed results describe reasonably well such observed load-displacement relations and rotation of the foundation.

Figure 10.28 shows the relation between the normalized vertical displacement (v/B) and the normalized horizontal displacement (h/B) of the foundation. The ratio of horizontal displacement to vertical displacement, h/v, is influenced by the stiffness of pile more than by the length of piles because there is not much difference between diagrams (a) and (b).

Figures 10.29 and 10.30 show the results under inclined loading (inclination angle $\alpha = 5°$). The results show that the load-displacement relations are almost the same as those for vertical loading, and small rotations occur

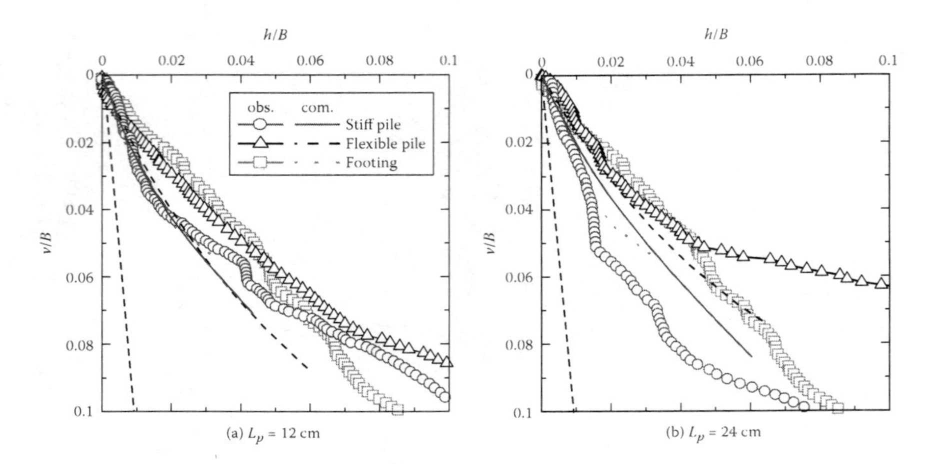

Figure 10.30 Vertical displacement versus horizontal displacement under inclined loadings.

in every foundation. However, the normalized vertical displacement (v/B) against the normalized horizontal displacement (h/B) of the foundation is rather large, even if the inclination angle is small ($\alpha = 5°$). We can see that the ratio of displacement, h/v, is larger than the ratio of horizontal load to vertical load (thin dotted straight line in the figure) under inclined loading. These observed behaviors of the foundation can be well simulated by the present analyses.

Chapter 11

Reinforced soils

11.1 REINFORCED FOUNDATION UNDER UPLIFT LOADING

Foundations with reinforcements stemming from the side of the foundation diagonally downward or horizontally were developed and constructed to increase the uplift bearing capacity of electric transmission towers and other similar structures (Matsuo and Ueno 1989; Tokyo Electric Power Company and Dai Nippon Construction 1990). To investigate the effect of the reinforcements, two-dimensional (2D) and three-dimensional (3D) model tests of the foundations using flexible reinforcements under vertical and inclined uplift loading conditions were carried out. The main objective was to investigate the mechanism of reinforcement and to find out the most effective arrangement under different loading conditions. Attention was paid particularly to the direction and position of the reinforcements. The numerical analyses were also carried out with finite element codes FEMtij-2D and FEMtij-3D. More detailed experimental and numerical results on this topic are described in Nakai et al. (2010).

11.1.1 Outline of model tests

Figure 11.1 shows a schematic diagram of the apparatus used for the 2D model tests. The model of the ground was 100 cm wide and 50 cm high. It was composed of the aluminum rod mass, for which the unit weight was 20.4 kN/m^3, as mentioned before. Figure 11.2 presents the schematic diagram and a photo of the apparatus used for the 3D model tests. The 3D model of the ground was 100 cm wide, 80 cm long, and 50 cm high. It was composed of alumina balls with unit weight of 21.5 kN/m^3. The foundation piles were set up before building the model of the ground for both 2D and 3D conditions. These piles were 23 cm long and penetrated 18 cm into the ground. The pile used in the 2D model tests was rectangular and 6 cm wide; the foundation used in the 3D model tests had the same dimensions but was cylindrical (see Figure 11.3).

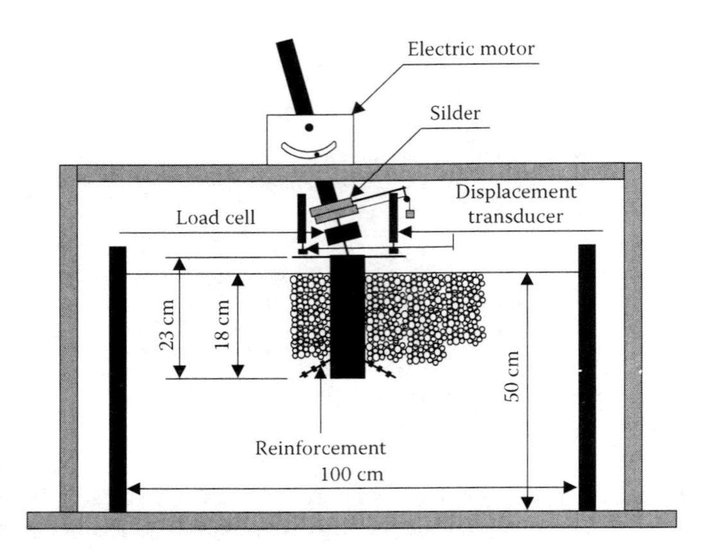

Figure 11.1 Schematic diagram of 2D model test apparatus (inclined uplift loading).

Figure 11.4 illustrates the arrangement of the reinforcements. Due to construction constraints in real cases, the length of reinforcement bars stemming from the sides of the foundation should be smaller than the caisson diameter. Therefore, these lateral bars were 5 cm long for the foundation diameter of 6 cm in the model tests. However, the reinforcements stemming from the bottom of the foundation can be longer. Two series of reinforcement patterns were investigated. The first series considered reinforcements stemming from a constant depth of 15 cm from the surface and inserted in three different directions: diagonally upward ($\beta = -30°$), horizontally ($\beta = 0°$), and diagonally downward ($\beta = 30°$). The second series analyzed the case where 10 cm long reinforcements stemmed vertically ($\beta = 90°$) and diagonally downward ($\beta = 60°$) from the bottom of the foundation, located at 18 cm from the ground surface.

Most tests used flexible aluminum plates as reinforcements. The reinforcements were 0.1 mm thick and this provides enough stiffness in tension, but negligible bending stiffness. In the tests of series I, reinforcement with 3 mm thickness was also used as stiff reinforcement having sufficient stiffness against tension and bending. In order to make the reinforcement surface rougher, 1.6 mm diameter aluminum rods spaced 1 cm from each other were glued along the upper and lower faces of the aluminum plates. The friction angle between the reinforcement and the ground was approximately 20° and 14.5° for the 2D and 3D model tests, respectively. Vertical and inclined upward pull out forces were imposed on the foundation continuously. The inclination angles used in the tests were $\alpha = 0°$, 15°, and 30°.

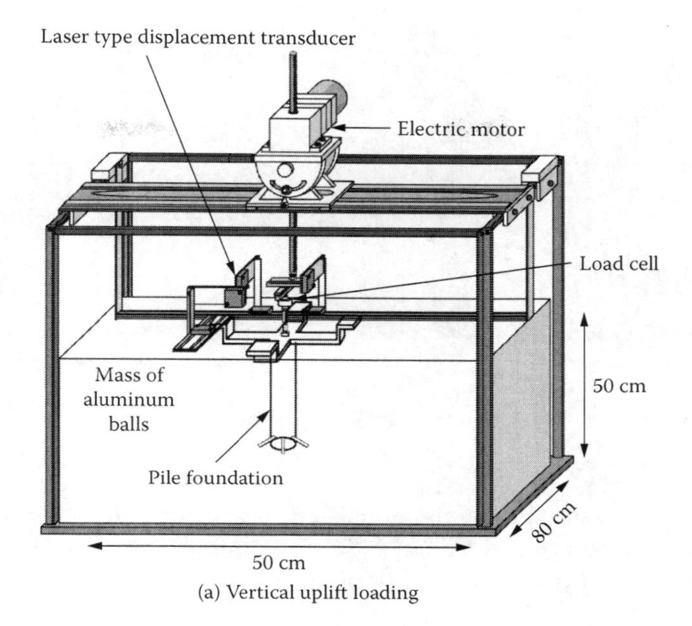

(a) Vertical uplift loading

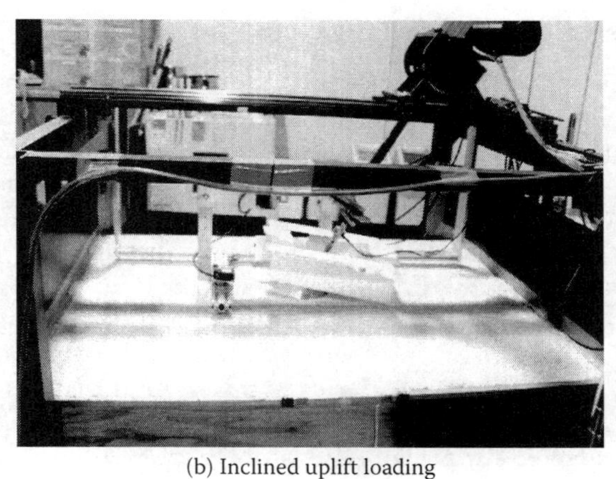

(b) Inclined uplift loading

Figure 11.2 Schematic diagram and picture of 3D model tests.

Figure 11.5 shows the direction of the uplift load and the plan view of the reinforcement arrangement for 3D conditions. During the tests, the displacements were controlled.

Displacement transducers and a load cell measured both the displacement and uplift loading of the foundation. In the 2D model tests, the rotation angle (θ) was measured from the displacements of the two edges of the loading plate. Digital photographs taken during the experiments recorded

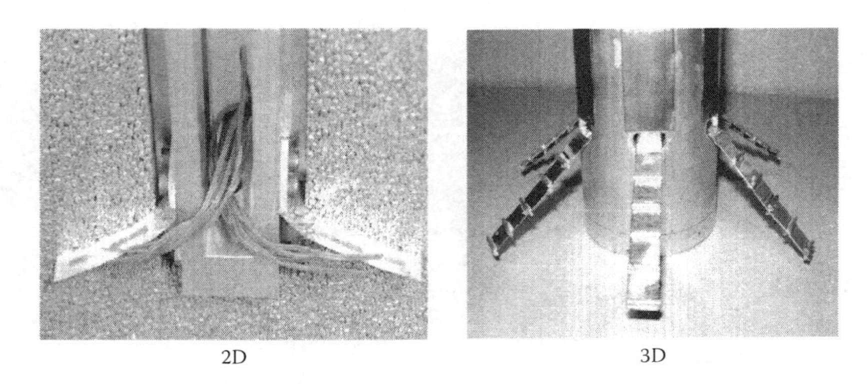

Figure 11.3 Pictures of piles with reinforcement (series I).

the ground movement under 2D conditions. In the 3D model tests, six laser transducers measured the foundation's displacements and rotations in 3D space. The reinforcement's axial force and bending moments were measured by strain gauges glued on both of its sides.

11.1.2 Outline of numerical analyses

The 2D finite element analyses considered plane strain and drained conditions, and they were carried out in the same scale as the 2D model tests. 3D finite element analyses were conducted for the 3D model tests as well. Figures 11.6(a) and (b) represent, respectively, 2D and 3D finite element meshes for the cases where the reinforcements stem diagonally downward from the bottom of the foundation. The initial state of the model ground was created by applying self-weight body forces to simulate 1D self-consolidation, which makes a ground at geostatic (K_0) conditions. Because the pile is a replacement pile, the numerical analyses do not account for installation procedures. The foundation was modeled assuming an elastic material with enough stiffness (the deformation of the pile can be neglected for the amount of load applied). In order to satisfy plane strain conditions, the reinforcements were simulated by beam elements in the 2D analyses.

On the other hand, the reinforcements were modeled using shell elements of a given width and thickness in the 3D analyses. For 2D analyses, axial stiffness and bending stiffness are assumed as $EA = 8.44 \times 10$ kN/cm and $EI = 7.04 \times 10^{-8}$ kNm²/cm for flexible reinforcements, and $EA = 2.95 \times 10^3$ kN/cm and $EI = 1.90 \times 10^{-3}$ kNm²/cm for stiff reinforcements, respectively. For 3D analyses, the values for axial and bending stiffness of each reinforcement were $EA = 7.03 \times 10$ kN and $EI = 5.86 \times 10^{-8}$ kNm², respectively. The frictional behavior between the wall and the

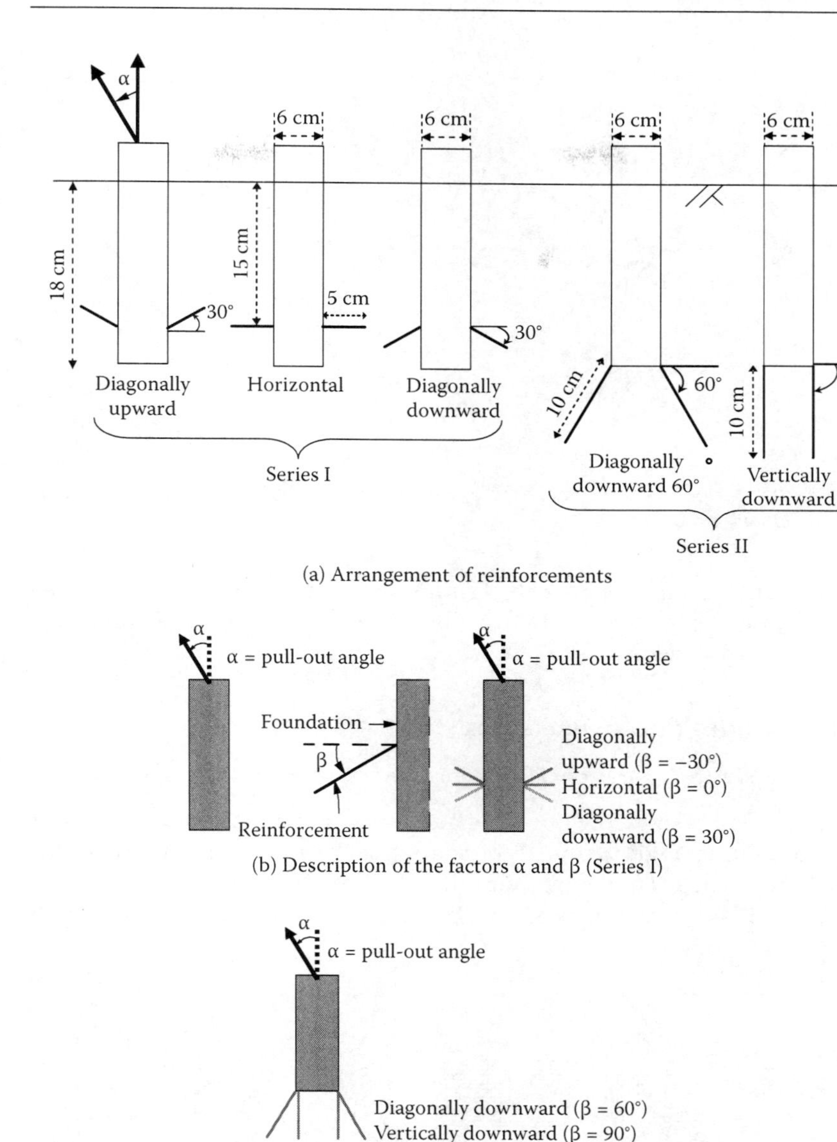

Figure 11.4 Explanations of reinforcements and applied loads.

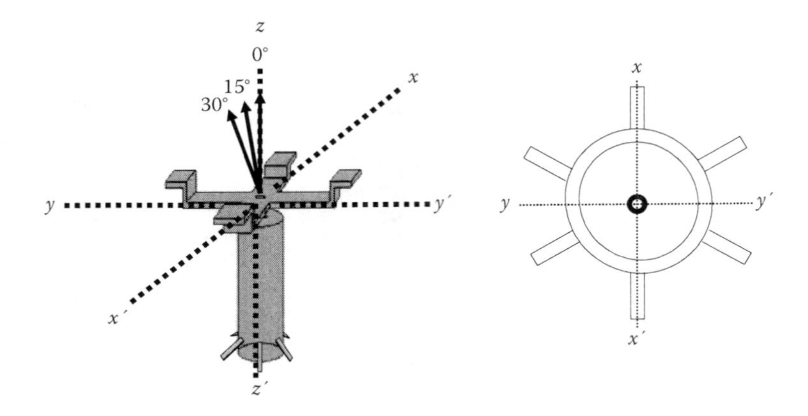

Figure 11.5 Direction of uplift load and plan view of the arrangement of reinforcements in 3D model tests.

ground was simulated using the elastoplastic joint elements. The friction angles of the joint elements used in the analyses were obtained from the laboratory experiments.

11.1.3 Results and discussions

11.1.3.1 Series I (influence of reinforcement direction)

Figures 11.7–11.9 show the observed and computed variations of the uplift load and rotation angle (θ) of the foundations versus their displacements for the cases in which the flexible reinforcements stem in different directions from the side of the foundation in 2D conditions. The total displacement

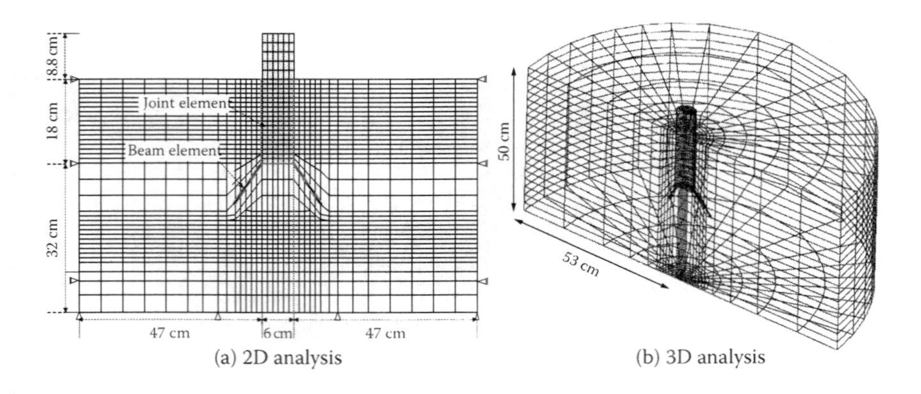

Figure 11.6 Finite element meshes (series II).

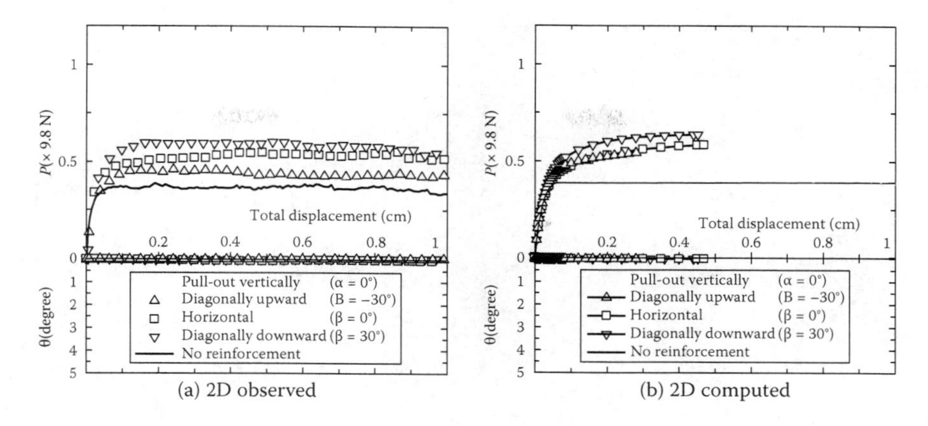

Figure 11.7 Load and rotation angle versus displacement relation under vertical uplift loading ($\alpha = 0°$): flexible reinforcements (2D: series I).

refers to the norm of the vector composed of the sum of the horizontal and vertical components. The factors α and β, mentioned in the captions, are described in Figure 11.4(b). Here, α is the uplift loading angle, measured from the vertical direction, while β denotes the reinforcement placement angle, measured from the horizontal direction.

In these figures, solid curves show the results obtained from a foundation without reinforcements. These figures indicate that reinforcements stemming diagonally downward are the most effective against vertical uplift loading. Flexible reinforcements worked against tensile forces. However, flexible reinforcements stemming from the side of the foundation at any inclination angle

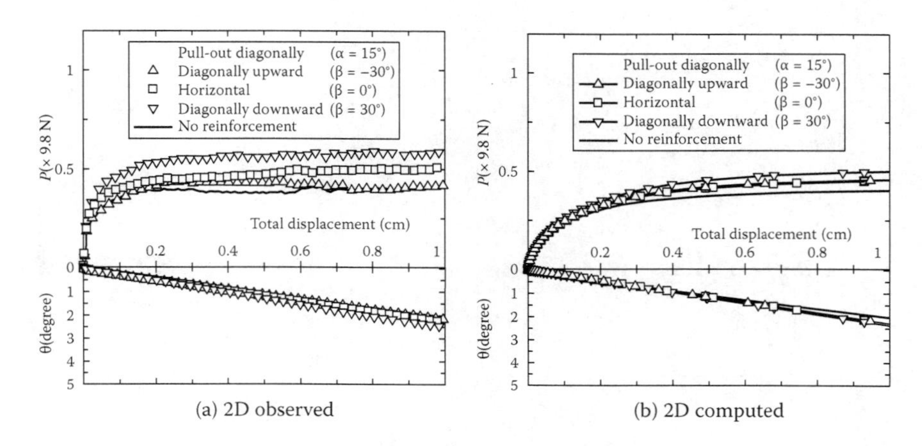

Figure 11.8 Load and rotation angle versus displacement relation under inclined uplift loading ($\alpha = 15°$): flexible reinforcements (2D: series I).

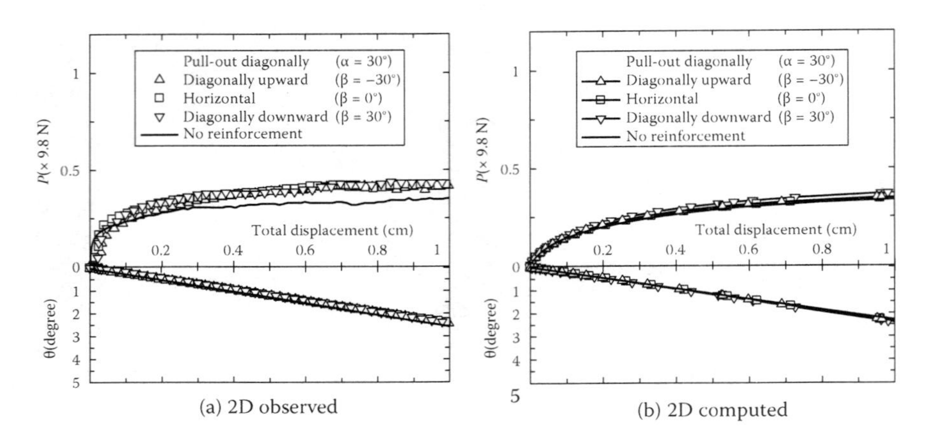

Figure 11.9 Load and rotation angle versus displacement relation under inclined uplift loading ($\alpha = 30°$): flexible reinforcements (2D: series I).

β were not effective against an inclined uplift load when the angle α was 30° (Figure 11.9). Figure 11.10 shows the results with stiff reinforcement when vertical uplift load is applied. It can be seen from Figures 11.7 and 11.10 that the reinforcements protruding diagonally downward are the most effective against vertical uplift loading in spite of the stiffness of the reinforcements. The observed behavior of the foundations is qualitatively and quantitatively described by the computed results with precision in every case.

Figures 11.11–11.13 show the observed and computed variations of uplift loading and rotation angle of the foundations versus the displacements of

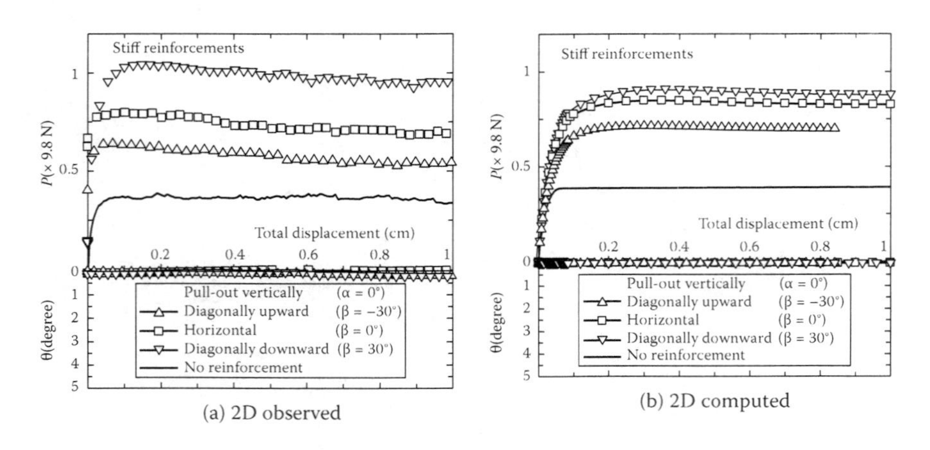

Figure 11.10 Load and rotation angle versus displacement relation under vertical uplift loading ($\alpha = 0°$): stiff reinforcements (2D: series I).

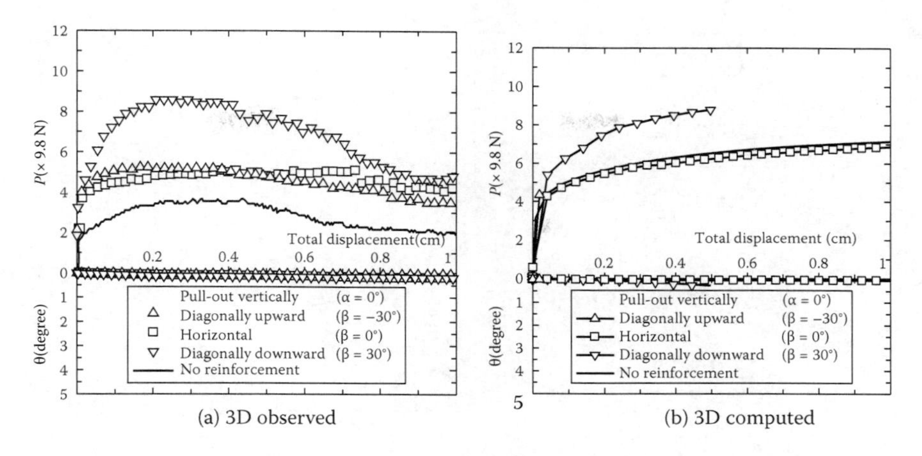

Figure 11.11 Load and rotation versus displacement under vertical uplift loading ($\alpha = 0°$): flexible reinforcements stemming from side (3D: series I).

the foundation in 3D conditions. The solid curves without marks show the results of a foundation without reinforcements in the same way as those in 2D conditions. Figure 11.11(a) shows that the reinforcements stemming from the side of the foundation diagonally downward are the most effective against vertical uplift loading. The results of the reinforcements stemming diagonally upward and horizontally are almost the same, but they are smaller than those of the reinforcements stemming diagonally downward. The results are also very similar to those for the 2D conditions (Figure 11.7).

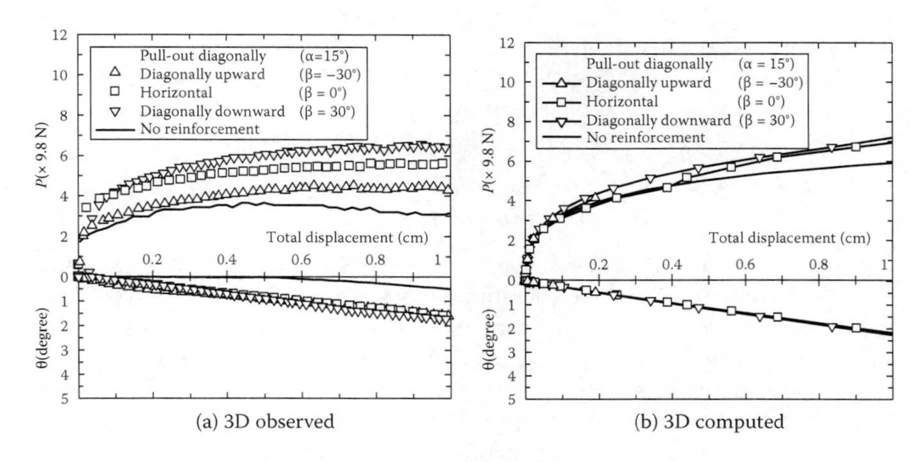

Figure 11.12 Load and rotation versus displacement under inclined uplift loading ($\alpha = 15°$): flexible reinforcements stemming from side (3D: series I).

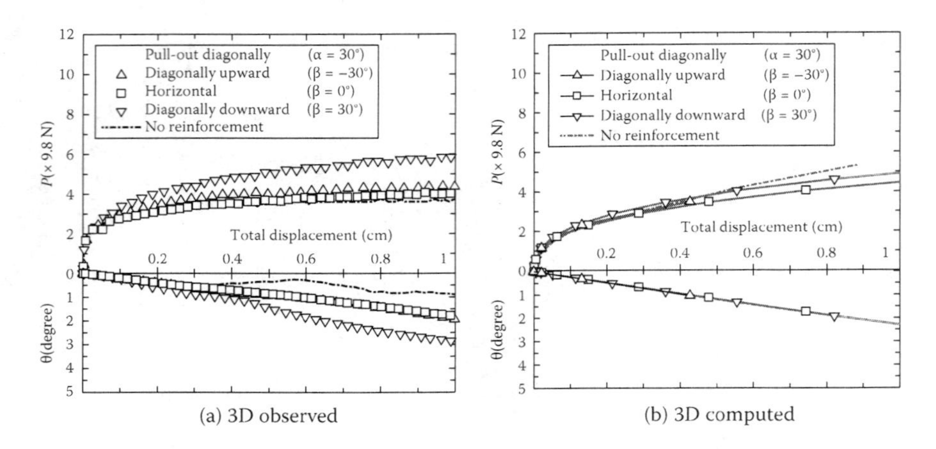

Figure 11.13 Load and rotation versus displacement under inclined uplift loading (α = 30°): flexible reinforcements stemming from side (3D: series I).

Figures 11.12 and 11.13 reveal that, unlike what was observed for vertical uplift loading, reinforcements stemming from the side of the foundation are not very effective against inclined loading in the same way as those in 2D conditions. However, reinforcements that stem diagonally downward are still better than the ones stemming horizontally and diagonally upward. In the case of the inclined uplift loading, the frictional force between the ground and the reinforcements in the loading side develops in the upward direction and acts as a negative resistance, which diminishes the positive resistance of the reinforcements. Therefore, the reinforcements become less effective as the inclination angle of the uplift loading increases. The computed results describe the observed behavior of the foundations. The observed and computed results are also in qualitative agreement with the results of the 2D observation (Figures 11.8 and 11.9).

11.1.3.2 Series II (reinforcements stemming from bottom of the foundation)

Figures 11.14 and 11.15 show the observed and computed results of uplift load and rotation angle of the foundation for cases in which the reinforcements stem from the bottom of the foundation in 2D conditions. Figure 11.4(c) illustrates the angles α and β, with β = 90° representing the case in which the reinforcement stems vertically downward. These types of reinforcements increase the uplift bearing capacity significantly against vertical uplift loading and inclined uplift loading. The computed results simulate well the observed load-displacement behavior and rotation of the foundation.

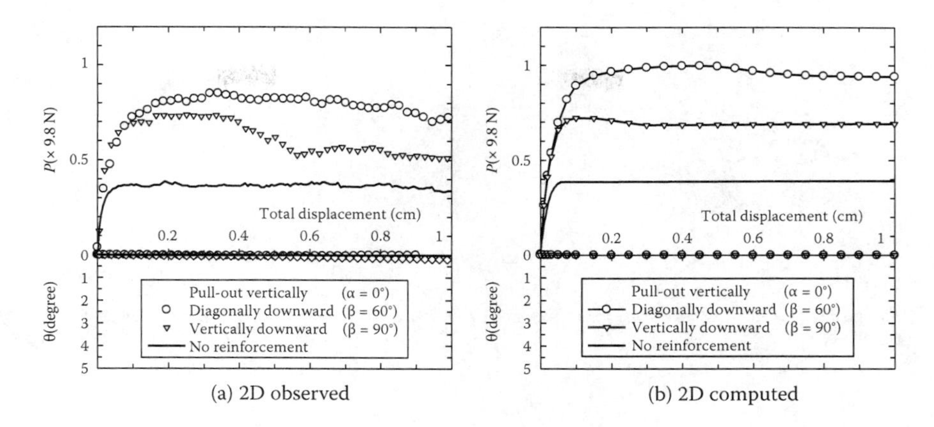

Figure 11.14 Load and rotation versus displacement under vertical uplift loading (α = 0°): flexible reinforcements stemming from bottom (2D: series II).

Figures 11.16 and 11.17 show the observed and computed movements of the foundation with flexible reinforcements after applying a displacement of 10 mm to the foundation. The uplift bearing capacity is much larger for reinforcements stemming from the bottom of the foundation, even though the deformed regions of ground do not spread as widely. Longer reinforcements stemming from the bottom are more effective against both inclined and vertical uplift forces.

Figures 11.18 and 11.19 show the observed and computed results of uplift loading and rotation angle of the foundation in 3D conditions. These figures indicate that these reinforcements significantly increase the uplift bearing capacity against vertical and inclined uplift loading, in accordance

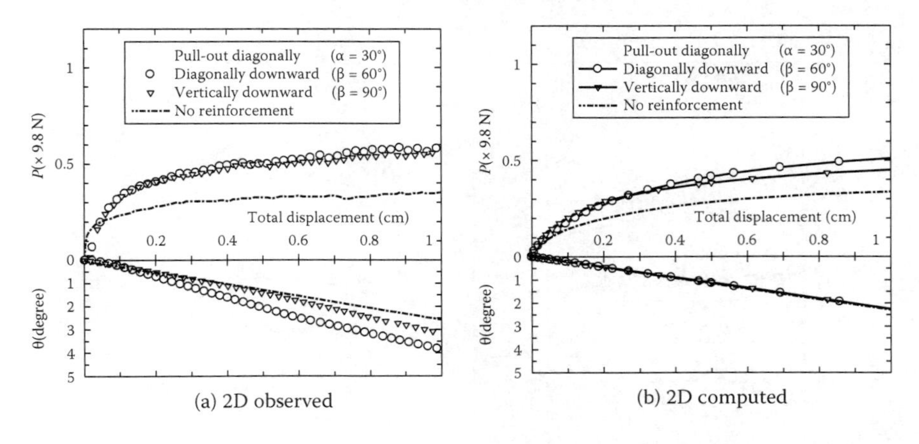

Figure 11.15 Load and rotation versus displacement under vertical uplift loading (α = 30°): flexible reinforcements stemming from bottom (2D: series II).

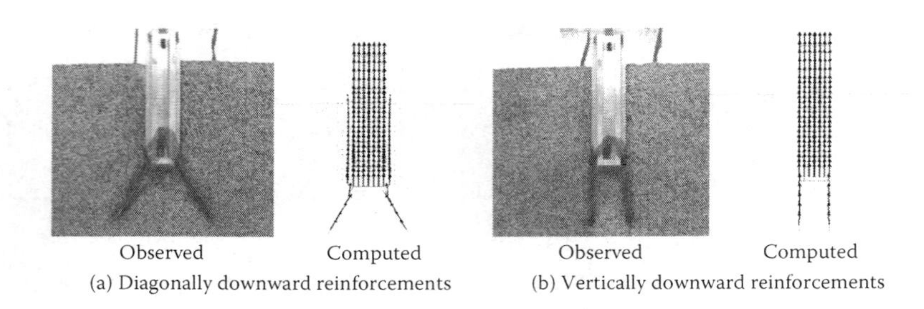

Figure 11.16 Ground movements under vertical uplift loading: flexible reinforcements (2D: series II).

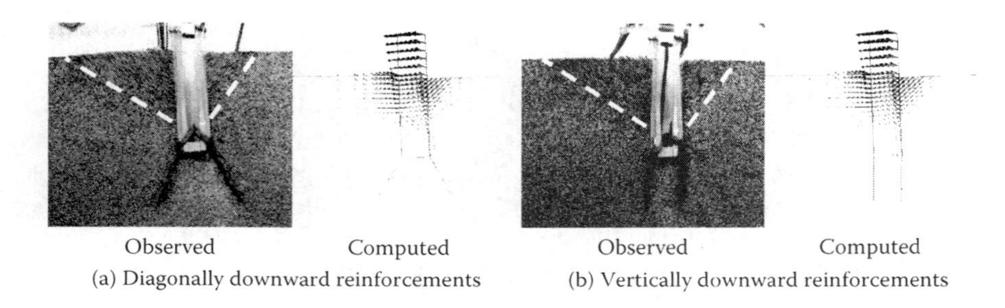

Figure 11.17 Ground movements under inclined uplift loading (α = 30°): flexible reinforcements (2D: series II).

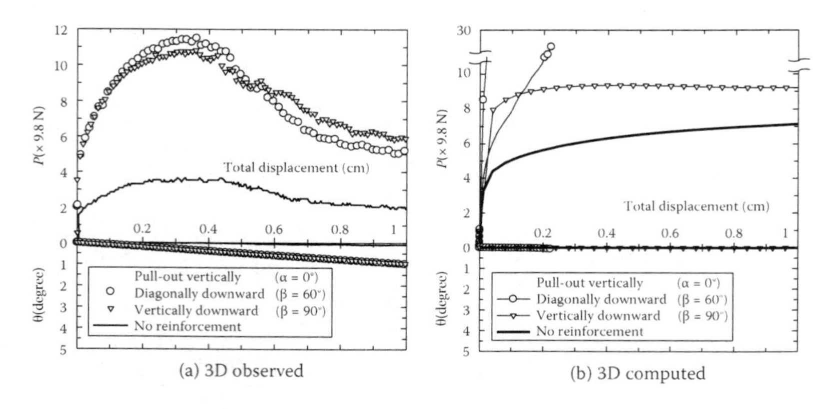

Figure 11.18 Load and rotation versus displacement under vertical uplift loading (α = 0°): flexible reinforcements stemming from bottom (3D: series II).

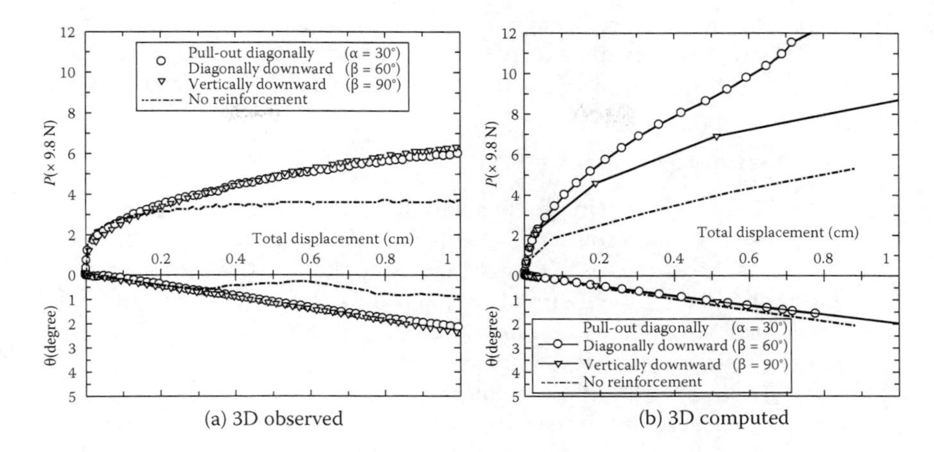

Figure 11.19 Load and rotation versus displacement under inclined uplift loading ($\alpha = 30°$): flexible reinforcements stemming from bottom (3D: series II).

with the results found under 2D conditions (Figures 11.14 and 11.15). Therefore, the 2D model tests can represent qualitatively the behavior of the 3D model tests in general 3D stress conditions.

Although the uplift bearing capacity for reinforcements stemming diagonally and vertically downward is almost the same in the model tests, the numerical analyses predict different results. They show that the uplift bearing capacity is slightly larger for the reinforcement placed diagonally downward than for the one set up vertically downward. However, there is a fair agreement between the results from the model tests and the numerical simulations. Under both loading conditions tested, the uplift bearing capacity of the reinforcements stemming from the bottom is much larger than the bearing capacity of those stemming from the side.

11.2 REINFORCING FOR INCREASING BEARING CAPACITY

To reduce ground deformation and earth pressure of the ground and to increase bearing capacity, the reinforced earth method is adopted using geosynthetics such as Geogrid and Geotextile. However, until now, the reinforcement mechanism for establishing a rational design method of reinforced-soil ground has not been fully understood. In this section, the bearing capacity problem of reinforced-soil ground is investigated and the reinforcing mechanism is clarified through 2D laboratory model tests and the corresponding numerical analyses. In the model tests, a mass of aluminum rods is used as ground, and tracing paper having almost no bending stiffness is used as reinforcement material. The numerical analyses using the FEMtij-2D program are also carried out. More detailed experimental and numerical results on this topic are described by Nakai, Shahin, et al. (2009).

11.2.1 Outline of model tests

Figure 11.20 shows a schematic diagram of a 2D model test apparatus. In the model tests, the aluminum rods mass is used as ground in the same way as in other 2D model tests. Vertical load is applied at the center of a flat foundation (width of $B = 12$ cm) using a motor. A slider, which permits the movement of the loading rod in the horizontal direction, and a load cell for measuring the magnitude of load are set in the loading system.

Vertical displacement and rotation of the foundation are measured with setting two displacement transducers at the two ends of the foundation. Tracing paper that has almost no bending stiffness is used as reinforcement material and is set underneath the foundation, changing the installation depths (D). Four sets of reinforcement lengths are considered: $L = 36$ cm ($L/B = 3$), $L = 24$ cm ($L/B = 2$), $L = 12$ cm ($L/B = 1$), and $L = 6$ cm ($L/B = 0.5$).

As shown in Figure 11.21, two kinds of frictional conditions of geosynthetics are employed. In the first type, aluminum rods with a diameter of

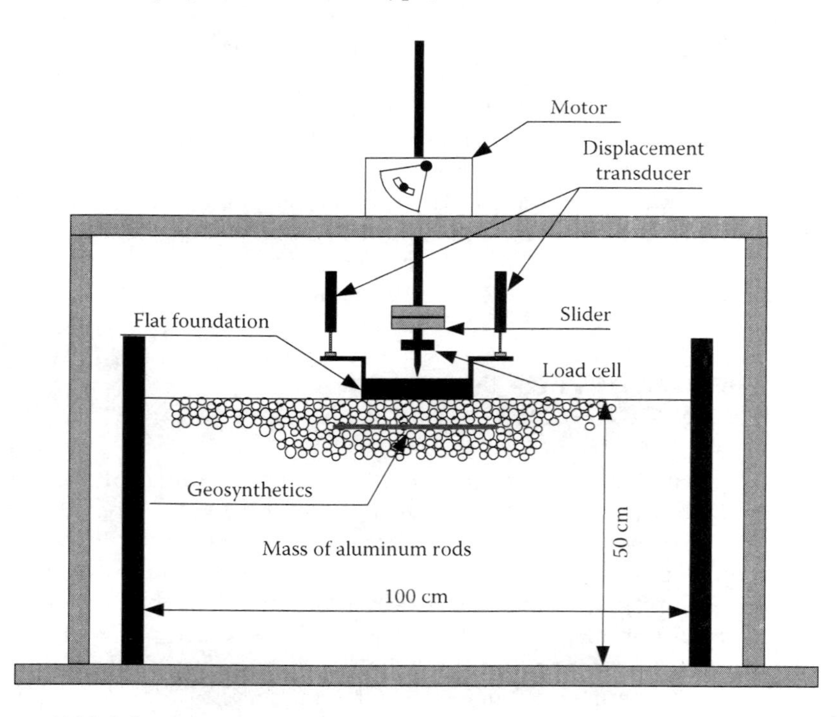

Figure 11.20 Schematic diagram of apparatus.

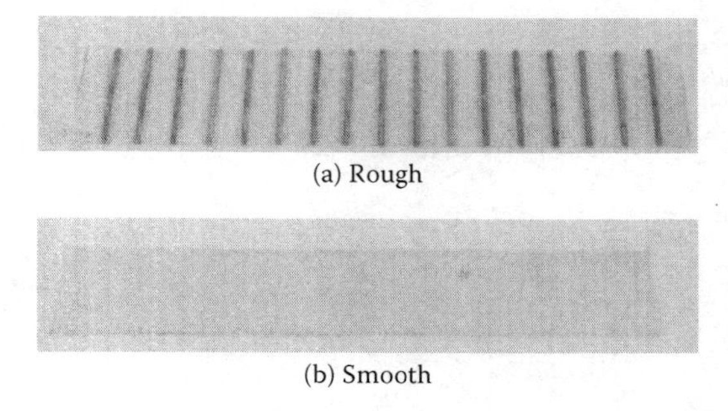

(a) Rough

(b) Smooth

Figure 11.21 Types of reinforcements.

1.6 mm are glued at both sides of the tracing paper to simulate rough conditions (frictional angle between the ground and the reinforcement $\delta = 20°$); in the second type, the smooth behavior of reinforcement is considered without gluing aluminum rods on the tracing paper (frictional angle between the ground and the reinforcement $\delta = 8°$). Changing the installation depth (D), the length (L), and the skin friction (δ) of the reinforcements, 13 test patterns are considered (Table 11.1). The deformation of the ground is captured by photographs of the model ground taken with a digital camera.

11.2.2 Outline of numerical analyses

Figure 11.22 represents the mesh used in the numerical analyses. Isoparametric four-noded elements are used for soil elements, and elastic beam elements are used to simulate reinforcements. The frictional behavior between the reinforcement and soil, and the foundation and soil, is modeled employing the elastoplastic joint element. The friction angle between reinforcement and soil is determined by the sliding tests ($\delta = 20°$ for rough reinforcements and $\delta = 8°$ for smooth reinforcements). The friction angle between foundation and soil is $\delta = 15°$. The stiffness parameters

Table 11.1 Test patterns

δ	*L/B*			*D/B*		
No reinforcement						
Rough reinforcement ($\delta = 20°$)	1/2			1/4		
	1			1/4		
	2			1/4		
	3	0	1/8	1/4	1/2	1
Smooth reinforcement ($\delta = 8°$)	2		1/8	1/4	1/2	1

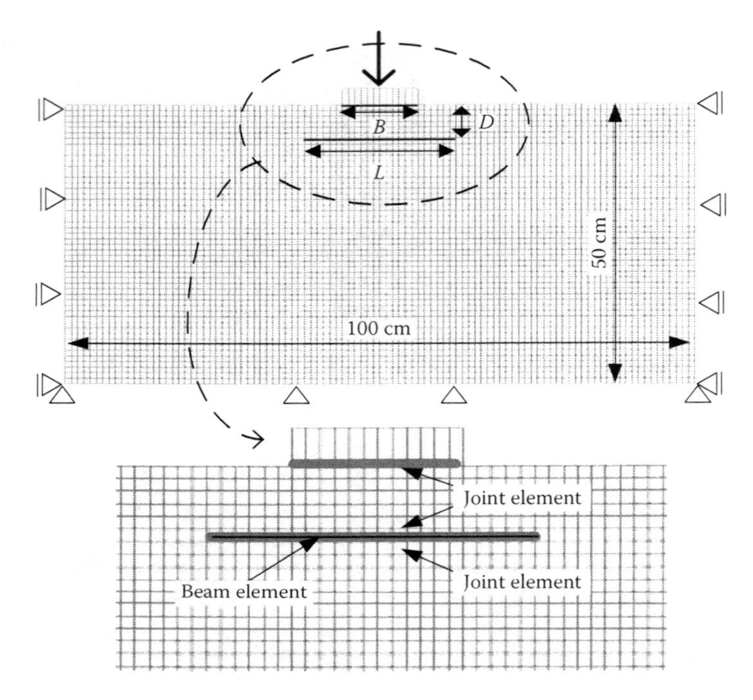

Figure 11.22 Finite element mesh.

of $EA = 1.31$ kN/cm to axial stiffness and $EI = 1.34 \times 10^{-10}$ kNm²/cm to bending stiffness were employed for the reinforcements.

11.2.3 Results and discussion

11.2.3.1 Influence of installation depth of reinforcement

Figure 11.23 shows the observed and computed bearing capacity for different installation depths where the reinforcement length is 24 cm. The vertical axes represent vertical load q_v, which is divided by $\gamma B/2$; abscissas show normalized vertical displacement that is normalized by the width of the foundation (B). The figures also show the results of no reinforcement. The influence of reinforcement cannot be expected when the installation depth is deeper than a certain depth (D/B is greater than 1/2). In addition, when the reinforcement is placed on the ground surface just below the foundation, there is almost no effect of the reinforcements except the friction coefficient of the foundation. The numerical analyses capture well the influence of the installation depth in the same way as the model tests.

Figure 11.24 represents the computed distributions of bending moment and axial force of reinforcements. It is seen that the bending moment is

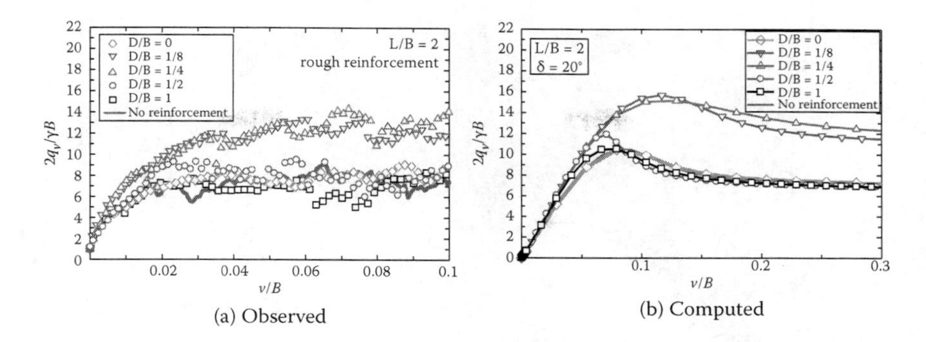

Figure 11.23 Vertical load versus vertical displacement for different installation depths of reinforcement.

almost zero in all cases, as expected. However, either a big range of larger axial force is distributed in the central part of the reinforcement ($D/B = 1/8$) or the value of axial force in the central part of the reinforcement becomes remarkably high where the effect of reinforcement exists ($D/B = 1/4$).

Figure 11.25 shows observed and computed deviatoric strain distributions of the ground while the load approaches peak strength (for the

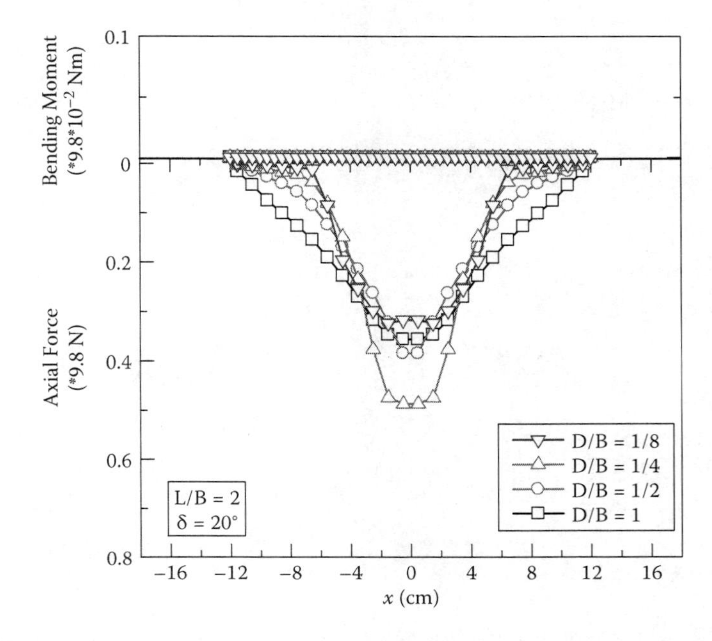

Figure 11.24 Computed bending moment and axial force for different installation depths of reinforcement.

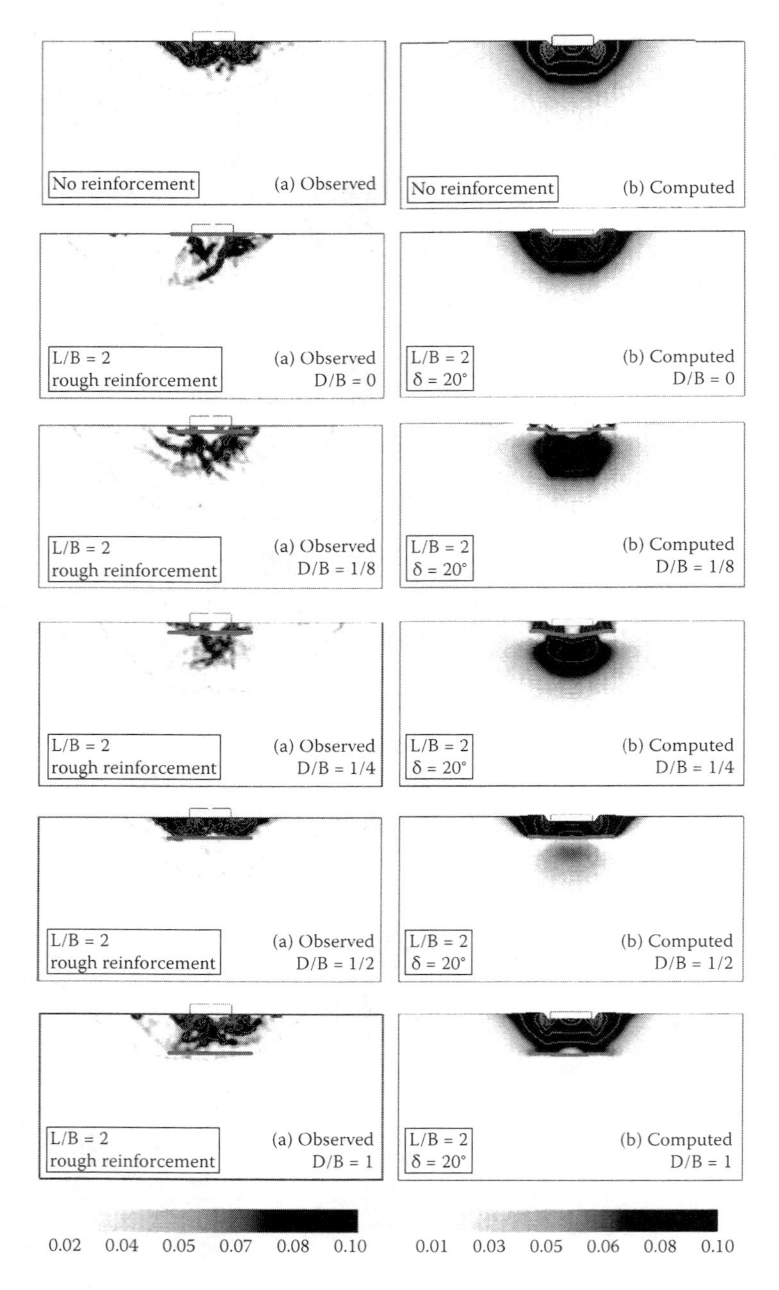

Figure 11.25 Deviatoric strain distribution for different installation depths of reinforcement.

model test, it is at $v/B = 0.04$, while it is at $v/B = 0.1$ for the analysis). The distribution of deviatoric strain of the model tests is obtained from the simulation of the PIV technique taking two photographs before and after loading. The topmost figure represents the results of the ground without reinforcement. It is seen that where the effect of reinforcement is the most ($D/B = 1/8$, $1/4$), a zone of large deviatoric strain spreads vertically below the foundation compared to no reinforcement. For $D/B = 1/2$ and 1, where there is almost no effect of reinforcement, the zone of large deviatoric strain does not spread vertically below the foundation. Especially for $D/B = 1.0$, the distribution of deviatoric strain is almost the same as the case of no reinforcement; in this case, reinforcement does not take part in ground improvement. The computed results describe well the differences of the deviatoric strain distribution as observed in the model tests.

11.2.3.2 Influence of reinforcement length

The influence of reinforcement is investigated for several reinforcement lengths keeping a constant installation depth ($D/B = 1/4$). The similar graphs illustrated in the previous section for different installation depths will be shown in this section. Figure 11.26 shows the relationships of load displacement for different reinforcement lengths. It is observed that even for $L = 6$ cm ($L/B = 1/2$), there are some effects of reinforcing on the bearing capacity of the ground. However, almost no differences of reinforcement effects on bearing capacity are observed for L/B greater than 1.0. The numerical analyses perfectly capture the results of the model tests.

Figure 11.27 represents distributions of bending moment and axial force of reinforcements. Almost the same distribution of axial force is seen for L/B greater than 1.0, which describes the mechanism of reinforcement

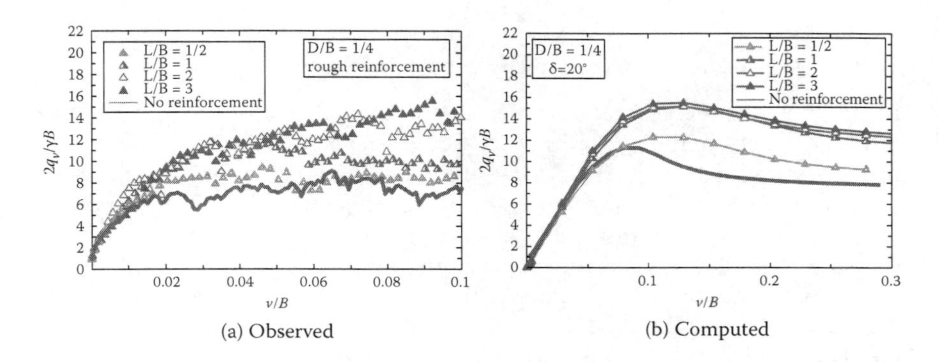

Figure 11.26 Vertical load versus vertical displacement for different lengths of reinforcement.

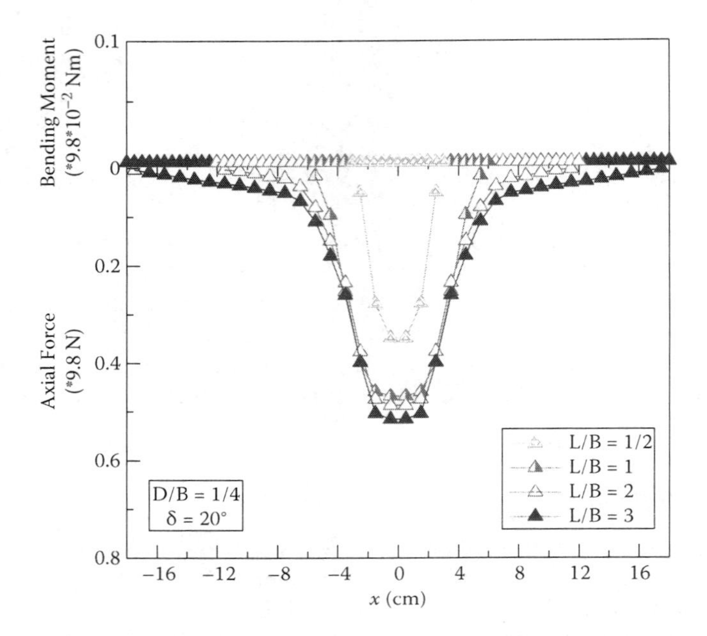

Figure 11.27 Computed bending moment and axial force for different lengths of reinforcement.

mentioned before. Figure 11.28 shows the distributions of deviatoric strain of the ground similar to Figure 11.25 (see the top ones in Figure 11.25 for the results of no reinforcement and $L/B = 2.0$). With the increase of reinforcement effect, a zone of large deviatoric strain spreads vertically below the foundation. The same effect of reinforcement can be expected when L/B is greater than 1.0.

11.2.3.3 Influence of friction angle of reinforcement

In the previous sections, the friction angle (δ) was 20°. In this section, the effects of friction angle will be discussed in the same condition illustrated in Figures 11.23–11.25 ($L = 24$ cm) except using a different friction angle ($\delta = 8°$). Here, the surface of reinforcement (tracing paper) is smooth; that is, there is no aluminum rod glued on the tracing paper. The relationship of load displacement is shown in Figure 11.29. Figure 11.30 illustrates the computed distribution of bending moment and axial force and Figure 11.31 represents the distributions of deviatoric strain in the ground (see Figure 11.25 for the results of no reinforcement).

Comparing Figure 11.29 with Figure 11.23 reveals that, in the smaller friction angle, the reinforcement effect decreases even for the same stiffness

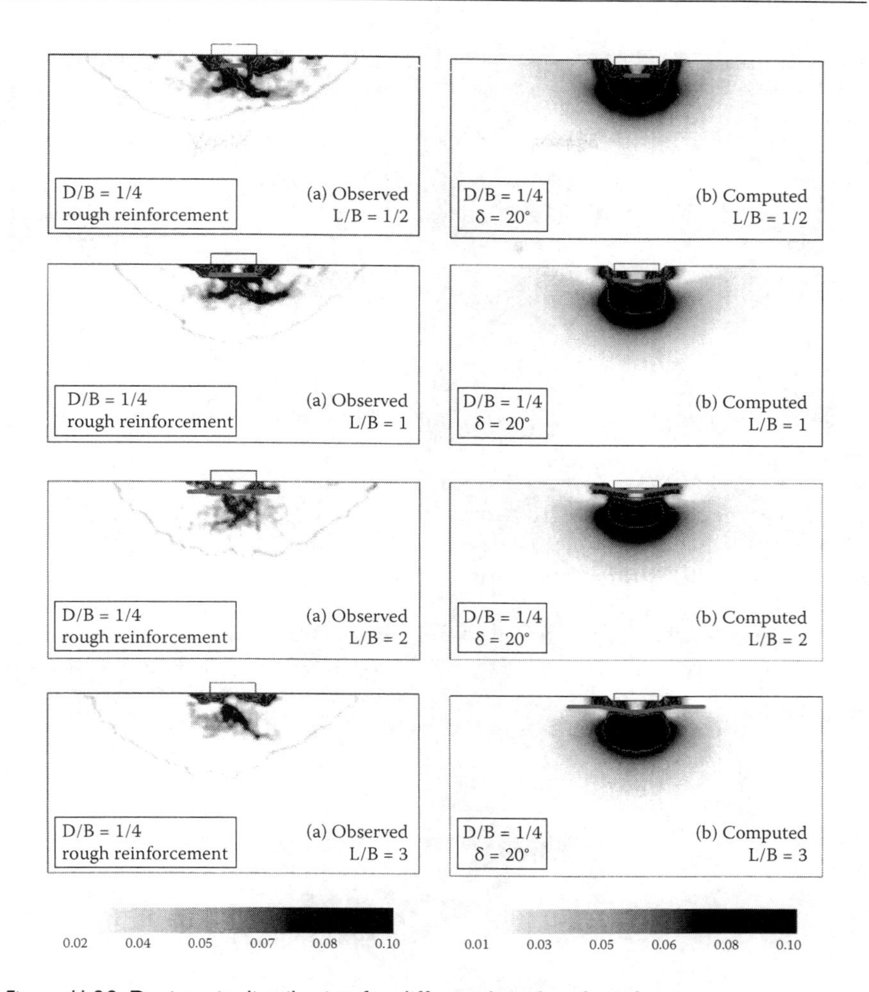

Figure 11.28 Deviatoric distribution for different lengths of reinforcement.

of the reinforcement. From Figure 11.29, it is seen that if the reinforcement is installed at an appropriate depth (for instance, $D/B = 1/8$), the bearing capacity of the ground increases even for the smooth surface of reinforcement. A reverse effect can also be seen (for example, $D/B = 1/2$) depending on the installation depth in the case of a smaller friction angle. Almost no reinforcing effect is observed in $D/B = 1$ for both types of friction angles. The numerical analyses depict well the phenomenon of reinforcing effect for different friction angles.

From Figure 11.30, it is found that the axial force for $D/B = 1/2$ is smaller than that for $D/B = 1$; therefore, a negative effect of reinforcement for $D/B = 1/2$ can be speculated. From Figures 11.24, 11.27, and 11.30, it

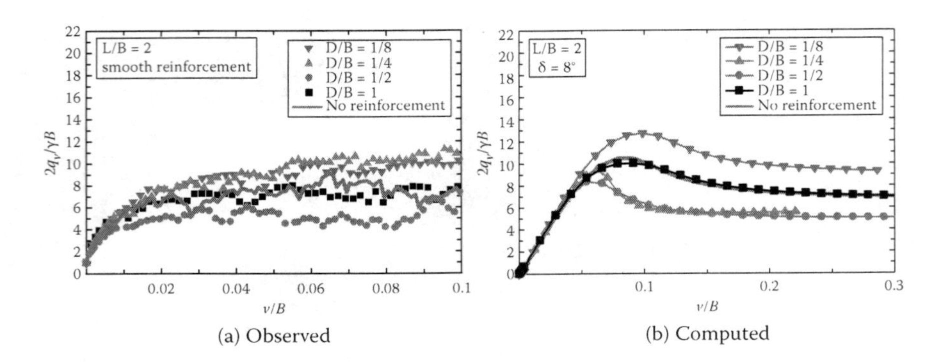

(a) Observed (b) Computed

Figure 11.29 Vertical load versus vertical displacement: smooth reinforcement.

is observed that though the axial force is smaller at the edges, it increases at the central part of the reinforcement for the cases where reinforcing effect exists. Deviatoric strain develops to a wider and deeper region even for the smooth reinforcement in $D/B = 1/8$, where a reinforcing effect exists (Figure 11.31).

On the other hand, for the smooth reinforcement in $D/B = 1/2$, where a negative reinforcing effect exists, the region of larger deviatoric strain is

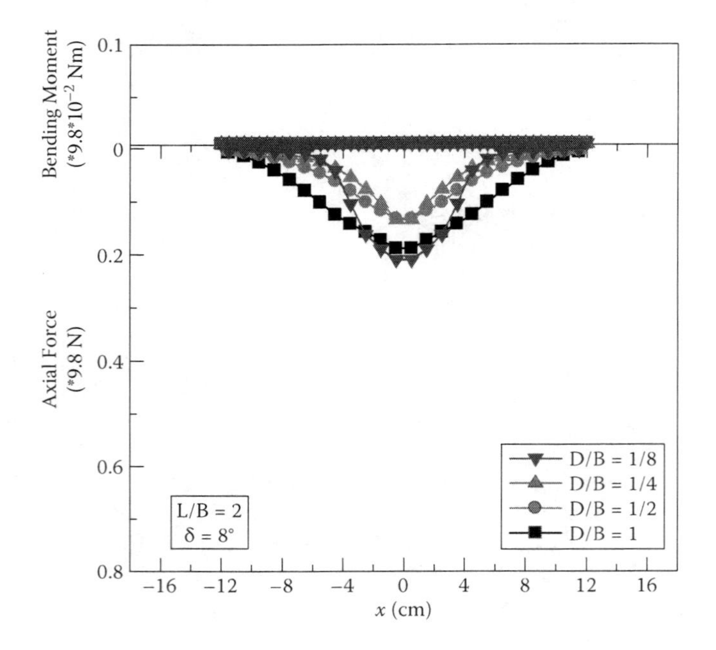

Figure 11.30 Computed bending moment and axial force: smooth reinforcement.

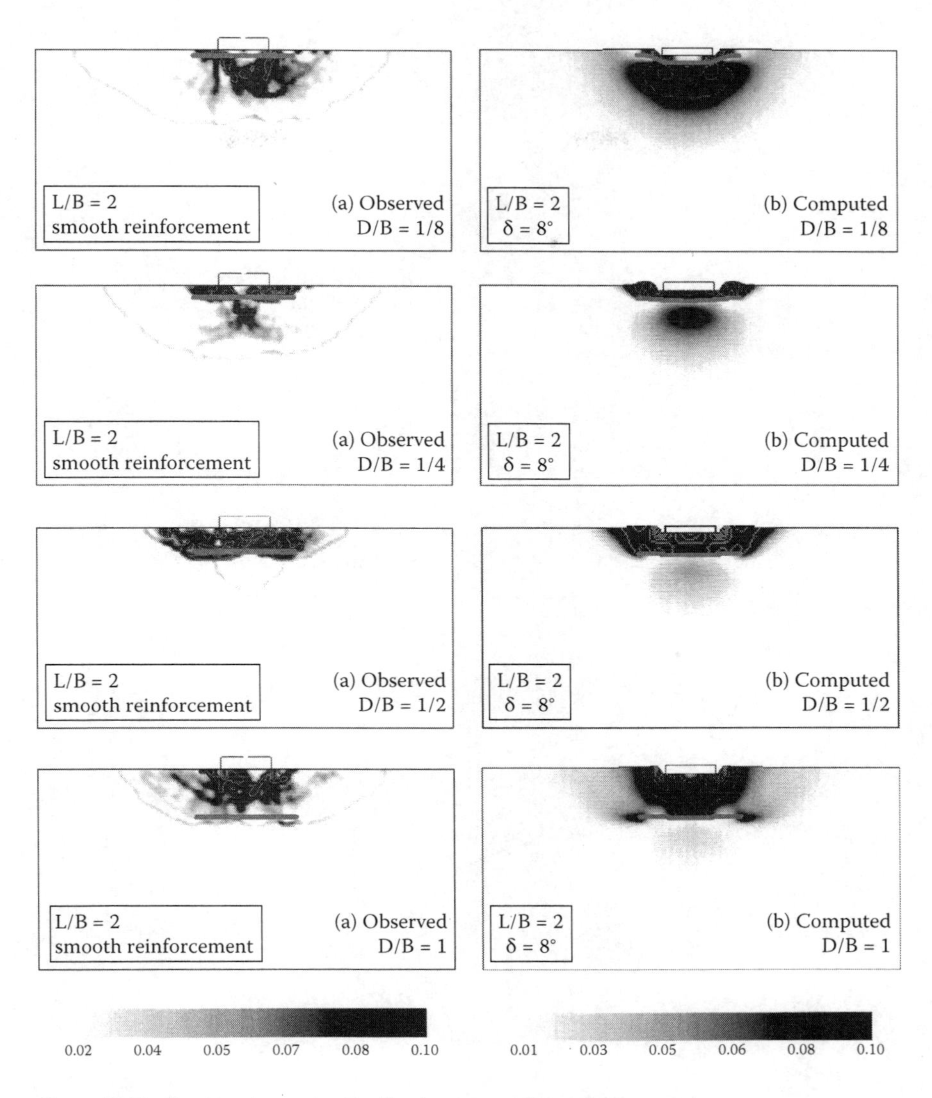

Figure 11.31 Deviatoric strain distribution: smooth reinforcement.

shallower than that for no reinforcement. This is because slippage occurs on the top surface of the reinforcement before developing a zone of shear strain vertically below the foundation. Hence, larger shear strain concentrates near the ground surface. The numerical analyses can reproduce well the mechanism of deviatoric strain distribution of the model tests.

Chapter 12

Localization and shear band development in element tests

12.1 INTRODUCTION

Localization of deformation is usually considered as a boundary value problem in numerical analyses. On the other hand, triaxial tests, plane strain tests, and other laboratory tests are generally simulated as a local (homogeneous) problem, integrating the stress–strain relation at a single point. This is the same as considering that deformations occur homogeneously over a finite element and as such will be referred to as the *ideal* test. In actual conditions, however, heterogeneous deformations occur in a sample as shear deformation develops; consequently, the localization of deformations should be considered as a form of shear banding.

In this chapter, shear banding is simulated numerically by finite element analyses using the advanced model of stage I (corresponding to the subloading t_{ij} model), considering localization of deformation as a boundary value problem. In much research, finite element simulations based on finite deformation theory have been used to reproduce shear bands numerically using soil-water coupled analyses (Yatomi et al. 1989; Asaoka and Noda 1995; Asaoka, Nakano, and Noda 1997; Oka, Adachi, and Yashima 1995). Nevertheless, it is possible to simulate localization using drained analysis. More detailed numerical results on this topic are described in Miyata et al. (2003).

12.2 OUTLINE OF ANALYSES

Analyses are carried out for drained compression tests under plane strain (2D) and triaxial (3D) conditions for both normally consolidated clay and overconsolidated clay (OCR = 10). Analyses based on infinitesimal deformation theory and finite deformation theory were performed as well. Moreover, 3D finite element analyses were also performed to compare with the results obtained under plane strain conditions.

Figure 12.1 shows the finite element meshes used in the current analyses. A rectangular specimen with a height of 10 cm and a width of 5 cm (and a

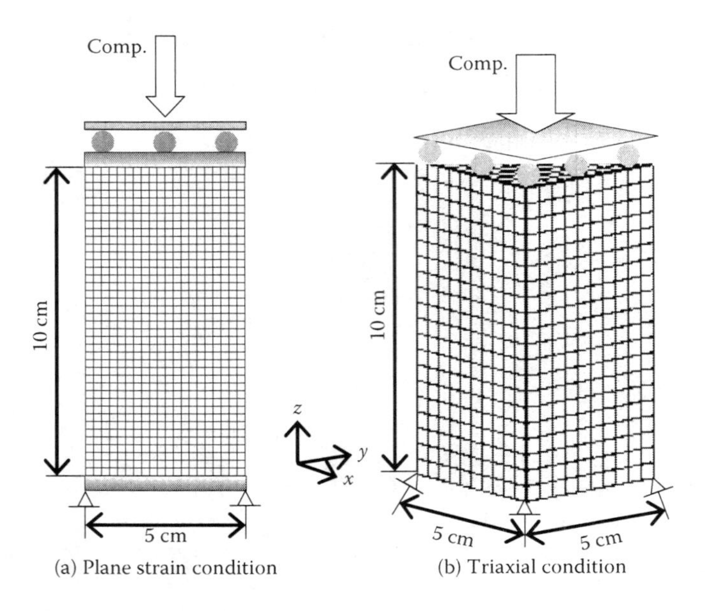

Figure 12.1 Finite element mesh.

depth of 5 cm in triaxial conditions) is considered. Figure 12.1(a) shows the mesh for a plane strain condition consisting of 800 elements (20 elements in width and 40 in height). In triaxial conditions, the mesh consists of 2,000 elements (20 elements in height, 10 elements in width, and 10 elements in depth). As for the boundary conditions, the lateral faces are free to move and have constant imposed stresses, the bottom face is kept perfectly fixed, and the top face is movable in the horizontal direction (in triaxial conditions, horizontal movement is kept free in the x-direction).

Vertical displacements are applied at the top face of the specimen to simulate loading, until a total axial strain of 20%. The initial stress state is isotropic with p_0 = 196 kPa for all elements. Fujinomori clay is assumed as the material (material parameters are shown in Tables 5.2 and 6.1). The *ideal* stress–strain behaviors of Fujinomori clay are described in Sections 5.3.3 and 6.2.3. The computations based on the finite deformation theory were carried out by the updated Lagrangian method using Jaumann's stress rate.

12.3 RESULTS AND DISCUSSION

The *average* behavior of the specimen is shown in Figure 12.2 (NC clay) and Figure 12.3 (OC clay) for plane strain conditions. The solid line shows the analytical solution for a single point, obtained integrating the stress–strain relation for given imposed strain increments (henceforth called *ideal* behavior).

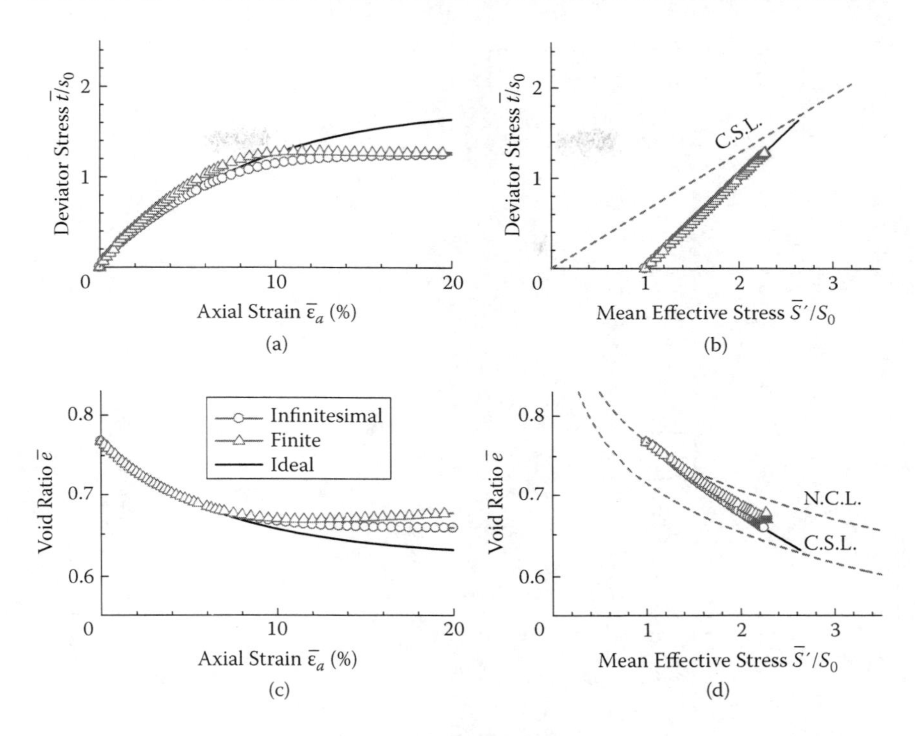

Figure 12.2 Computed stress–strain relations as a mass –2D (OCR=1).

The upper bar (–) sign with stress, strain, and void ratio is used to indicate the *average* result in the specimen as a whole. The average axial stress is obtained by dividing the total vertical force, computed from the nodal reactions at the top face, by the cross-sectional area of the sample. The void ratio as a whole is the average value of the void ratios of all elements in the specimen.

From Figures 12.2 and 12.3, it can be seen that the average stress–strain behavior is the same as the ideal behavior in the early stages of shearing. However, after reaching peak strength, the average deviatoric stress departs from the *ideal* behavior and either decreases or becomes constant (see Figure 12.3 in particular). In all cases, the average peak deviator stress occurred at smaller strains and reached lower values than those computed for the *ideal* curve. With respect to the strain theory, there is not much of a difference between the results of the finite deformation and infinitesimal deformation theories for both normally and overconsolidated clay.

Figures 12.4 and 12.5 show the distributions of deviatoric strain for different values of the axial strain, after the average behavior started to deviate from the ideal behavior. The displacement pattern shows the top face moving to the left-hand side for both conditions. In the case of normally consolidated clay, the localized deviatoric strains distributed over a wider

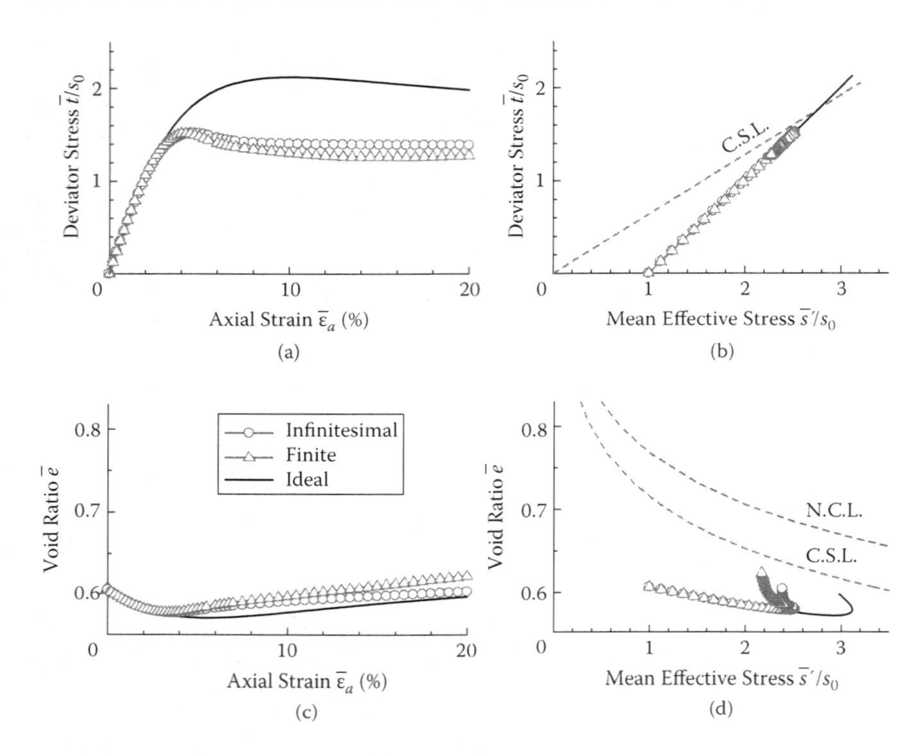

Figure 12.3 Computed stress-strain relations as a mass –2D (OCR=10).

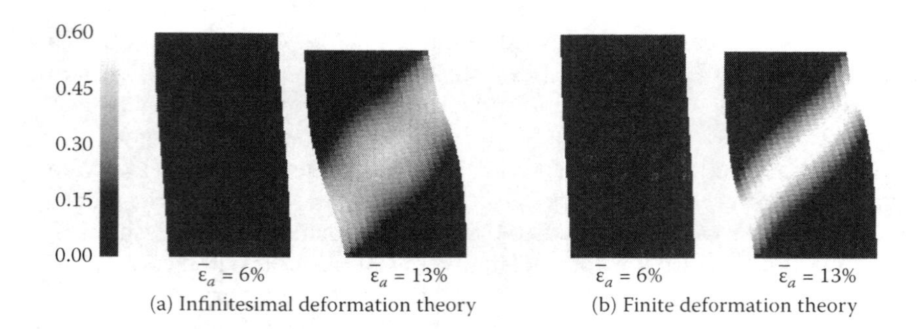

Figure 12.4 Distributions of deviatoric strain –2D (OCR=1).

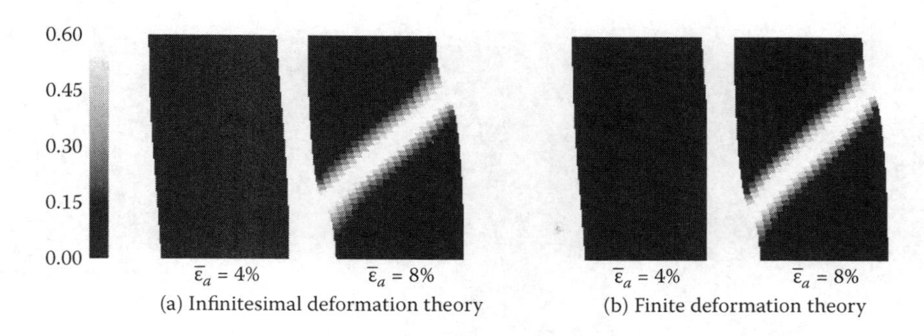

Figure 12.5 Distributions of deviatoric strain –2D (OCR=10).

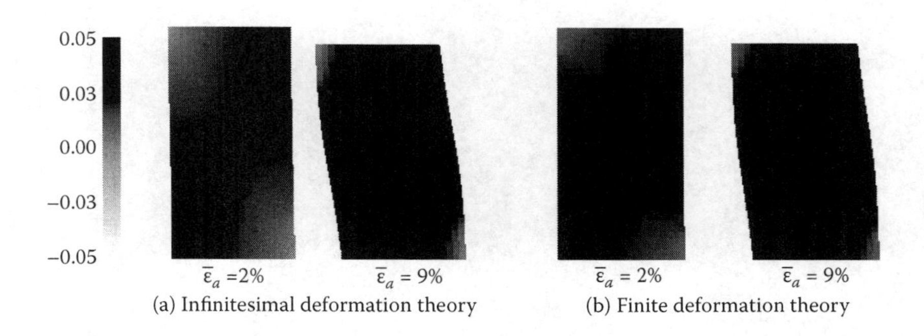

Figure 12.6 Distributions of volumetric strain –2D (OCR=1).

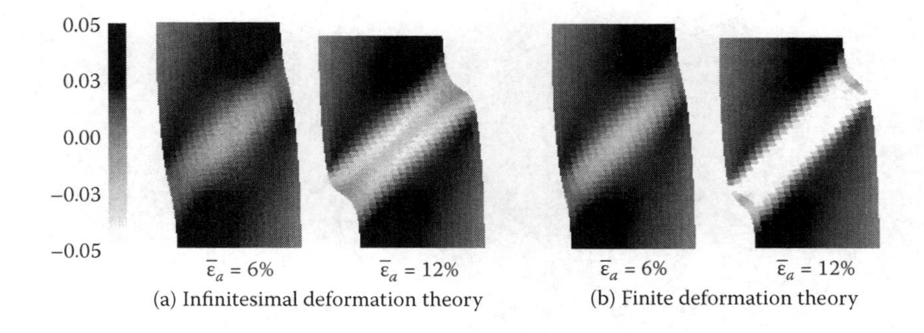

Figure 12.7 Distributions of volumetric strain –2D (OCR=10).

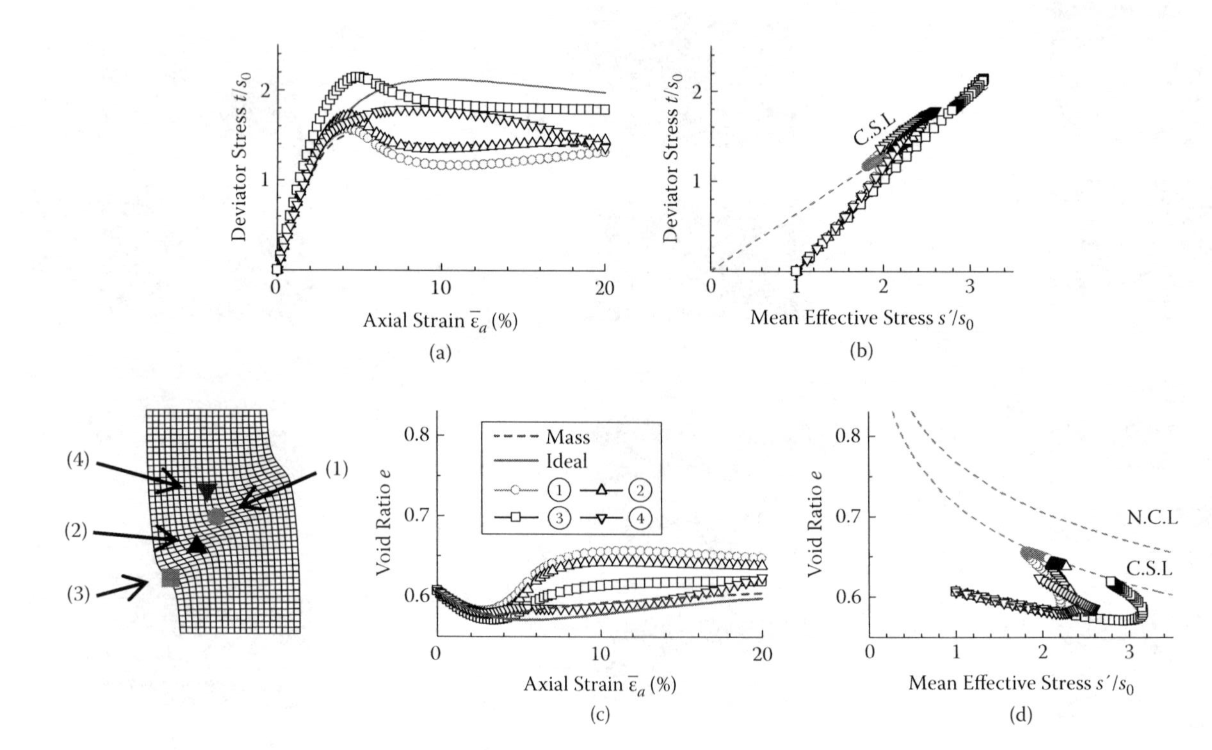

Figure 12.8 Stress–strain void ratio relation of each element –2D (infinitesimal deformation theory, OCR=10).

region and the shear band cannot be seen as clearly as in the case of overconsolidated clay with infinitesimal deformation theory. With finite deformation theory, the deviatoric strains occur locally for both normally and overconsolidated clays.

Figures 12.6 and 12.7 show the distributions of volumetric strain for NC and OC clays, respectively. For normally consolidated clay, large volumetric contraction is observed over most of the specimen except at the upper left and the lower right parts (Figure 12.6) because normally consolidated clays show negative dilatancy. On the other hand, for overconsolidated clay, the regions of concentrated shear stresses and volumetric expansion are almost the same. Although the elements in the shear band expand uniformly for finite deformation theory, the elements in the edge of the shear band expand more than the elements in the center of the shear band for infinitesimal deformation theory.

The mechanism of development of the shear band is examined using the results for the overconsolidated clay, for which the shear band is clearer. Four elements in different positions inside the specimen are

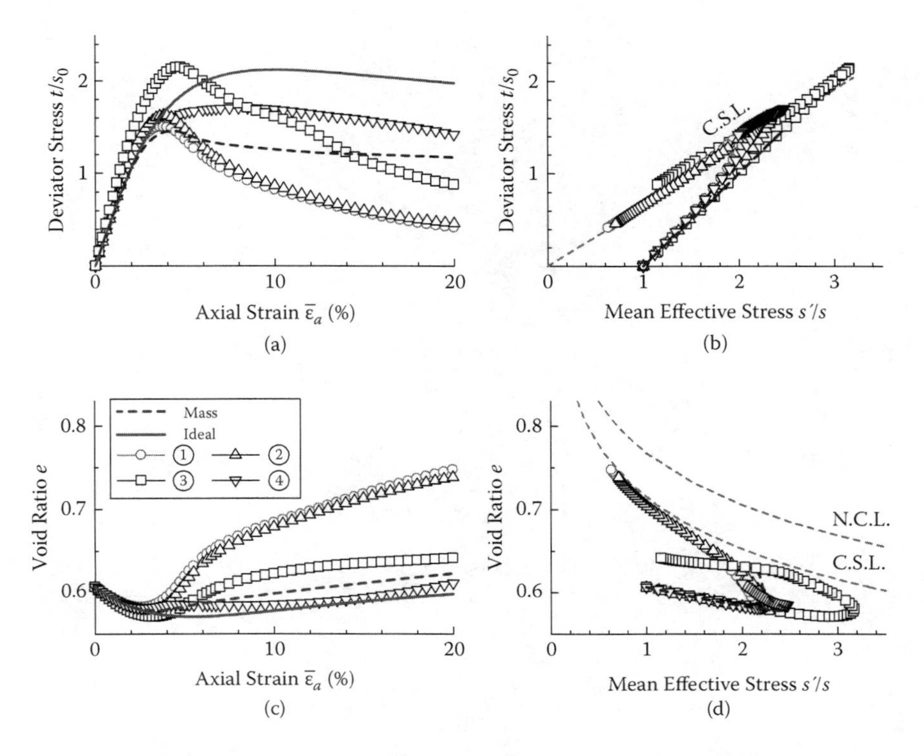

Figure 12.9 Stress–strain void ratio relation of each element –2D (finite deformation theory, OCR=10).

analyzed: element (1) in the center of the specimen and the shear band, (2) in the shear band, (3) at the end of the shear band, and (4) outside the shear band. The stress–strain relation of each element is shown in Figure 12.8 (infinitesimal deformation theory) and Figure 12.9 (finite deformation theory). Element (1) in the center of the specimen reaches peak strength and then softens earlier than the others. Elements (2) and (3) show more or less the same tendency as element (1). Therefore, the element at the center of the specimen softens first and acts as a trigger. After that, the strain-softening domain expands toward the borders of the specimen, forming a shear band.

Figures 12.10 and 12.11 show the result of the 3D finite element analysis based on infinitesimal deformation theory. From the stress–strain relation in 12.10(a) and 12.11(a), it can be noticed that the average curve shows some hardening and then deviates from the ideal behavior after some stage of shearing, similarly to what was observed under plane strain conditions. The change of void ratio is almost the same as that

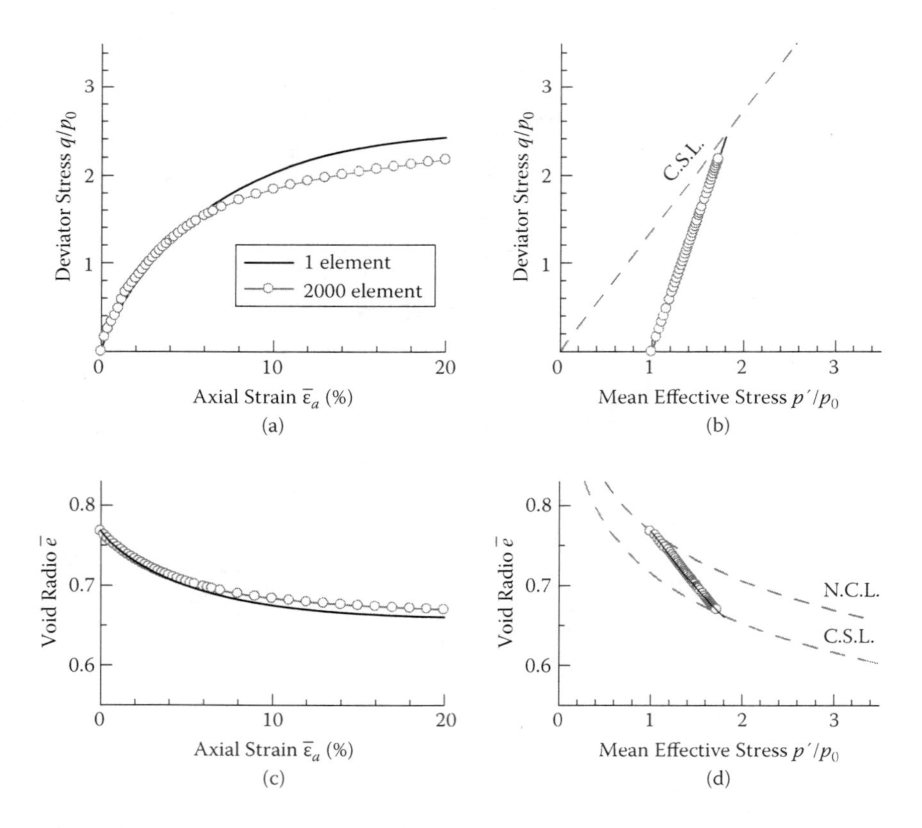

Figure 12.10 Stress–strain void ratio relation as amass –3D (OCR=1).

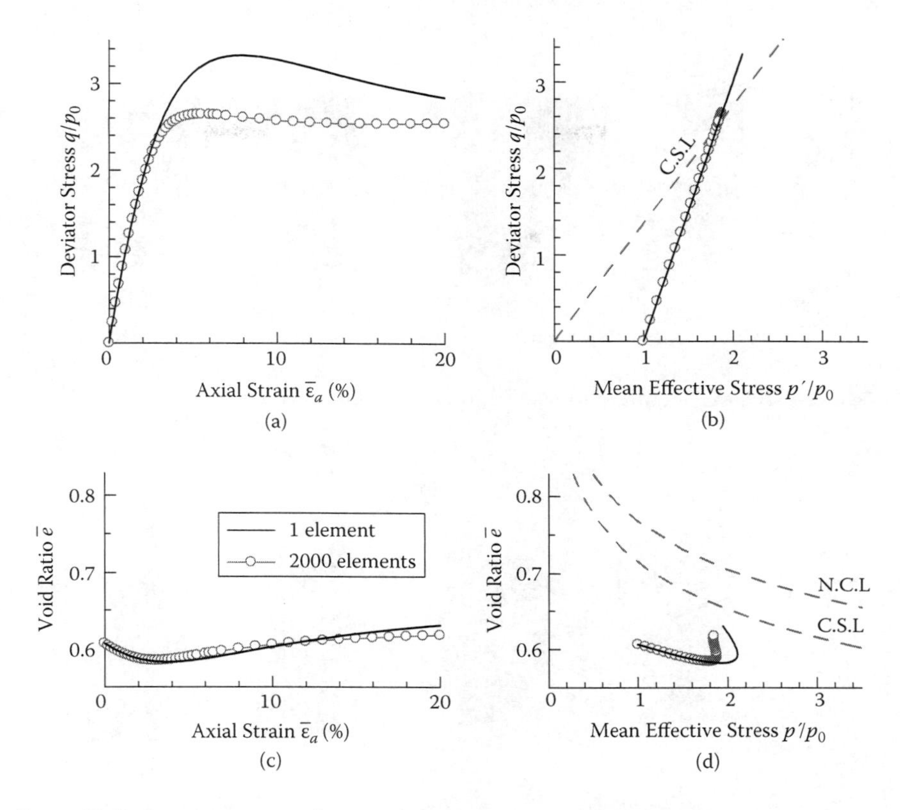

Figure 12.11 Stress–strain void ratio relation as a mass –3D (OCR=10).

of the ideal behavior. The same general tendency is observed for the 3D and plane strain simulations, with the average curve failing earlier and reaching lower peak stresses than those computed for the ideal curve at a single point.

Figures 12.12 and 12.13 show the distributions of deviatoric strain for normally consolidated and overconsolidated clays. Figures 12.12(a) and 12.13(a) show the distribution of deviator strain in the middle section of the specimen, and Figures 12.12(b) and 12.13(b) show the same in a section near the lateral face in the first quarter section of the specimen. An inclined region where deviatoric strains are concentrated can be seen in these figures, similarly to what was observed under plane strain conditions. However, deviatoric strains are not as localized as in the plane strain case, and a shear band is not clearly formed. Comparing Figures 12.12(a) and 12.12(b) or Figures 12.13(a) and 12.13(b), it is seen that the distribution of deviatoric strain is almost the same in every section of the specimen.

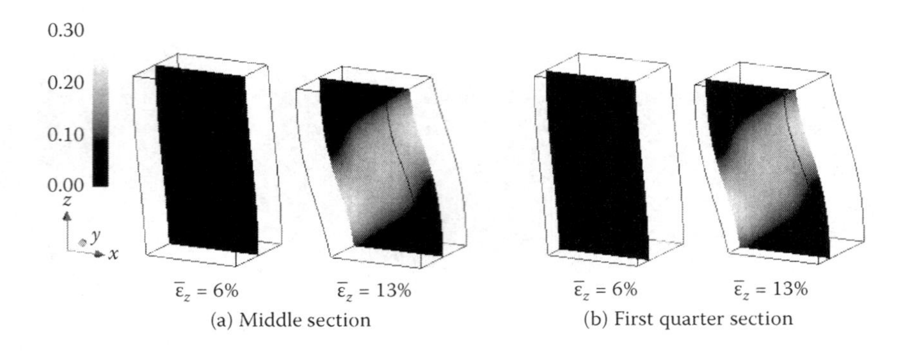

Figure 12.12 Distributions of deviatoric strain –3D (OCR=1).

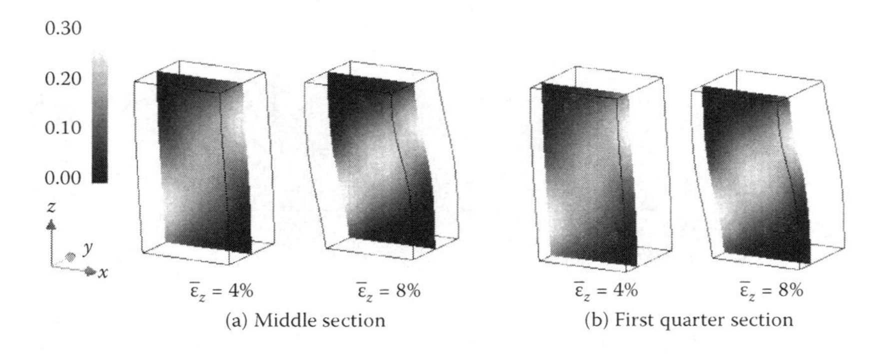

Figure 12.13 Distributions of deviatoric strain –3D (OCR=10).

In every analysis, the initial condition of the specimen is assumed to be homogeneous. Nevertheless, the computed deformation of the specimen becomes not only heterogeneous but also dissymmetric with respect to the vertical center line; finally, a shear band is formed because a very small error in numerical calculation becomes a trigger, and the constitutive model used can describe properly the influence of confining pressure and/or density on the soil behavior.

Chapter 13

Conclusions of Part 2

Constitutive models for soils should be developed in order to apply them to the practical design of various geotechnical problems with satisfactory accuracy as their final goal. It is then necessary to know the applicability of the constitutive models for practical uses. For this purpose, in Part 2, numerical simulations (numerical modeling) of various geotechnical problems using the constitutive model presented in Part 1 and the corresponding model tests (physical modeling) have been carried out. Two-dimensional (2D) and three-dimensional (3D) model tests were performed under earth gravity (1 g tests).

The numerical simulations using the finite element codes FEMtij-2D and FEMtij-3D were also carried out under the same stress level and boundary conditions as those in the physical model tests. In every model test, the same geomaterial was used: aluminum rod mass for 2D model tests and alumina balls for 3D model tests. Therefore, the common material parameters were used in the numerical analyses of these model tests.

Although the title of each chapter in Part 2 is fundamental and popular in various geotechnical engineering problems, most of the contents discussed in each chapter cannot be considered properly and/or cannot be explained reasonably by conventional design methods. The main results obtained in each chapter are summarized here.

In Chapter 8, after describing the purpose of Part 2 and giving a brief introduction of each chapter, the details of geomaterials used in the 2D and 3D model tests were shown. Here, the material parameters of 2D model ground were determined from biaxial tests on aluminum rod mass and those of 3D model ground from triaxial tests on alumina balls. These materials show stress–strain behavior similar to that of a medium dense sand, so the advanced model at stage I ("subloading t_{ij} model") is used. To describe precisely the interface behavior between soil and structure in the numerical simulations, the formulation of elastoplastic joint elements was also presented.

In Chapter 9, 2D model tests of trapdoor problems and the corresponding numerical simulations were carried out for the investigation of fundamental variables during tunnel excavations. The influences of soil cover and construction sequence on settlement and earth pressure of the ground obtained by the model tests were simulated accurately by the numerical analyses. To simulate the sequential tunneling process, 3D model tests and numerical analyses were carried out under the same soil covers as those in 2D model tests. There was also good agreement between the test results and the computed results.

Comparing these results with the results in 2D conditions, the effects of 3D excavation on surface settlement and earth pressure were discussed. Particularly, it becomes obvious that in real 3D tunneling, the earth pressure of the tunnel is changed due not only to the arching effects in the transverse direction to the tunneling axis but also to the arching effects in the longitudinal direction.

More realistic 2D model tests and numerical simulations were carried out to investigate the surface settlement and the earth pressure generated during excavation of circular tunnels. The influence of excavation patterns and the influence of existing building loads on earth pressure and ground movements were investigated in detail. Particularly, it was shown through the model tests and numerical simulations that maximum surface settlement may not occur above the tunnel crown, but rather underneath the building load, and that the pattern of settlement of the existing building depends on the type of the foundation and the relative distance between tunnel and foundation. The computed results described well all the test results not only qualitatively but also quantitatively.

In Chapter 10, 2D model tests and numerical simulations of active and passive earth pressure problems with retaining walls were carried out. Using the apparatus in which the deflection process and deflection mode of the wall are changed, the influence of these factors on the earth pressures and the ground movement were investigated. The corresponding computed results describe well the observed influences. For example, the earth pressure on the wall is not much influenced by the wall deflection mode, but rather depends on the wall deflection process, when wall deflection is applied progressively from the upper part of the wall to the lower part, as shown in real diaphragm walls.

Two-dimensional model tests and the corresponding numerical simulations of braced open excavation problems were carried out, and the influence of nearby existing structure loads on the wall and the ground, as well as the influence of the construction sequence, was discussed. The computed results described all the phenomena observed during the model tests comprehensively, including the fact that the maximum settlement occurs at the position of the existing structure and its magnitude and pattern depend on the type of foundation.

To investigate the settlement and bearing capacity of different types of foundation (strip foundation and piled raft), 2D model tests and the corresponding numerical simulations were carried out for three different types of loading conditions (concentric, eccentric, and inclined loadings). There was good agreement between the computed results and the test results not only qualitatively but also quantitatively. Therefore, it is considered that the finite element code (FEMtij) using the present constitutive models is applicable to bearing capacity problems as well as excavation problems such as tunneling and open excavation.

In Chapter 11, 2D and 3D model tests and the numerical simulations were carried out to investigate the mechanism of reinforced foundations (some reinforcement bars stem from the pile foundation) under various uplift loading conditions. It is experimentally and numerically shown that the highest uplift bearing capacity of the foundation (e.g., electric transmission tower) is obtained when the bars stem diagonally downward regardless of their bending stiffness. It is also shown that such reinforcement bars are not so effective when the uplift load becomes more inclined. On the other hand, the reinforcements stemming downward from the pile bottom are effective under both vertical and inclined uplift loads.

The bearing capacity of soil-reinforcement using geosynthetics under vertical loading was investigated by means of 2D model tests and the corresponding numerical simulations, in which the installation depth of reinforcement, the length of reinforcement, and the skin friction of reinforcement are changed. The results obtained from the numerical simulations show very good agreement with the results of the model tests in all patterns investigated. It was also found that, although reinforcements with lower skin friction have a moderate reinforcing effect, such reinforcements have a reverse effect depending on the installation depth.

In Chapter 12, to check the applicability of the present numerical method to the analysis of failure problems in geotechnical engineering, finite element analyses of plane strain tests (2D) and triaxial compression tests (3D) based on infinitesimal deformation theory and finite deformation theory were carried out for simulating the development of shear bands. Normally consolidated and overconsolidated Fujinomori clays (OCR = 1 and 10, respectively) were assumed. In spite of assuming simple drained analyses without soil-water coupled effects, the computed results describe well the phenomenon of strain localization including the shear band developments in normally and overconsolidated clays. It is seen through the analyses in this chapter that, by employing the constitutive model, which can describe typical soil features, including the influence of confining pressure (density), not only finite deformation analysis but also infinitesimal analysis can simulate the localization of deformation in geotechnical problems, although the computed results using finite deformation theory are clearer.

In Part 2, the model tests and the corresponding finite element analyses were carried out under low stress levels compared to the prototype stress levels. However, the material parameters of the present constitutive models are independent of the confining pressure and density as described in Part 1, so the present numerical method can be applied to the analyses of practical geotechnical problems. Finite element codes (FEMtij-2D and FEMtij-3D) were applied to the prediction of deformation and failure in real grounds, the validation of construction methods, and others (Hinokio et al. 2004; Nakai, Farias et al. 2007; Konda, Shahin, and Nakai 2010; Shahin et al. 2012).

References

Aboshi, H. (1973): An experimental investigation on the similitude in the consolidation of a soft clay, including the secondary creep settlement. *Proceedings of 8th International Conference on Soil Mechanics and Foundation Engineering,* Moscow, 4.3:88.

Adachi, T., Kimura, M. and Osada, H. (1993): Interaction between multitunnels under construction. *Proceedings of 11th Southeast Asian Geotechnical Conference,* Singapore, 51–60.

Adachi, T. and Oka, F. (1982): Constitutive equation for normally consolidated clays based on elasto/viscoplasticity. *Soils and Foundations* 22 (4): 57–70.

Adachi, T., Tamura, T., Kimura, M. and Aramaki, S. (1994): Experimental and analytical studies of earth pressure. *Proceedings of 8th International Conference on Computer Methods and Advances in Geomechanics,* Morgantown, 3:2417–2422.

Adrain, R. J. (1991): Particle imaging techniques for experimental fluid mechanics, *Ann. Rev. Fluid Mechanics*, 23, 261–304.

Akai, K. and Tamura, T. (1978): Numerical analysis of multidimensional consolidation accompanied with elastoplastic constitutive equation. *Proceedings of JSCE* 269:95–104 (in Japanese).

Alonso, E. E., Gens, A. and Josa, A. A. (1990): A constitutive model for partially saturated soils. *Geotechnique* 40 (3): 405–430.

Asaoka, A (2003): Consolidation of clay and compaction of sand—An elastoplastic description. *Proceedings of 12th Asian Regional Conference on Soil Mechanics and Geotechnical Engineering,* Singapore, 2:1157–195.

Asaoka, A., Nakano, M. and Noda, T. (1994): Soil-water coupled behavior of saturated clay near/at critical state. *Soils and Foundations* 34 (1): 91–105.

———. (1997): Soil-water coupled behavior of heavily overconsolidated clay near/at critical state. *Soils and Foundations* 37 (1): 13–28.

———. (2000a): Superloading yield surface concept for highly structured soil behavior. *Soils and Foundations* 40 (2): 99–110.

———. (2000b): Elastoplastic behavior of structured overconsolidated soils. *Journal of Applied Mechanics, JSCE* 3:335–342 (in Japanese).

Asaoka, A. and Noda, T. (1995): Imperfection-sensitive bifurcation of Cam clay under plane strain compression with undrained boundary. *Soils and Foundations* 35 (1): 83–100.

Baldi, G., Hueckel, T., Piano, A. and Pellegrini, R. (1991): Developments in modeling of thermo-hydro-geomechanical behavior of Boom clay and clay-based buffer materials. Report ER 13365, Commission of the European Communities, Nuclear Science and Technology.

Bishop, A. W. (1959): The principle of effective stress. *Teknisk Ukeblad* 39:859–863.

Bjerrum, L. (1967): Engineering geology of Norwegian normally consolidated marine clays as related to settlements of buildings. *Geotechnique* 17 (2): 81–118.

Bolton, M. D. and Lau, C. K. (1993). Vertical bearing capacity factors for circular and strip footings on Mohr–Coulomb soil. *Canadian Geotechnical Journal* 30 (6): 1024–1033.

Boudali, M., Leroueil, S. and Srinivasa Murthy, B. R. (1994): Viscous behavior of natural clays. *Proceedings of 13th International Conference on Soil Mechanics and Foundation Engineering,* New Delhi, 1:411–416.

Chowdhury, E. Q. and Nakai, T. (1998): Consequence of the t_{ij} concept and a new modeling approach. *Computers and Geotechnics* 23 (3): 131–164.

Chowdhury, E. Q., Nakai, T., Tawada, M. and Yamada, S. (1999): A model for clay using modified stress under various loading conditions with the application of subloading concept. *Soils and Foundations* 36 (6): 103–116.

Dafalias, Y. (1982): Bounding surface elastoplasticity–viscoplasticity for particulate cohesive media. *Proceedings of IUTAM Conference on Deformation and Failure of Granular Materials,* Delft, 97–107.

Dafalias, Y. and Popov, E. P. (1975): A model of nonlinearly hardening materials for complex loading. *Acta Mechanica* 23:173–192.

El-Shohby, M. A. (1969): Deformation of sands under constant stress ratio. *Proceedings of 7th International Conference on Soil Mechanics and Foundation Engineering,* Mexico, 1:111–119.

Eriksson, L.G. (1989): Temperature effects on consolidation properties sulphide clays, *Proceedings of 12th International Conference on Soil Mechanics and Foundation Engineering*, Rio de Janeiro, 3: 2087–2090.

Farias, M. M., Pedroso, D. M. and Nakai, T. (2009): Automatic substepping integration of the subloading t_{ij} model with stress path dependent hardening. *Computers and Geotechnics* 36 (4): 537–548.

Farias, M. M., Pinhiero, M. and Cordao Neto, M. P. (2006): An elastoplastic model for unsaturated soils under general three-dimensional conditions. *Soils and Foundations* 46 (5): 613–628.

Fredlund, D. G. and Morgenstern, N. R. (1977): Stress state variables for unsaturated soils. *Canadian Geotechnical Journal* 15 (3): 313–321.

Goodman, R. E., Taylor, R. L. and Brekke, T. L. (1968): A model for the mechanical jointed rock. *Proceedings of ASCE* 94 (SM3): 637–659.

Graham, J., Crooks, J. H. A. and Bell, A. L. (1983): Time effects on the stress–strain behavior of natural soft clays. *Geotechnique* 33 (3): 327–340.

Green, G. E. (1971): Strength and deformation of sand measured in an independent stress control cell. *Stress–Strain Behavior of Soil, Proceedings of Roscoe Memorial Symposium,* 285–323.

Hansen, J. B. (1961): A general formula for bearing capacity. Bulletin 11, Danish Geotechnical Institute, Copenhagen, Denmark, 38–46.

Hashiguchi, K. (1980): Constitutive equation of elastoplastic materials with elastoplastic transition. *Journal of Applied Mechanics, ASME* 102 (2): 266–272.

———. (2009): *Elastoplasticity theory*. In Lecture notes in applied and computational mechanics. New York: Springer, 42.

Hashiguchi, K. and Okayasu, T. (2000): Time-dependent elastoplastic constitutive equation based on the subloading surface model and its application to soils. *Soils and Foundations* 40 (4): 19–36.

Hashiguchi, K. and Ueno, M. (1977): Elastoplastic constitutive laws for granular materials, constitutive equations for soils. *Proceedings of Specialty Session 9, 9th International Conference on Soil Mechanics and Foundation Engineering*, Tokyo, 73–82.

Henkel, D. J. (1960): The relationship between effective stresses and water content in saturated clays. *Geotechnique* 10 (2): 41–45.

Hicher, P. Y. and Shao, J. F. (2008): *Constitutive modeling of soils and rocks* (English version). New York: Wiley.

Hill, R. (1948): A variation principle of maximum plastic work in classical plasticity. *Quarterly Journal of Mechanics and Applied Mathematics* 1:18–28.

———. (1950): *The mathematical theory of plasticity*. Oxford, England: Clarendon Press.

Hinokio, M., Nakai, T., Hoshikawa, T. and Yoshida, H. (2001): Dilatancy characteristics and anisotropy of sand under monotonic and cyclic loadings. *Soils and Foundations*, domestic ed. 41 (3): 107–124 (in Japanese).

Hinokio, M., Sato, H., Suzuki, M. and Nakai. T. (2004): Numerical analyses on the behavior of single piles under horizontal loading in clayey and sandy grounds. *Journal of Applied Mechanics, JSCE* 7:703–712 (in Japanese).

Honda, M. (2000). Prediction method of mechanical behavior on unsaturated ground. Dissertation for PhD in engineering, Kobe University, 125–126 (in Japanese).

Hueckel, T. and Baldi, G. (1990): Thermoplasticity of saturated clays: Experimental constitutive study. *Proceedings of ASCE* 116 (GE12): 1778–1796.

Hueckel, T. and Pellegrini, R. (1991): Thermoplastic modeling of undrained failure of saturated clay due to heating. *Soils and Foundations* 31 (3): 1–16.

Iwata, N., Nakai, T., Zhang, F., Inoue, T. and Takei, H. (2007): Influences of 3D effects, wall deflection process and wall deflection mode in retaining wall problems. *Soils and Foundations* 47 (4): 685–699.

Jennings, J. E. B. and Burland, J. B. (1962): Limitations to the use of effective stresses in partly saturated soils. *Geotechnique* 12 (2): 125–144.

Kaneda, K. (1999): A modeling of elastoplastic behavior of structured soils and analysis of time-dependent-like behavior of saturated soils by soil-water coupled computation. Dissertation for PhD in engineering, Nagoya University, 125–126 (in Japanese).

Karube, D. and Kato, S. (1989): Yield functions of unsaturated soil. *Proceedings of 12th International Conference on Soil Mechanics and Foundation Engineering*, Rio de Janeiro, 1:615–618.

Katona, M. G. (1984): Evaluation of viscoplastic cap model. *Journal of Geotechnical Engineering, Proceedings of ASCE* 110 (8): 1106–1125.

Kawai, K., Karube, D., Kato, S. and Kado, Y. (2000): Behavior of unsaturated soil and water characteristics in undrained shear. Research paper of Research Center for Urban Safety and Security, Kobe Univ., 4:231–239 (in Japanese).

Khalilia, N. and Khabbaz, M. H. (1998): A unique relationship for the determination of the shear strength of unsaturated soils. *Geotechnique* 48 (2): 1–7.

Kikumoto, M., Kyokawa, H. and Nakai, T. (2009): Comprehensive modeling of the water retention curve considering the influences of suction histories, void ratio and temperature. *Journal of Applied Mechanics, JSCE* 12:343–352 (in Japanese).

Kikumoto, M., Kyokawa, H., Nakai, T. and Shahin, H. M. (2010): A simple elastoplastic model for unsaturated soils and interpretations of collapse and compaction behaviors. *Proceedings of the 5th International Conference on Unsaturated Soils* (UNSAT 2010), Barcelona, 849–855.

Kikumoto, M., Kyokawa, H., Nakai, T., Shahin, H. M. and Ban, A. (2011): An elastoplastic constitutive model for soils considering temperature dependency. *Proceedings of 60th National Congress of Theoretical and Applied Mechanics,* Tokyo, OS02-16 (in Japanese)

Kikumoto, M., Nakai, T. and Kyokawa, H. (2009): New description of stress-induced anisotropy using modified stress. *Proceedings of 17th International Conference on Soil Mechanics and Geotechnical Engineering,* Alexandria, Egypt, 1:550–553.

Kohgo, Y., Nakano, M. and Miyazaki, T. (1993): Theoretical aspects of constitutive modeling for unsaturated soils. *Soils and Foundations* 33 (4): 49–63.

Konda, T., Shahin, H. M. and Nakai, T. (2010): Numerical analysis for backside ground deformation behavior due to braced excavation. *Journal of Materials Science and Engineering* (IOP Publishing), 10, doi:10.1088 /1757-899X/10/1/012010.

Kyokawa, H. (2010): Elastoplastic constitutive model for saturated and unsaturated soils considering the deposited structure and anisotropy. Dissertation for PhD in engineering, Nagoya Institute of Technology.

Kyokawa, H., Kikumoto, M., Nakai, T. and Shahin, H. M. (2010): Simple modeling of stress–strain relation for unsaturated soil. *Experimental and applied modeling of unsaturated soils* (geotechnical special publication no. 202), ASCE, 17–25.

Ladd, C. C., Foott, R., Ishihara, K. and Poulos, H. G. (1977): Stress-deformation and strength characteristics. *Proceedings of 9th International Conference on Soil Mechanics and Foundation Engineering,* Tokyo, 1:421–494.

Lade, P. V. and Duncan J.M. (1973): Cubical triaxial tests on cohesionless soil. *Proceedings of ASCE 99* (SM10): 793–812.

———. (1975): Elastoplastic stress–strain theory for cohesionless soil. *Proceedings of ASCE 101* (GT10): 1037–1053.

Lade, P. V. and Musante, H. M. (1978): Three-dimensional behavior of remolded clay. *Proceedings of ASCE 104* (GT2): 193–209.

Leonards, G. A. and Girault, G. A. (1961): A study of the one-dimensional consolidation test. *Proceedings of 5th International Conference on Soil Mechanics and Foundation Engineering,* Paris, 1:213–218.

Leroueil, S., Kabbaj, M., Tavenas, F. and Bouchard, R. (1985). Stress–strain–strain rate relation for the compressibility of sensitive natural clays. *Geotechnique* 35 (2): 159–180.

Leroueil, S. and Marques, M. E. S. (1996): Importance of strain rate and temperature effect in geotechnical engineering. *Measuring and Modeling Time Dependent Soil Behavior,* Geotechnical Special Publication (61), ASCE, 1–60.

Maeda, K., Hirabayashi, H. and Ohmura, A. (2006): Micromechanical influence of grain properties on deformation—Failure behaviors of granular media by DEM. *Proceedings of Geomechanics and Geotechnics of Particulate Media,* Yamaguchi, 173–179.

Mair, R. J., Taylor, R. N. and Bracegirdle, A. (1993): Subsurface settlement profiles above tunnels in clays. *Geotechnique* 43 (2): 315–320.

Matsuo, M. and Ueno, M. (1989): Development of ground reinforcing type of foundation. *Proceedings of 12th International Conference on Soil Mechanics and Foundation Engineering,* Rio de Janeiro, 2:1205–1208.

Matsuoka, H. (1974): Stress–strain relationship of sand based on the mobilized plane. *Soils and Foundations* 14 (2): 47–61.

Matsuoka, H. and Nakai, T. (1974): Stress-deformation and strength characteristics of soil under three different principal stresses. *Proceedings of JSCE* 232:59–70.

Meyerhof, G. G. (1963): Some recent research on the bearing capacity of foundations. *Canadian Geotechnical Journal* 1 (1): 16–26.

Mitchell, J. K. and Soga, K. (2005): *Fundamentals of soil behavior,* 3rd ed. New York: Wiley.

Miyata, M., Nakai, T., Hinokio, M. and Murakami, K. (2003): Numerical analysis for localized deformation of clay based on infinitesimal and finite deformation theories. *Journal of Applied Mechanics, JSCE* 6:455–465 (in Japanese).

Modaressi, H. and Laloue, L. (1997): A thermo-viscoplastic constitutive model for clays. *International Journal for Numerical and Analytical Methods in Geomechanics* 21:313–335.

Mroz, Z., Norris, V. A. and Zienkiewicz, O. C. (1981): An anisotropic critical state model for soils subjected to cyclic loading. *Geotechnique* 31 (4): 451–469.

Muir Wood, D. (1990): *Soil behavior and critical soil mechanics.* Cambridge, England: Cambridge University Press.

Murayama, S. (1964): A theoretical consideration on a behavior of sand. *Proceedings of IUTAM Symposium on Rheology and Soil Mechanics,* Grenoble, 146–159.

———. (1990): *Theory of mechanical behavior of soils.* Tokyo: Giho-do (in Japanese).

Murayama, S. and Matsuoka, H. (1971): Earth pressure on tunnels in sandy ground. *Proceedings of JSCE* 187:95–108 (in Japanese).

Nakai, T. (1985): Finite element computations for active and passive earth pressure problems of retaining wall. *Soils and Foundations* 25 (3): 98–112.

———. (1989): An isotropic hardening elastoplastic model for sand considering the stress path dependency in three-dimensional stresses. *Soils and Foundations* 29 (1): 119–137.

———. (2007): Modeling of soil behavior based on t_{ij} concept. *Proceedings of 13th Asian Regional Conference on Soil Mechanics and Geotechnical Engineering,* Kolkata, 2:69–89.

Nakai, T., Farias, M. M., Bastos, D. and Sato Y. (2007): Simulation of conventional and inverted braced excavation using subloading t_{ij} model. *Soils and Foundations* 47 (3): 597–612.

Nakai, T., Fujii, J. and Taki, H. (1989): Kinematic extension of an isotropic hardening model for sand. *Proceedings of 3rd International Symposium on Numerical Methods in Geomechanics,* Niagara Falls, 36–45.

Nakai, T. and Hinokio, T. (2004): A simple elastoplastic model for normally and overconsolidated soils with unified material parameters. *Soils and Foundations* 44 (2): 3–70.

Nakai, T. and Hoshikawa, T. (1991): Kinematic hardening model for clay in three-dimensional stresses. *Proceedings of 7th International Conference on Computer Methods and Advances in Geomechanics*, Cairns, Australia, 655–660.

Nakai, T., Kyokawa, H., Kikumoto, M. and Zhang, F (2009): Elastoplastic modeling of geomaterials considering the influence and density and bonding. *Proceedings of Prediction and Simulation Methods for Geohazard Mitigation*, Kyoto, 367–373.

Nakai, T. and Matsuoka, H. (1980): A unified law for soil shear behavior under three-dimensional stress condition. *Proceedings of JSCE* 303:85–77 (in Japanese).

———. (1983): Shear behaviors of sand and clay under three-dimensional stress condition. *Soils and Foundations* 23 (2): 26–42.

———. (1986): A generalized elastoplastic constitutive model for clay in three-dimensional stresses. *Soils and Foundations* 26 (3): 81–98.

———. (1987): Elastoplastic analysis of embankment foundation. *Proceedings of 8th Asian Regional Conference on Soil Mechanics and Geotechnical Engineering*, Kyoto, 1:473–476.

Nakai, T., Matsuoka, H., Okuno, M. and Tsuzuki, K. (1986): True triaxial tests on normally consolidated clay and analysis of the observed shear behavior using elastoplastic constitutive models. *Soils and Foundations* 26 (4): 67–78.

Nakai, T. and Mihara, Y. (1984): A new mechanical quantity for soils and its application to elastoplastic constitutive models. *Soils and Foundations* 24 (2): 82–94.

Nakai, T. and Muir Wood, D. (1994): Analysis of true triaxial and directional shear cell tests on Leighton Buzzard sand. *Proceedings of International Symposium on Pre-failure Deformation Characteristics of Granular Materials*, Sapporo, 419–425.

Nakai, T., Shahin, H. M., Kikumoto, M., Kyokawa, H., Zhang, F. and Farias, M. M. (2011a): A simple and unified one-dimensional model to describe various characteristics of soils. *Soils and Foundations* 51 (6): 1129–1148.

———. (2011b): A simple and unified three-dimensional model to describe various characteristics of soils. *Soils and Foundations* 51 (6): 1149–1168.

Nakai, T., Shahin, H. M., Watanabe, A. and Yonaha, S. (2009): Reinforcing mechanism of geosynthetics on bearing capacity problems—Model tests and numerical simulations. *Proceedings of 17th International Conference on Soil Mechanics and Geotechnical Engineering*, Alexandria, Egypt, 1:917–920.

Nakai, T., Shahin, H. M., Zhang, F., Hinokio, M., Kikumoto, M., Yonaha, S. and Nishio, A. (2010): Bearing capacity of reinforced foundation subjected to pull-out loading in 2D and 3D conditions. *Geotextiles and Geomembranes* 28 (3): 268–280.

Nakai, T., Taki, H. and Funada, T. (1993): Simple and generalized modeling of various soil behaviors in three-dimensional stresses. In *Modern approaches to plasticity*, ed. D. Kolymbas. New York: Elsevier, 561–584.

Nakano, M., Asaoka, A. and Constantinescu, D. T. (1998): Delayed compression and progressive failure of the assembling of crushed mudstone due to slaking. *Soils and Foundations* 38 (4): 183–194.

Nova, R. (1982): A viscoplastic constitutive model for normally consolidated clay. *Proceedings of IUTAM Conference on Deformation and Failure of Granular Materials,* Delft, 287–295.

Nova, R. (2010): *Soil mechanics* (English version). New York: Wiley.

Oda, M. (1972): The mechanism of fabric changes during compressional deformation of sand. *Soils and Foundations* 12 (2): 1–18.

———. (1993): Inherent and induced anisotropy in plasticity theory of granular soils. *Mechanics of Materials* 16 (1–2): 35–45.

Ohta, H. (1971): Analysis of deformations of soils based on the theory of plasticity and its application to settlement of embankments. Dissertation for PhD in engineering, Kyoto University.

Oka, F., Adachi, T. and Yashima, A. (1995): A strain localization analysis using a viscoplastic softening model for clay. *International Journal of Plasticity* 11 (5): 523–545.

Oka, F., Kodaka, T., Kimoto, S., Ishigaki, S. and Tsuji, C. (2003): Step-changed strain rate effects on the stress–strain relations of clay and a constitutive modeling. *Soils and Foundations* 43 (4): 189–202.

Oshima, A., Ikeda, Y. and Masuda, S. (2002). Effects of load increment ratio on step loading consolidation test of clay. *Proceedings of 37th Annual Meeting on JGS* 1:289–290 (in Japanese).

Peck, R. B. (1969): Deep excavations and tunneling in soft ground. *Proceedings of 7th International Conference on Soil Mechanics and Foundation Engineering,* Mexico City, 3:225–290.

Pedroso, D. M., Farias, M. M. and Nakai, T. (2005): An interpretation of subloading t_{ij} model in the context of conventional elastoplasticity theory. *Soils and Foundations* 45 (4): 61–77.

Perzyna, P. (1963): The constitutive equations for rate sensitive plastic materials. *Quarterly Applied Mathematics* 20 (4): 321–332.

Roscoe, K. H. and Burland, J. B. (1968): On the generalized stress–strain behavior of wet clay. In *Engineering plasticity,* ed. J. Heyman and F. A. Leckie. Cambridge, England: Cambridge University Press, 535–609.

Roscoe, K. H., Schofield, A. N. and Thurairajah, A. (1963): Yielding of clays in states wetter than critical. *Geotechnique* 13 (3): 211–240.

Satake, M. (1982): Fabric tensor in granular materials. *Proceedings of IUTAM Conference on Deformation and Failure of Granular Materials,* Delft, 63–68.

———. (1984): Anisotropy in ground and soil materials. *Tsuchi to Kiso* 32 (11): 5–12 (in Japanese).

Schofield, A. N. and Wroth, C. P. (1968): *Critical state soil mechanics.* London: McGraw–Hill.

Sekiguchi, H. (1977): Rheological characteristics of clays. *Proceedings of 9th International Conference on Soil Mechanics and Foundation Engineering,* Tokyo, 1:289–292.

———. (1984): Theory of undrained creep rupture of normally consolidated clay based on elasto-viscoplasticity. *Soils and Foundations* 24 (1): 129–147.

———. (1985): Macrometric approaches—static—intrinsically time-dependent. *Proceedings of Discussion Session on Constitutive Laws of Soils, 11th International Conference on Soil Mechanics and Foundation Engineering,* San Francisco, 66–98.

Sekiguchi, H. and Ohta, H. (1977): Induced anisotropy and time dependency in clays. *Proceedings of Specialty Session 9, 9th International Conference on Soil Mechanics and Foundation Engineering,* Tokyo, 229–238.

Sekiguchi, H. and Toriihara, M. (1976): Theory of one-dimensional consolidation of clays with consideration of their rheological properties. *Soils and Foundations* 16 (1): 27–44.

Shahin, H. M., Nakai, T., Hinokio, M., Kurimoto, T. and Sada, T. (2004): Influence of surface loads and construction sequence on ground response due to tunneling. *Soils and Foundations* 44 (2): 71–84.

Shahin, H. M., Nakai, T., Hinokio, M. and Yamaguchi, D. (2004): 3D effects on earth pressure and displacements during tunnel excavations. *Soils and Foundations* 44 (5): 37–49.

Shahin, H. M., Nakai, T., Zhang F., Kikumoto, M., Ito, F. and Baba, S. (2012): Investigation of failure mechanism during slope excavation and its countermeasure by elastoplastic FE analysis. *Proceedings of 11th International Symposium on Landslides,* Banff (submitted).

Shahin, H. M., Nakai, T., Zhang, F., Kikumoto, M. and Nakahara, E. (2011): Behavior of ground and response of existing foundation due to tunneling. *Soils and Foundations* 51 (3): 395–409.

Shibata, T. (1960): On the volume changes of normally consolidated clays. *Annals of Disaster Prevention Research Institute,* Kyoto University, 6:128–134 (in Japanese).

Shibata, T. and Karube, D. (1965): Influence of the variation of the intermediate principal stress on the mechanical properties of normally consolidated clays. *Proceedings of 6th International Conference on Soil Mechanics and Foundation Engineering,* Montreal, 1:359–363.

Sivakumar, V. (1993): A critical state frame work for unsaturated soil. PhD dissertation, University of Sheffield.

Suklje, L. (1957): The analysis of the consolidation process by the isotaches method. *Proceedings of 4th International Conference on Soil Mechanics and Foundation Engineering,* London, 1:200–206.

Sun, D. A., Sheng, D. and Xu, Y. (2007): Collapse behavior of unsaturated compacted soil with different initial densities. *Canadian Geotechnical Journal* 44:673–686.

Sung, E., Shahin, H. M., Nakai, T., Hinokio, M. and Yamamoto, M. (2006): Ground behavior due to tunnel excavation with existing foundation. *Soils and Foundations* 46 (2): 189–207.

Tanaka, H., Udaka, K., and Nosaka, T. (2006): Strain rate dependency of cohesive soils in consolidation settlement. *Soils and Foundations* 46 (3): 315–322.

Tatsuoka, F. (1978): Stress–strain behavior of an idealized granular material by simple elastoplastic theory. *Proceedings of United States–Japan Seminar on Continuum Mechanical Statistical Approaches of Granular Materials,* Sendai, Japan, 301–320.

Taylor, D. W. (1948): *Fundamentals of soil mechanics.* New York: Wiley.

Terzaghi, K. (1943): *Theoretical soil mechanics.* New York: Wiley.

Terzaghi, K. and Peck, R. P. (1948): *Soil mechanics in engineering practice.* New York: Wiley.

Tokyo Electric Power Company and Dai Nippon Construction (1990): Report on bearing capacity of caisson type foundation with reinforcing bars, 4, (in Japanese).

Tschebatarioff, G. P. (1951): *Soil mechanics: Foundations and earth structure.* New York: McGraw–Hill.

van Genuchten, M.T. (1980): A closed form equation for predicting the hydraulic conductivity of unsaturated soils. *Soil Science of American Journal* 44: 892–898.

Vesic, A. S. (1975): Bearing capacity of shallow foundations. In *Foundation engineering handbook*. eds. H. F. Winterkom and H. Y. Fang. New York: van Nostrand Reinhold, 121–147.

Wan, R. G. and Guo, P. J. (1999): A pressure and density dependent dilatancy model for granular materials. *Soils and Foundations* 39 (6): 1–11.

Watabe, Y., Emura, T., Yamaji, T., Udaka, K. and Watanabe, Y. (2009). A consideration on upper limit of applied load in long term consolidation tests. *Proceedings of 64th Annual Meeting on JSCE,* III-011, 21–22 (in Japanese).

Watabe, Y., Udaka, K. and Morikawa, Y. (2008): Strain rate effect on long-term consolidation of Osaka Bay clay. *Soils and Foundations* 48 (4): 495–509.

Wheeler, S. J. and Sivakumar, V. (1995): An elastoplastic critical state framework for unsaturated soil. *Geotechnique* 45 (1): 35–53.

Yamaguchi, H. (1969). *Soil mechanics.* Tokyo: Giho-do (in Japanese).

Yashima, A., Leroueil, S., Oka, F. and Guntaro, I. (1998): Modeling temperature and strain rate dependent behavior of clays—One dimensional consolidation. *Soils and Foundations* 38 (2): 63–73.

Yatomi, C., Yashima, A., Iizuka, A. and Sano, I. (1989): General theory of shear bands formulation by a non-coaxial Cam clay mode. *Soils and Foundations* 29 (3): 41–53.

Yong, R. N. and Mckyes, E. (1971): Yield and failure of clay under triaxial stresses. *Proceedings of ASCE* 97(SM1): 159–176.

Yoshikuni, H., Moriwaki, T., Ikegami, S. and Tajima, S. (1990): The effect of stress history on the secondary compression characteristics of overconsolidated clay. *Proceedings of 25th Annual Meeting on JGS* 1:381–382 (in Japanese).

Yu, H. S. (2006): *Plasticity and geotechnics.* New York: Springer.

Zhang, F., Kimura, M., Nakai, T. and Hoshikawa, T. (2000): Mechanical behavior of pile foundations subjected to cyclic lateral loading up to the ultimate state. *Soils and Foundations* 40 (5): 1–18.

Zhang, F., Yashima, A., Nakai, T., Ye, G. L. and Aung, H. (2005): An elasto-viscoplastic model for soft sedimentary rock based on t_{ij} concept and subloading surface. *Soils and Foundations* 45 (1): 65–73.

Zhang, S. and Zhang, F. (2009): A thermo-elasto-viscoplastic model for soft sedimentary rock. *Soils and Foundations* 49 (4): 583–595.

Zienkiewicz, O. C. and Taylor, R. L. (1991): *The finite element method,* vol. 2, 4th ed. New York: McGraw–Hill.

Index

D

P

S

U

V

W

Y

Z